Modelling Coastal and Marine Processes

2nd Edition

Modelling Coastal and Marine Processes

2nd Edition

Phil Dyke

University of Plymouth, UK

ICP

Imperial College Press

Published by

Imperial College Press
57 Shelton Street
Covent Garden
London WC2H 9HE

Distributed by

World Scientific Publishing Co. Pte. Ltd.
5 Toh Tuck Link, Singapore 596224
USA office: 27 Warren Street, Suite 401-402, Hackensack, NJ 07601
UK office: 57 Shelton Street, Covent Garden, London WC2H 9HE

Library of Congress Cataloging-in-Publication Data
Dyke, P. P. G.
  [Modeling coastal and offshore processes]
  Modeling coastal and marine processes / Phil Dyke (University of Plymouth, UK). -- 2nd edition.
    pages cm
  Includes bibliographical references.
  ISBN 978-1-78326-769-9 (hardcover : alkaline paper) --
  ISBN 978-1-78326-770-5 (paperback : alkaline paper)
  1. Coastal engineering--Mathematical models. 2. Coastal engineering--Environmental aspects.
  3. Coastal ecology--Mathematical models. 4. Marine ecology--Mathematical models.
  5. Oceanography--Mathematical models. 6. Climatic changes--Mathematical models. I. Title.
  II. Title: Modelling coastal and marine processes.
  TC209.D95 2015
  627'.58--dc23
                                   2015035758

British Library Cataloguing-in-Publication Data
A catalogue record for this book is available from the British Library.

In-house Editors: Catharina Weijman/Dr. Sree Meenakshi Sajani

Typeset by Stallion Press
Email: enquiries@stallionpress.com

Printed in Singapore

To My Parents

Violet Susan Dyke (1914–1972)

and

George Dyke (1906–1980)

without whom this book would not have been possible.

Preface to Second Edition

It has been nine years since the first edition and some updating was certainly required. There have been some gradual changes over the past years. First of all, the software available is more wide ranging and depends on a larger variety of methods. In 2006, finite differences dominated as the numerical method used, almost to the exclusion of all others apart from a few that used finite elements. Now, differences still dominate as the most commonly used method, but within this method modellers widely use more sophisticated methods such as flux corrected transport and finite volume methods using unstructured grids. Recently these unstructured grids have become adaptive, meaning that they change as the time integration proceeds. The applications of this are still in their infancy, but nevertheless the principles are important to understand and there is a section on them in this revised text. Another change has less to do with technicalities than with use and design. More of the better software has become free to use and modular. Research students now routinely use these tools to analyse particular problems and they can change, reprogram, even redesign bits of the code themselves. This is due to the non-commercial sourcing which has grown as the complexity has grown. The teams of people who design such software are normally drawn from across universities and government funded research institutes, so the products are in the public domain, and there is a team of people adding modules, maintaining the codes and redesigning the output.

The opportunity has been taken to re-order some of the rest of the material. The first edition was a bit guilty of using the last big chapter as a catch-all for all topics not mentioned elsewhere. Therefore, estuaries and trapped waves now have their own chapters. There is also a separate chapter on tides surges and sea-level modelling as this has assumed greater importance in the last few years in particular, the modelling of tsunamis is now included. The chapter on numerical methods has become two chapters, one on the methods and the other on applications. It is still the case, however, that although the technical content has increased a little, the emphasis remains on describing what is going on rather than letting mathematics do the talking. Climate change has not gone away, but it is fair to say that fear of its onset over the past ten years has not been quite as intense as it used to be, and this has thrown a lifeline to the climate change sceptics. The section in what is now Chapter 10 has been changed to capture the latest research. The fundamental message is the same; mankind needs to stop using so much fossil fuel. To reflect the change in emphasis, the title of the book has also changed from Modelling Coastal and Offshore processes to Modelling Coastal and Marine Processes. The first three chapters are as before with minor changes. Chapter 4 (Numerical methods) has become two chapters, the first methods and the second applications of these methods. Chapter 6 is the new one on tides etc. Old Chapters 5, 6 and 7 are now Chapters 7, 8 and 9 and all three have had minor changes. Old Chapter 8 (now Chapter 10) has had bits removed: to Chapter 6 (tides etc.), to new Chapter 11 (estuaries) and to new Chapter 12 (trapped waves). The material in both Chapters 11 and 12 is extended compared to that in the 2007 edition. What is left in Chapter 10 are climate change and ocean scale modelling, and these too have been changed to reflect recent developments. The final chapter is still a conclusion, but now containing a brief description of one of these newly designed pieces of software, NEMO (see NEMO, 2012).

It is a pleasure to thank all those who have helped me improve the text, not least the students who I have taught over the years. Recently these have become more concentrated in Civil and Coastal Engineering and this has probably led to a slight change in emphasis in favour of the quantitative model. Thanks also to my son Adrian, who has stepped in to help with networking and other software difficulties.

Professor Phil Dyke
University of Plymouth
April 2015

Preface

It has been over 15 years that the author has now been responsible for the delivery of a modelling module on a graduate studies degree in marine science; latterly the class has included coastal engineering students and this has influenced both the contents and the style of delivery. Two earlier texts, Dyke (1996), Dyke (2001), are unashamedly marine science; this one certainly leans more towards coastal engineering. It has definitely benefitted from the existence of these two previous books, and the content has some overlap.

The motivation for studying how to model coastal sea processes has not really changed since the publication of the second of these books in 2001; it is still by and large environmental protection, although the need for this has, if anything, moved slightly from the pollution and direct interference by man to global warming and climate change. Even more today is the need to express what are very complicated ideas to important people who lack a technical background. The students for whom this book is intended will by and large have some technical facility but upon graduation these students will almost certainly need to explain the modelling outcomes to journalists, politicians and the like, so the emphasis in this text is on understanding the processes themselves. It has been assumed that readers will have a knowledge of elementary calculus and algebra at least equivalent to that met in the first stages of most physical science or engineering degrees. Even if this is not the case, there are only a few places where not being familiar with this mathematics will inhibit understanding. At all times in this textbook pains have been taken to explain what is happening in terms of plain English and not to rely on equations to do the talking; over the years this author has taught too many biologists and geographers to know not to do this.

The first three chapters contain the background material. This runs from the modelling and techniques of Chapters 1 and 2, to the derivation of the fundamental equations in Chapter 3. In the previous texts, this derivation has been avoided, but although this chapter can be skipped if absolutely necessary, the author has taken pains to explain precisely what is going on as expressions are being derived, so it is worth at least some effort to understand. The numerical procedures are explained in Chapter 4, and again this is fuller than in the author's previous texts on this subject. Once again, the emphasis is on explanation of what is happening. After this, the book is about applying modelling techniques. Chapter 5 is on diffusion, Chapter 6 is about modelling waves, sediments and coasts and is definitely core coastal engineering modelling. Chapter 7 is a little different. In the last ten or so years, ecosystem and biological modelling has come of age. In this chapter, it is believed for the first time, a systematic development is given that is aimed at the modeller whose principal motivation is understanding physical processes but who recognises the important interaction between this and biological processes. No previous biology is assumed (indeed, it is difficult to see how biological knowledge can be much less than that possessed by the author). The final chapter sweeps up various other topics, including modelling climate change, modelling storm surges and other flows usually too large for the coastal engineer, but vital for those concerned with shallow sea dynamics.

It is a pleasure to acknowledge all my students over the years for the feedback on the lecture material which has helped a great deal in the selection of the subject matter for this book. Thanks also to the participants at the biennial JOint Numerical Sea MODelling Group (JONSMOD) who have done much to keep me abreast of developments over the past 25 years as the subject has quickly evolved; I single out Alan M Davies of the Proudman Oceanographic Laboratory who has always been extremely helpful, especially in explaining to me the application of finite elements to oceanographic modelling. Particular thanks also to colleagues at the Plymouth Marine Laboratory for the indoctrination into biological modelling.

Professor Phil Dyke
University of Plymouth
September 2006

Contents

Chapter 1

Modelling Preliminaries

1.1 Introduction

In this chapter we shall introduce the modelling process. This is a semi formal process by which modelling takes place. In order to do any real modelling however it is still useful to understand a little mathematics and statistics. Precisely how much is still controversial. With software these days it is tempting to think that very little of either is required, however this is like a car driver who knows absolutely nothing of what goes on under the bonnet. Fine as long as everything works, but what if it breaks down? True, we call the experts to fix it, but this text is about model building so it is necessary to be a bit of a mechanic. Basic mathematics is therefore good to have although it is possible to do some modelling without it, see Dyke (1996). In this chapter, we cover the basic mathematics that will enable you to understand the derivations of the fundamental equations in Chapter 3.

The statistics part has a slightly different function. As well as its use in building models, statistics is used for making sense of data, in particular inferring from data. It is difficult to see when this will ever be unnecessary so an introduction to statistics will always be an extremely useful inclusion in a book such as this. There is nothing in this chapter that should be beyond anyone who has studied mathematics in the year they turned 18, and most of it should be accessible to those who dropped mathematics after age 16. The prime motivator for learning anything new is its usefulness, and rest assured everything in this chapter is useful at least once in the rest of the text. Also, when the mathematics and statistics are introduced, it will be done in a way to highlight this motivation, that is whenever possible through the use of relevant examples.

1.2   Introduction to Modelling

To the majority, the word modelling still means something to do with photography or, if they have a scientific background, the building of scaled-down replicas that ought to mimic real life situations. In this latter category one thinks of Civil Engineering consultants building models of harbours with attendant breakwaters and jetties, and then subjecting them to a particular wave climate. The way this is done is to build a physical model, usually in a large area reminiscent of an aircraft hanger. In this model, the area of coast or river or estuary (whatever) is built from materials such as concrete, sand and cement. Of course there is a scale, perhaps 1:20 or even larger, which needs to be considered when examining results. If waves are of interest, then there has to be a paddle mechanism included in order to generate them. Exactly how the scale factors can be calculated is the subject of Chapters 2 and 3, but suffice it to say that measurements of quantities such as wave height, current speed and direction, the force on pier or jetty can be made on the model. Appropriate scale factors are then applied and an estimate of the real life wave height, current speed and direction, force on the pier or jetty or whatever can then be made. Up to 30 or so years ago virtually all modelling in coastal engineering or oceanography referred to this kind of activity. These days, modelling invariably means use of the computer and the big, once national, facilities (e.g. Hydraulics Research in the UK Delft Hydraulics in The Netherlands — both now privatised) now have much scaled down (no pun intended) the facilities for these physical models but have many sophisticated computer models to replace them. Many would prefer the word enhance rather than replace, as there is still the place for a physical model where the carefully placed strain gauge can give information to reinforce the output from a mathematical model. In most cases, the results from a mathematical model implemented via software on a computer will tell the same story as the results from a physical model, but if there are contradictory results, neither should automatically be believed. Perhaps they are both wrong and the situation is more complicated that either model builder thought. Areas where physical modelling is still dominant are in the building of bridges and in the design of spacecraft. In both of these areas, the final costs are so huge that the expense of building a physical model is less critical than, for example, estimating dilution rates of a dissolved substance in an environmentally sensitive estuary. In this latter case, mathematical models are now almost always used.

We shall not be discussing physical models in this text. The parallels will be explored a little further in Chapter 2 where this is natural, but thereafter physical models are left behind. The academic discipline termed "Mathematics" is correctly thought of as operating within a very well-defined set of axioms using well-defined techniques to give precise answers to well-defined problems. Pure mathematicians tend to hold this view; applied mathematicians are more liberal. This rigid view of mathematics is not well suited to contribute to the description of a practical science such as oceanography or coastal engineering. Research papers that are very mathematical, which may be very interesting in their own right, often have only a tenuous link with reality. At the other end of the spectrum there are some very simple mathematical models that embody the essence of oceanographic truth, and we will definitely be meeting some of these. It is in this blending of mathematics with the knowledge of oceanographic processes where the art of successful mathematical modelling lies.

There is no doubt whatever that mathematical modelling has been greatly assisted by the rapid advances in computer power. Nevertheless, mathematical modelling can still certainly take place without it, it is just that computing increases the scope of the modelling. The emphasis in this text is definitely not on computing, which may be thought of as a tool that enables modelling to take place. Instead, the concentration is on the correct mathematical description of the ocean and coastal physics and, to some extent, biology. Here, we shall treat the terms coastal oceanography, coastal marine physical science and coastal engineering as synonymous. Coastal oceanography is a science that has grown through painstaking observation and progressed through scientists making judgements and deductions from these observations. There are several distinctive features that, although not peculiar to coastal engineering, oceanography, meteorology and earth science, in general make it particularly amenable to the relatively new art of mathematical modelling. Yes, although there is a great deal of scientific method and rigour in mathematical modelling, it still remains in many ways very much an art. First, coastal oceanography as an applied science has to incorporate aspects of Physics, Chemistry and Biology. Indeed, it may be argued that the sea provides an ideal vehicle for the study of some, but certainly not all, of the fundamentals of these basic sciences. In order to understand some of the processes that go on in the sea, it is therefore necessary to simplify some aspects and ignore others. This is what occurs in modelling. Secondly, there is a very important aspect to modelling called validation. In most sciences and engineering, validation means trying out the model and comparing it with the real situation. In coastal oceanography, the entire history of the science, from the accumulated wisdom of fishermen through the voyages of discovery to modern day scientific expeditions, is centred around observations and provides, in some respects, an ideal scenario for validation. Conditions are however not controlled, as the day has not arrived where weather can be prescribed, therefore there is no control over at least one input. This is certainly a disadvantage in some respects, although it does encourage the continuation of the lively debate between the modeller and the observer.

Outside the coastal environment modelling has had a large part to play in our understanding of how the ocean currents are driven. Ocean circulation modelling dates back to just after the Second World War, and although this text does not dwell on global ocean modelling, the understanding of global ocean physics forms the basis for models that help us understand global climate models and models of El Niño and so has its place in this text.

1.2.1 *Environmental issues*

In the last 60 years or so, many new substances have been developed, from those used in foodstuffs, packaging, building materials to paint additives and the many plastics used everywhere in our daily lives. With the industrial processes that are used to produce these substances has come an awareness that care has to be taken about the disposal of the byproducts of the manufacturing. The environmental lobby has become particularly strong in the last 30 or so years, and it is really only now that we are at last beginning to become aware of the lasting effect of all the foreign material mankind seems to be continually pumping into the Earth's environment. Of course, much research needs to be done before our understanding is anywhere near adequate, but now, whenever there is a new manufactured chemical or the production of a hitherto unsuspected byproduct, there is in most countries strict legislation governing what can and cannot be allowed into the environment. Unquestionably, sometimes the environmental lobby prevents what is an innocent process taking place, but this is much less worrying than permitting a pollution that may unwittingly cause widespread environmental damage.

Some worrying case histories have come to light following the collapse of the Soviet Union. In Poland, Czechoslovakia (now split into the Czech Republic and Slovakia) and other former Warsaw Pact countries, cumulative environmental effects have shortened the life expectancy of people living within the range of rivers polluted by the unthinking discharge of industrial waste, including that from nuclear industries. Major rivers in Poland were so polluted that they were useless even for industry let alone recreational use. Using them for drinking water was completely out of the question. Before 1991, East Germany (as it was then called) was encouraged to increase industrial production regardless of any impact this might have on the environment. In December 1990, drinking water in Brandenberg did not meet European Union (EU) quality standards. Tankers had to be used to supply drinking water in some areas. Nitrate levels were up to 25 times EU legal limits. Samples from the Havel river contained high levels of phosphate, ammonium, benzol and zinc. Purification ensued, but even then water was found still to be contaminated with oil, even phenol. Post 1991, the West has become horrifically aware of dead rivers and dead lakes with toxic levels of similar chemicals. The unified Germany is doing its best to clean up the mess left by its communist predecessor, but it takes both time and money. It is a cruel irony that Germany now comprises two pre-1991 countries that were at the opposite ends of the environmental spectrum as far as cleanliness is concerned. A cruelty that has hit the ex West German where it hurts; in his pocket. Some of these more legal and political issues are outlined in Section 1.2.4.

There are many ways in which chemicals can affect the environment. Most of these are local effects and come under the name pollution. There are two effects however that have had an impact over the whole globe since the mid-1980s. In order to reach the public attention these days requires the adoption of a "sound bite". The sound bites in question are "global warming" and "ozone depletion". Ozone depletion has dropped out of the headlines; in the 1990s and early part of the century there were the holes in the polar ozone layers that first brought this to attention of the public, which led to targeting chlorofluorocarbons (CFCs) as the prime cause and the subsequent banning of most aerosols, see below. The public might be forgiven for thinking that they have disappeared; far from it. The polar holes are bigger than ever and there are now holes in other places, for example over Tibet. Maybe one problem is public confusion with global warming that has in some sense now monopolised the environmental agenda. If the public can get things wrong it usually does, and these effects, although very, independent, have been first confused then — as far as ozone depletion is concerned — overlooked. To sum up global warming, this is the increase of probably man-made gaseous discharge of chemicals such as carbon dioxide, methane and sulfurous oxides into the atmosphere, which alters the balance between incoming and outgoing radiation, decreasing the latter so that the Earth as a whole warms up. That the Earth's albedo (as this balance between incoming and outgoing radiation is called) is changing is beyond question, but whether it is through industrial pollution, the eradication of huge swathes of equatorial rainforests or some of both or is all part of natural oscillations is still controversial. More is said about climate modelling in Chapter 10. In the last few years though, almost everyone, but not the USA, seems convinced that global warming is largely due to man-made greenhouse gases. When the first edition of this book was being written, the southern USA was suffering hugely from the after effects of hurricane Katrina. Although hurricanes happen every year and there have been similar sized ones in the past, thankfully missing crucial vulnerable areas, questions will no doubt be asked. Do hurricanes contain more energy on average than they used to? If the answer is yes, could this be yet another manifestation of global warming. In fact, hurricane activity goes in cycles of around 30 to 40 years, and we are ending a quiet period just now — the 1960s was the last active period. This more energetic hurricane activity is coinciding with the enhanced media awareness of global warming, so the temptation to put two and two together and get five is overwhelming, especially for those who like to give the USA a wake up call.

There is less controversy about ozone depletion. This is the name given to the disappearance of that layer enveloping the Earth, largely comprising ozone, that is responsible for shielding us from the harmful effects of the Sun's ultraviolet rays. The aerosol can is less than 100 years old, and by the 1980s it had mushroomed in use until the gas used for the propulsion of the spray, CFC or chlorofluorocarbons, was found to persist long enough to attack and destroy parts of the ozone found in the upper parts of the atmosphere. As has been said briefly above, this ozone protects us from the ultraviolet rays arising out of the Sun and overexposure to which can lead to skin cancers. This whole ozone depletion problem has been brought to the attention of the public by the publication of colourful pictures of the increasingly large hole in the ozone layer over the Antarctic and a smaller one discovered over the Arctic. The problem in fact has been known for some time; the depletion of stratospheric ozone was first observed as long ago as 1979, but not reported until 1985. A hole which is largest in the springtime, was observed in the ozone layer, but at this time no culprit was identified. It was perhaps an oddity that would go away. Not so. By the mid 1980s it was recognised that

CFCs had an important role and the Montreal Protocol was signed: 100 countries signed a protocol in 1988 to discuss ways of arresting the depletion of ozone in the upper layers of the atmosphere. By 1990 and 1991 there were two successive years of severe ozone depletion; it was not going away and CFCs were confirmed as being linked to the problem. Although CFCs, first introduced via aerosols in the 1930s, were contributing directly to ozone depletion, there was additionally a positive feedback mechanism at work. The CFC-derived chlorine lowered air temperatures, which in itself made CFC-derived chlorine more effective in depleting ozone. By 1993, ozone had depleted to 21% of "normal" values. In 1996, the hole was as big as USA and Canada combined and by 1998 it extended over an area nearly twice the size of the Antarctic continent itself. The increase in atmospheric carbon dioxide and methane interacts with this ozone depletion problem by cooling the upper atmosphere even though the lower atmosphere is warming, which leads to the maintenance of the hole. All this is a salutary lesson in unforeseen environmental consequences due to man-made chemicals. Besides UV rays leading to the enhanced risk of skin cancer as well as other health problems, other consequences, as yet unknown, may still await us. Not before time, in December 1995 signatories to the Montreal Protocol met in Vienna and agreed limits on ozone depleting substances: methyl bromide and low density chemicals, the former to be phased out by 2010, the latter to stabilise at 1995–1998 levels by 2002. Even though ozone depletion has receded from being newsworthy, the hole in the Antarctic ozone layer is likely to be there beyond 2050 no matter what environmental measures are adopted now.

These global problems have helped, and continue to help, to focus public attention on the environment and how important it is to protect it, even though these two large problems are perhaps beyond the individual to influence to any measurable extent. Nevertheless, people are now aware that they must "do their bit" to protect the environment, whether this is by being careful about waste disposal — who had heard of a bottle bank or biodiesel 35 years ago — or by using their cars less. Nowadays, using unleaded fuel in the family car or supporting so-called organically grown foodstuffs is recognised good practice. This text will only touch on such difficult worldwide environmental problems; instead the focus will be on smaller scale modelling, and these large problems form the context in which many of the smaller models are embedded. In the next section let us look at the modelling process itself.

1.2.2 *The modelling process*

In many books on mathematical modelling, the starting place is the description of some kind of idealised modelling process using as a vehicle some equally idealised problem. The trouble with this is that both students and experienced practitioners alike find this less than convincing. The element of trial and error that seems to be involved in the classical modelling process is unrealistic to the practicing oceanographer, whilst to students of ocean science the whole process looks too ideal, not related to actuality. However, it is the heuristic trial and error side of modelling that makes it so successful in its mimicry of real life. We therefore must design our modelling process particularly with the coastal ocean scientist in mind.

The singular most important aspect of modelling that has led to its recent popularity is the ready availability of cheap but increasingly powerful computers. In the whole of the 1960s and 1970s and the first half of the 1980s, in order to use these computers it was necessary to be able to program them. In order to program them it was necessary to learn a high level computer programming language such as ALGOL in the early days, FORTRAN, PASCAL, C++ (now C#) or any one of a heap of object-oriented programming languages. The details of the programming in turn demand a detailed knowledge of mathematics and the numerical methods that are used to translate the mathematics into the discrete mathematics that computers can use. By their very nature, these programs — utilising as they do powerful computers — are complicated and are based on sophisticated rather than simple mathematics. A requirement for those involved in marine modelling was therefore some knowledge of mathematics, including the calculus that is used to describe the dynamic balances in a fluid and the transport of heat and salt, and the techniques of discretising that in turn demand knowledge of numerical methods. Much of this kind of modelling is still of course going on, but unquestionably it is no longer mandatory to be as close to the mathematics.

Many marine scientists are concerned with models on computers because they wish to answer engineering or environmental impact questions, but they lack the mathematical background to formulate and then program models on computers themselves. Software, the name for a computer program that is commercially available, is only a successful product if it accessible to the majority of likely users. Of course, it must also be useful in terms of producing meaningful results. It is widely recognised that not enough marine scientists have the mathematical background to comprehend the details of today's marine models, and it is indeed fortunate that this is no longer necessary. The very computer power that enables the models themselves to be complex also enables so-called "front ends" to be incorporated into models. These front ends act as an interface between the program and the lay user and enable the lay user to use the program constructively without the need to get involved in the programming itself. One common method is to use English commands to enquire of the user what features he or she wishes to incorporate into a model. In effect the user operates with a series of menus, choosing from a set or answering yes or no to simple questions. In this way, a particular problem can be solved by adapting a complex program through menus and without detailed programming knowledge. This is the general philosophy behind what used to be called "expert systems" but now are simply part of decision support that are these days quite widespread, especially in the medical field.

The general philosophy behind modelling in marine systems can be expressed succinctly in a flow chart of the type shown in Figure 1.1. In the language of systems, ocean science might be thought of as a mixture of soft system and hard system. A soft system is one that is ill-posed and usually involves humans, whereas a hard system is one that is controllable, obeys well formed laws and is by and large amenable to exact mathematical solution. Perhaps a more natural division for marine science is into three classes: natural systems, e.g. biological organisms; artificially designed but

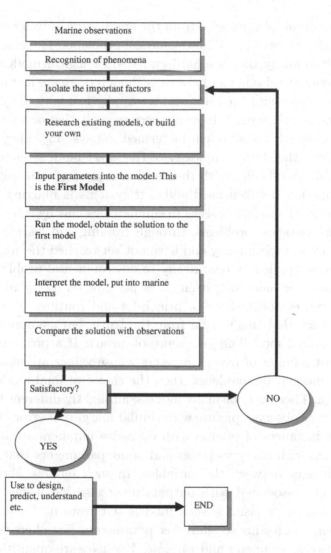

Fig. 1.1 A marine modelling process flow chart.

physical systems e.g. engineering devices; and artificially designed but abstract systems e.g. economic models and models involving scheduling. The kind of modelling indicated by the flow chart of Figure 1.1 mostly fits into the first of these categories, however models of the physics of the sea are very different from biological models. Most models of ocean physics arise from the application of well established laws and lead to well posed mathematical problems. Therefore, in systems methodology they are hard systems. Biological models related to the ocean are also in fact hard as, although there are no universally recognised laws with the same stature as physical laws, they are well posed and solutions exist. It is only very recently that ocean scientists have brushed with what might be termed soft systems. They are only soft because the sheer complexity of the latest models renders exact predictability difficult. With the development of genetic algorithms, the distinction between hard and soft systems is blurring. Related neural network modelling seeks to simulate systems by emulating the way humans tackle problems, through training and learning. These notions are in their infancy and have not yet reached the stage where they can be applied systematically to environmental problems.

Thinking of modelling itself, it is perhaps tempting to describe it as a well established set of principles and routines. It would be wrong to say that this was far from the truth, but as more variables are considered modelling gets more of an art. If a proposed model has a vast number of free parameters whose values are more or less at the behest of the modeller, then the choice of final model is also very large. There may be many models, all slightly different but all of which fit the observed picture with similar margins of error. The most common instances of models with excessive numbers are models of ecosystems with many variables and many parameters that describe the exchanges between the variables. In such models, the level of uncertainty associated with outputs must also inevitably increase. This is not a criticism of such models, far from it. It is merely a consequence of having so many free parameters. Note here the difference between *parameter* and *variable*. *Variables* are quantities whose behaviour we wish to model, whereas *parameters* are quantities that can be estimated either directly or indirectly and are there as a direct consequence of the modelling process. Parameters are not natural quantities, variables are. Of course if there is a perceived fault in the outcome of a model, it is usually a matter of conjecture whether any fault lies in what has been left out of the model, the method the software uses or the interpretation of the output. Perhaps it is the observations themselves that are wrong and the model is actually working. Usually, experience alone tells us the most likely source of the error, and one of the aims of this book is to help you to gain some experience through the eyes of the author.

1.2.3 *Engineering projects and consultancy*

The prime reason engineering projects are of interest in a book such as this is because they produce waste. The environmental scientist views a factory on the shore as a potential hazard. In contrast, the economist will view the same factory as a potential for growth and employment. Most large factories will use water, perhaps as a coolant or more actively in a chemical process, and the only way to ensure that the factory is economically viable is to discharge the water into the nearby river or estuary. In recent times, the recycling of waste has increased due to advances in recycling technologies. Nevertheless, waste products are still produced, which emerge from the factory carried by the waste water. When a new factory is proposed, it is common for various parts of its construction to be put out to tender. This means that various sets of experts will be paid to build sections of it. One section will undoubtedly be the design of the waste disposal, which commonly takes the form of a diffuser to make sure that any waste water remains legal. The more technical aspects of how discharges behave are dealt with in Chapter 7 where diffusers are described. Here, we are concerned with ensuring that any potentially hazardous dissolved substance present in the waste may remain within environmentally acceptable values, as laid down by legislature. In order to do this, the team involved in designing the diffuser often contains someone who is capable of building or acquiring an appropriate mathematical model. This model should be capable of simulating the worst case scenario whereby the levels of the hazard are in some sense maximal. If even this maximum level is legally acceptable, then normally the go ahead for the manufacture of the diffuser as designed can be given. Some of the more technical aspects of plumes are covered in Chapter 7.

The environmental questions that arise from not only the building of factories but many other activities such as sport, the modification of harbours, housing developments, etc. has meant that environmental consultancy is now big business. As the population increases, and man explores and exploits more of the Earth's surface, the environmental scientist assumes the role of a guardian. It is now certainly not optional to consider most carefully environmental questions every time industrial or other unnatural activity impacts on the world around us. Some of the things that were done in the name of industrial advancement in the UK are now seen as very wrong and make us wince. Unfortunately, like activities are still going on in other parts of the world. The framework for the control of this is the law, which forms part of the next section.

1.2.4   *Legal issues and public perception*

When man entered the industrial age back in the early 19th century, he started releasing chemicals into the environment. However, back then the quantities were not large and the sociology and knowledge of environmental chemistry at this time also meant that any discharges were disregarded. More recently, things have certainly changed. The accidental or deliberate release of chemicals into the environment is now a very live issue. When a chemical is so released a whole host of processes ensue. As far as the activity of man is concerned, there are legal processes which will be outlined later. To give a flavour of the scientific processes, first of all the introduced chemical will interact with naturally occurring chemicals in the environment. This interaction could be just mixing, but could also involve chemical reactions with important consequences for the environment. Then there is the physics of how the introduced substance interacts with the environment. In water, here a river, estuary or sea, this could involve diffusion, sedimentation or simply the migration of the substance with local currents. In addition, the substance could be buoyant and contribute to a surface slick or sink and interact with the sediment. Finally the introduced chemical could interact with the biology: the plankton, fish, aquatic reptiles, amphibians, mammals and ultimately man. It is this interaction, almost always the most difficult to forecast, that of course is the most worrying for the public and lies behind the heightened interest in environmental protection. We have to be so very careful as biological consequences are the most difficult of all to predict. To cite a non marine example, who would have thought that an insecticide (DDT) would cause birds that accidentally consumed it to lay eggs that had abnormally thin shells. The birds of prey were more interested in the small mammals and birds that ate the insects that absorbed the DDT, so this was a chain. These shells could not withstand the weight of the incubating adult, hence the population of peregrine falcons, in particular, was decimated. In the medical field, we all now know the consequences of a pregnant woman taking the anti morning sickness drug thalidomide. It is this kind of worry that is behind the reaction of the general public — or do I mean the popular UK press — to genetically modified (GM) crops. A public, it must be remembered, which is still smarting and facing uncertainty over bovine spongiform encephalopathy (BSE) and its relationship to new variant CJD (Creutzfeldt–Jakob disease) which, although numbers seem to have peaked in 2003, is still potentially a serious worry. Then of course there is the use of the word "nuclear". A technology that has tremendous potential but the name of which has associated with it the erroneous millstone of being linked with bombs and destruction as well as the not so erroneous problem of radioactive waste disposal.

In recent times, the control and management of any waste products of industry have become increasingly governed by legislation. Pollution laws, as they are colloquially known, tend to be different in different regions of the world. In the USA, the laws are usually very strict in terms of health, but less so in environmental protection terms. In Singapore, all environmental laws are very strict indeed. In the old eastern block countries, notoriously East Germany as it tried to build its industry up from the ravages of the Second World War, environmental legislation was virtually non-existent and ignored even when it was there. Even in the UK, pollution laws can be different in England and Wales than in Scotland and Northern Ireland. Legal principles are there to protect the interests of those who can be threatened or damaged by pollution. The legal process is notoriously expensive and often tortuous, therefore it is in everyone's interest, apart perhaps from the lawyers, not to go to court. Most cases are thus settled by insurers. Some large cases however need a national forum and usually arise because they are in some way new. Obvious examples are the large tanker accidents the Torrey Canyon (in 1967) and the Amoco Cadiz (in 1978) and more recently the Exxon Valdez, 1989 and Gulf of Mexico blow-out, 2010, where liability was contested, not surprisingly given the extent of the environmental disaster in each case. On a smaller scale, the liability for minor slicks caused by the flushing out of "empty" tanks lies squarely with the captain and the only problem is catching him (or her). More relevant to modelling is the legislation that exists to prevent more than particular concentrations of certain chemicals in the sea. This is the maritime equivalent to monitoring lead in car emissions, which led to the development of lead-free petrol and the catalytic converter. Examples of what is meant in the maritime context are the pollution caused by painting boats with anti fungal paint (tributyl-tin), or the control of heavy metals that arise from chemical process factories — lead (again), mercury and cadmium enhanced levels of which can occur in the waste water which is discharged into the nearby river or estuary. Experts in the biochemical effects of toxic chemicals usually formulate levels of chemicals that are deemed reasonable to tolerate, and if these are exceeded, prosecutions ensue. As new processes are developed, these are scrutinised and the legislation is modified accordingly. The enforcers of the legislation vary. Sometimes it is a national body such as the Environmental Agency River Authority (in England and Wales) or the River Purification Boards (in Scotland), sometimes the transgressor is in breach of a law such as the Environmental Protection Act in which case the police can be involved. Successful prosecution can result in the closing of a factory and the fining or imprisonment of offenders. As authorities become convinced that industrial concerns can consistently meet minimum standards, they are able to issue some kind of licensing agreement. This grants the licensee to manufacture.

In the international sphere, there are international treaties that all signatories obey. Examples of this include the treaty that preserves Antarctica for scientific study and the Treaty of Rome, which set up the fundamental legislature for the European Union. Over the past ten years, the law has become more complex in Europe as EU legislation increases. In the maritime field there have always been complications due to offences taking place in international waters or there being at least three nations involved; one which owns the offending material, usually oil, one which owns the container oil tanker, or the country of origin of the captain and the country to which the territorial waters in which the act occurred belong. To this complex picture needs to be added the question of market forces. It is still broadly true that the small guy fighting the multi-million pound company loses.

In the UK case, law still plays an important part. That is, the court is a very powerful body, and if a case sets a precedent — perhaps in the level of compensation paid by a company that has caused bodily harm for example — this is cited in similar future cases. Some kind of convergence then occurs if there are enough cases. This is the power of "common law".

The big question of global warming is causing meetings with high profile politicians, starting with Kyoto 1997 and Rio 1998, then Montreal 2005; in Copenhagen in 2009 once more there was no general agreement, but a statement saying that the increase in temperature should be kept to under 2°C. This figure was arrived at from a compromise dating back to 1990 but has no real scientific basis. At 2°C the scientific community say that by 2050 the world would be seriously compromised. In South Africa in 2012 a conference announced that it would take at least 10 more years to reach a robust agreement; this sounds optimistic. The reason no treaty worth the name has emerged is due to the partisan interests of some of the delegates, contradictory differences between environmental desires and local political reality; they are irreconcilable. A few years ago, the representatives of the Florida Keys community joined forces with other island communities around the world (Fiji, Samoa, etc.) in the hope of providing a convincing lobby to the USA industries who are widely recognised as main contributors to the industrial gases that cause global warming. Then there are the two large emerging economies of China and India that are altering the balance of the world economy and have to be part of any global warming legislation. If the sea level rises continue, then island communities will be no more in a century or two. Despite the current preoccupation with terrorism, global warming is probably our biggest long term problem and it is a very difficult one to solve.

Given the very long time scales, the signs that mankind can reverse the effects of global warming, or the effects of ozone depletion for which there is more agreement, are not good. There are some very recent encouraging signs, but not enough. One convincing strategy that will help is to provide accurate models, and to do this the underlying processes have to be understood. This is what this textbook is about.

1.3   Mathematical Preliminaries

The basis of most models lies in the mathematical description of the laws obeyed by the thing being modelled. For coastal engineers and marine scientists this is most often the sea, though it could be a marine ecosystem (see Chapter 9). The sea is a fluid (salty water) and so the equations obeyed by the sea are those of fluid mechanics. This presents us with a problem as the mathematical description of a fluid is not something one meets outside quite advanced courses. Moreover, the sea is a fluid with dissolved substances, changes in temperature plus, most of the time, turbulence. So, what we attempt to do here is difficult, but not impossible. It amounts to giving some mathematical background in order that the derivation of the equations in Chapter 3 can be understood. The kind of mathematics required comes under the heading of "vector calculus" or sometimes "advanced calculus". This presupposes that everyone knows about calculus of course, which might not be the case.

1.3.1   *Calculus*

Calculus was developed in the 17th century simultaneously — being diplomatic — by Newton in England and Leibniz in Germany. It is Leibniz' description that is followed nowadays. It is a powerful tool for studying things that change; the word "calculus" derives from the Latin for "a small stone" which were used to help in calculations in

ancient times. The question is how much do we need to know about
calculus in general. One useful definition is that of the derivative.
The usual way to define a derivative is graphically in terms of tan-
gents and this is alluded to when we outline numerical methods in
Chapter 4. However, we need calculus to describe rates of change, so
it is this definition that wins here. In mathematics one uses letters
to stand for the values of quantities. If a quantity remains the same
everywhere and for all time, it is a constant. One can think of g the
acceleration due to Earth's gravity as a constant, usually taken as
9.81 in SI units (ms^{-2}). On the other hand, if a quantity varies it is
called a *variable* and it is also given a letter. This letter is sometimes
followed by parentheses containing the quantities it varies with. For
example, if a quantity is labelled $u(x, y, z, t)$ it depends on where it is,
the co-ordinates (x, y, z) and when it is time t, and it will in general
change with time and space. The letter u usually denotes the current
in an easterly direction, and it needs no leap of imagination to realise
that this current will change in time even if we consider a fixed point
in space. Think of the current due to the tide at a particular location
as an example. The easterly component of this current will also be
different depending on where it is measured. There are thus different
rates of change for u; four in fact, one each for x, y, z and t. We cater
for this in words by writing "the rate of change of u *with respect to*
t", or x or whatever. The rate of change of u at a particular fixed
location with respect to t is the derivative of u with respect to t and
is written

$$\frac{\partial u}{\partial t}.$$

It has the definition

$$\frac{\partial u}{\partial t} = \lim_{\Delta t \to 0} \left\{ \frac{u(x, y, z, t + \Delta t) - u(x, y, z, t)}{\Delta t} \right\}.$$

Note that only t is varying; this is the reason for the curly style of
the "d" and it means that everything except t is being kept constant.
In books on calculus, which are normally dauntingly large, there are
usually a lot of practice examples and exercises on finding the rates
of change, first of functions like $u(t) = t^2$ where the "d" really is

just d and not curly as there is only the one variable. This is the derivative section of the calculus of one variable and usually comes first. Later — usually much later — there will be examples such as $u(x, y, z, t) = x^2 + y^2 + z^2 - t^2$. Although it is useful to know how to find rates of change of various functions like polynomials, exponential and trigonometric functions, this is not the point here. This section cannot hope to duplicate large texts on the calculus. If the value t^2 is actually substituted for u in the right-hand side, the limit takes the form

$$\lim_{\Delta t \to 0} \left\{ \frac{(t + \Delta t)^2 - t^2}{\Delta t} \right\} = \lim_{\Delta t \to 0} \left\{ \frac{(\Delta t)^2 + 2t\Delta t}{\Delta t} \right\} = 2t.$$

Taking this kind of limit gives all the basic rules of the calculus found in the textbooks and now available on your PC or local area network, provided you have paid for the software MAPLE or MATHEMATICA, which are very powerful, but there are small calculators that can do symbolic calculus and algebra. These are usually banned from examinations, but this is a book not an examination. So, we have introduced what is called differentiation — that is taking this limit and finding rates of change — and it is the meaning of the partial derivative — a rate of change of a quantity with respect to a single variable, the others remaining constant — that is crucial to the understanding of the balances derived in Chapter 3.

The inverse of addition is subtraction; the inverse of multiplication is division; so the inverse of differentiation is integration. The mechanics of subtraction are more difficult than the mechanics of addition; the mechanics of division are more difficult than the mechanics of multiplication; so the mechanics of integration are more difficult than the mechanics of differentiation. Indeed, sometimes it is impossible to find certain integrals whereas it is always possible to find derivatives, provided they exist — there are strange functions that do not have derivatives, but we shall not encounter them very much in this text. Do we need to bother about integration for our modelling? The answer is not very much; we really only need to understand the symbols, not to find actual integrals. Although, actually doing the integration is a good aid to understanding. It is the difference between playing with a dog (doing) and seeing its picture (just looking). The symbol for the integral of a function $u(t)$ is

$$\int u(t)dt.$$

The symbol derives from the letter "S", which is elongated and stylised but stands for "summation". There are, unfortunately, several different kinds of integration, though they are all the inverse of differentiation in some sense, and all involve summing over some domain or other. The integral of $2t$ will therefore be t^2 although we add a constant c, as the derivative of a constant is zero. The derivative of $t^2 + c$ is $2t$, no matter what the value of c. The physical interpretation of the first type of integration usually encountered by students involves determining the area under a curve. This is the easiest to understand, but unfortunately it is not the one we need here. In this text, we take a curve and divide it up into infinitesimally small straight bits, then sum all these infinitely many bits. Consider, as an example, the density of the sea, given the standard symbol $\rho(x, y, z)$ and ignoring for the moment that it might also depend on time as well as position. The integral of this density from one end of a given path to the other is then the mass of this worm-like solid and the summed quantity called a *line integral*. The integrals we actually need in Chapter 3 involve summing over surfaces and volumes rather than just lengths. The area or volume is divided into small chunks and summed over, but the principle is the same and this time the result of the integral is the mass of the surface or volume. The actual rigorous definition stems from the one-dimensional definition of the line integral. The integral written

$$\int_{P_1}^{P_2} \rho(x, y, z)ds$$

can be interpreted as follows. Suppose a thin wire is in the form of a curve, and this curve starts at point P_1 and ends at point P_2. The ds represents a very small (infinitesimal) arc length which is so tiny that the density is constant along it. There is a limiting process at work here, and in this limit we can say that ds has in fact zero length so that, at each *point* along the wire, it has a density $\rho(x, y, z)$ that depends upon x, y and z and is therefore a function of position given by the Cartesian co-ordinates (x, y, z). The integral is the total mass of the wire. If $\rho = $ constant, then the mass is simply this constant times the length of the wire, so the integral is adding up all the little arc lengths ds. In this case the mass is the density times the length as expected. In reality, we need to multiply by the very small and constant cross-sectional area of the wire to achieve "mass = volume × density" of course. Rather than specify the end points as P_1 and P_2, the curve is often represented by the letter C and the integral is written

$$\int_C \rho(x, y, z) ds.$$

If the curve is closed, that is it is a loop without an end, there is a special symbol

$$\oint_C \rho(x, y, z) ds.$$

Now you may be wondering how on earth such objects are evaluated. That is, given some functional form of ρ, say $x^2 + y^2 + z^4/a^2$ or some such, how do we find these integrals? Fortunately, we do not really need to know, but let us just play with the dog a little and do a single simple example. The trick used involves working on the equation of the shape that describes the curve of the wire, then parameterising it in terms of mathematics. For example, a circle of unit radius lying in the (x, y) plane, centre the origin, can be parameterised by $x = \cos\theta$, $y = \sin\theta$, $z = 0$ with θ taking the range $0 \le \theta < 2\pi$. If this last bit is gobbledygook, it does not really matter. If you really want to work through this example, just spend a little time with a textbook to get the appropriate technical details. It is reinforced that such technical details are not actually essential here. Just feel happy if you know about such things, because they do help with analogy and insight. The actual calculation for this example is as follows. If the density is $x^2 + y^2 + z^4/a^2$ on the wire itself, this can be written in terms of θ; the notation is displayed as Figure 1.2. Of course $z = 0$, but we can substitute $x = \cos\theta$ and $y = \sin\theta$. Additionally the arc length of the unit circle is simply $ds = 1 \cdot d\theta$. The limits of the integral will be $\theta = 0$ and $\theta = 2\pi$, geometrically the same point of course,

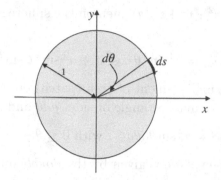

Fig. 1.2   The unit circular shaped wire.

but then the wire is a closed curve, a unit circle. The integral (mass of the wire) is thus

$$\int_0^{2\pi} (\cos^2 \theta + \sin^2 \theta)d\theta,$$

which actually has the value 2π using $\sin^2 \theta + \cos^2 \theta = 1$, which follows as Pythagoras' theorem once the triangle definitions of sine and cosine are used. If the curve was a little more twisted or had the odd kink or was actually three dimensional like a coiled spring, then you can see that the technicalities of evaluating such integrals can get overwhelming. The important feature to glean is that the density is evaluated *on the curve* and that it is the *shape of the curve*, and not the density, that determines the limits of the integral. This is useful to remember in Chapter 3 and in particular when Finite Elements are described in Chapter 4. So, in this way, it is possible to calculate this kind of integral. What about integrals that find the weight of surfaces and volumes? The principle is exactly the same. For a surface, we integrate twice and for a volume we integrate three times, the number of times coinciding with the dimension of the object. In mathematical terms, a surface needs *two* parameters to describe it, so when the density is expressed as $\rho(x, y, z)$, it becomes $\rho(\theta, \phi)$ *on the surface* because the surface itself will be described by a set of equations that relate x, y and z to θ and ϕ. For example, the surface of a sphere $x^2 + y^2 + z^2 = a^2$ is parameterised by $x = a \cos \phi \sin \theta$, $y = a \sin \phi \sin \theta$ and $z = a \cos \theta$. This means that if $\rho(x, y, z)$ retains the previous algebraic form of $x^2 + y^2 + z^4/a^2$ then on the sphere $x^2 + y^2 + z^2 = a^2$ and we have

$$\rho(x, y, z) = \rho(\theta, \phi) = a^2 \sin^2 \theta \cos^2 \phi + a^2 \sin^2 \theta \sin^2 \phi + a^2 \cos^4 \theta.$$

Additionally, the small element of area (patch) on the surface of the sphere is an infinitesimal rectangle of sides $ad\theta$ and $a \sin \theta d\theta$. Hence,

$$dS = a^2 \sin \theta d\theta d\phi \quad \text{with } 0 \leq \theta \leq \pi.$$

So, the mass of the sphere is given by the *double* integral

$$\int \int \rho(\theta, \phi) a^2 \sin \theta d\theta d\phi,$$

which is

$$\int_0^\pi \int_0^{2\pi} (a^4 \sin^3 \theta \cos^2 \phi + a^4 \sin^3 \theta \sin^2 \phi + a^4 \cos^4 \theta \sin \theta) d\theta d\phi.$$

To avoid cancellation due to spurious negative densities, we take the integral over only $1/8$ of the sphere's surface, that corresponding to positive values of x, y and z, then multiply by eight using symmetry

$$8 \int_0^{\pi/2} \int_0^{\pi/2} (a^4 \sin^3 \theta \cos^2 \phi + a^4 \sin^3 \theta \sin^2 \phi + a^4 \cos^4 \theta \sin \theta) d\theta d\phi.$$

This will not be evaluated, but a few comments are worth making. The first is about notation. The initial limits of integration ensure that the surface is swept out just once. Think of the sphere as the Earth; then θ is the co-latitude (90-latitude) so the range of θ sweeps out a line of longitude from north to south pole (180° or π radians). Still with the first double integral, the inner integral's limits take this semicircle and sweep it around the globe a full circle (360° or 2π radians) to form the sphere. The second comment is that the order of integration is dictated by the order in which the surface is described. The convention is to work from the left, so we integrate with respect to θ first (holding ϕ constant), then we integrate with respect to ϕ. By the second integration, all the θs have disappeared. This order is also indicated by the order of the $d\theta$ and $d\phi$ at the end of the expression. We could evaluate this now of course, but by not doing so emphasises that it is not the mathematical technicalities that are important but understanding what the notation means. In fact, for this integral the ϕ integral is very easy as we use $\cos^2\phi + \sin^2\phi = 1$, and for those interested, the final result is $44\pi a^4/15$. The extension to three dimensions is straightforward; we have three integrals to evaluate. Here it is hardly worth parameterising, and integrating first with respect to x then y then z can be done directly. In general, this will look like

$$\int_{f(x_0,y,z)}^{f(x_1,y,z)} \int_{g(y_0,z)}^{g(y_1,z)} \int_{z_0}^{z_1} \rho(x,y,z)\,dxdydz.$$

Again the first integral is with respect to x holding y and z constant. Once the integration has been done and the limits (the f functions) have been inserted, all xs have disappeared and only y and z remain. The second integral is with respect to y, and similarly when the integration is done and the limits inserted (the g functions) only z remains. Finally, the third integration takes place and the limits are z_0 and z_1 and we get a number answer. The order is the order of dx, dy and dz. Some books work the integrals from the inside out, but this is considered old fashioned and poor practice by mathematicians these days. That is enough calculus.

1.3.2 *Vectors*

The other branch of mathematics that needs our attention is vectors. A vector is the name given to a quantity that has direction as well as magnitude. Quantities that only have magnitude and do not have a direction are called scalars. Here are some examples of vectors: force, ocean current and wind. All of these have both direction and magnitude. On the other hand, density, temperature and salinity are scalars because they only have magnitude. In the above discussion of calculus we only integrated scalars, specifically the density, so what about vectors? It turns out that they too can be differentiated and integrated, but first we need notation. Vectors are indicated in one of three ways. If a single letter is to indicate a vector then it is either underlined, $\underline{a}$, or it is written in boldface, **a**. Underlining vectors really belongs in the classroom, so we shall use boldface characters to indicate vector quantities. Geometric vectors are denoted by $\overrightarrow{AB}$, the vector joining point A to point B, but these are not used in this text. The third way of indicating vectors is by giving its three components. Components are scalar quantities and represent how the vector quantity is made up. For example, we might write $\mathbf{u} = (u, v, w)$ to indicate that the current $\mathbf{u}$ has components u, v and w. The double use of the letter u is not confusing as one is a scalar (the component) and the other a vector. Components are simply the proportion of the vector in each of three perpendicular directions. Commonly, x is east, y is north and z is up. So a wind can be specified by how much is due east, how much is due north and how much is vertical (usually very small). Therefore a north east wind conventionally coming from the north–east will have equal x and y components, and a zero z component. In mathematical terms the letters $\mathbf{i}, \mathbf{j}$ and $\mathbf{k}$ are used to denote vectors in the x, y and z directions with magnitude one, called unit magnitude; the vectors of unit magnitude are called *unit vectors* and are a very handy device. We certainly need them in Chapter 3 when deriving the Coriolis term due to the rotation of the Earth. It means that we have an alternative and easier way of writing vectors in terms of their components. For example,

$$\mathbf{u} = (u, v, w) = u\mathbf{i} + v\mathbf{j} + w\mathbf{k}$$

and the right-hand side can be manipulated more easily than the middle triple, provided we know some basic rules for adding and multiplying the unit vectors. To add or subtract two vectors, we simply add or subtract the components; nothing could be more straightforward. Here, is a simple example

$$\mathbf{i} + 2\mathbf{j} + 3\mathbf{k} + \mathbf{i} + 3\mathbf{j} - \mathbf{k} = (1+1)\mathbf{i} + (2+3)\mathbf{j} + (3-1)\mathbf{k} = 2\mathbf{i} + 5\mathbf{j} + 2\mathbf{k}.$$

Again, there is no time to really explore the algebra of vectors here; they can be used very fruitfully in geometry and a modern application is in visualization and computer animation. To multiply vectors is less straightforward. In fact, there are two different types of product: the scalar product that gives rise to a scalar and the vector product that gives rise to a vector. The definitions here are given in terms of components for two reasons. First they are the only ones used, second they are the easiest to grasp.

The scalar or "dot" product of two vectors $a_1\mathbf{i} + a_2\mathbf{j} + a_3\mathbf{k}$ and $b_1\mathbf{i} + b_2\mathbf{j} + b_3\mathbf{k}$ is $a_1b_1 + a_2b_2 + a_3b_3$. In other words (symbols in fact),

$$\mathbf{a} \cdot \mathbf{b} = a_1b_1 + a_2b_2 + a_3b_3.$$

One usually does lots of elementary examples in much the same way as when one meets quadratic equations for the first time, but this is not done here for reasons of space as well as relevance. For us, it is more important to know the definition and one or two important facts. One of these is that if the dot product of a pair of vectors is zero then either one of the vectors is zero or they are at right angles to each other. This is called *orthogonality* and the scalar product is a special case of an *inner product*. More is made of these concepts once vectors have been dealt with. The vector or "cross" product is written $\mathbf{a} \times \mathbf{b}$ and has the definition

$$\mathbf{a} \times \mathbf{b} = \begin{vmatrix} \mathbf{i} & \mathbf{j} & \mathbf{k} \\ a_1 & a_2 & a_3 \\ b_1 & b_2 & b_3 \end{vmatrix}.$$

For those who are not familiar with determinants, the right-hand side multiplies out to

$$(a_2b_3 - b_2a_3)\mathbf{i} + (a_3b_1 - b_3a_1)\mathbf{j} + (a_1b_2 - b_1a_2)\mathbf{k}.$$

The determinant is merely a convenient notation that displays the symmetry. There are only a few important properties of cross products needed here. First of all $\mathbf{a} \times \mathbf{b} = -\mathbf{b} \times \mathbf{a}$, secondly the direction of $\mathbf{a} \times \mathbf{b}$ is perpendicular to both $\mathbf{a}$ and $\mathbf{b}$ so as to form a right-handed system. Using the right hand, if the thumb is aligned with $\mathbf{a}$ and the forefinger aligned with $\mathbf{b}$ then the cross product $\mathbf{a} \times \mathbf{b}$ is in the direction of the second finger if this is held at right angles to thumb and forefinger. The other properties and all the algebraic examples usually associated with this branch of mathematics will be glossed over. Finally, mention needs to be made of the triple products. The scalar triple product of three vectors $\mathbf{a}$, $\mathbf{b}$ and $\mathbf{c}$ is

$$\mathbf{a} \cdot (\mathbf{b} \times \mathbf{c}) = \begin{vmatrix} a_1 & a_2 & a_3 \\ b_1 & b_2 & b_3 \\ c_1 & c_2 & c_3 \end{vmatrix},$$

and the result is preserved if either **a**, **b** or **c** are cyclicly permuted; so the above result is the same as **b**·(**c** × **a**) and **c**·(**a** × **b**). It represents the area of a parallelepiped with three adjacent sides as the three vectors; a parallelepiped is a solid whose plane faces are parallelograms. The final object needed in this whistle stop tour of vector algebra is the vector triple product, and this is best introduced through the formula

$$\mathbf{a} \times (\mathbf{b} \times \mathbf{c}) = (\mathbf{c} \cdot \mathbf{a})\mathbf{b} - (\mathbf{a} \cdot \mathbf{b})\mathbf{c}.$$

With this vector product not only is the order important, but the placement of the parentheses also changes its value; mathematicians say that the expression is not associative. The only time this vector triple product is met is if the centripetal acceleration due to the Earth's rotation ever needs to be calculated (see Chapter 3). The term is $\mathbf{\Omega} \times (\mathbf{\Omega} \times \mathbf{r})$ where $\mathbf{\Omega}$ is the angular velocity of the Earth, and **r** is the vector pointing out perpendicularly from the Earth's axis to the surface. Using the formula above, and that the dot product of two vectors that are at right angles is zero this gives

$$\mathbf{\Omega} \times (\mathbf{\Omega} \times \mathbf{r}) = -\Omega^2 \mathbf{r}.$$

So we have gone through calculus and skated through vector algebra. The mathematics required to fully understand Chapter 3 combines these in the form of vector calculus. As a simple example, the integral

$$\int_C \mathbf{u} \cdot d\mathbf{r}$$

represents the (infinitesimal) quantity **u**.$d\mathbf{r}$ summed along the curve C. In fluid mechanics this quantity is called the circulation. It combines the scalar product and the notion of a line integral. The evaluation of such objects can be simplified through the use of vector identities that relate the three vector derivatives called "grad", "div" and "curl". These are defined now. First we have grad ϕ or, in full, the gradient of ϕ; this is

$$\nabla\phi = \mathbf{i}\frac{\partial\phi}{\partial x} + \mathbf{j}\frac{\partial\phi}{\partial y} + \mathbf{k}\frac{\partial\phi}{\partial z}$$

and is a vector. The operator ∇ is called the gradient operator and calculates a spatial gradient of a scalar such as temperature or density. A little thought will tell you that such gradients must have a direction as well as magnitude, so that they are vectors is not surprising. The next definition is for "div", or the divergence; this is

$$\nabla \cdot \mathbf{u} = \frac{\partial u}{\partial x} + \frac{\partial v}{\partial y} + \frac{\partial w}{\partial z},$$

where we have used the components (u, v, w) of the vector $\mathbf{u}$. The physical meaning of this is as follows. The vector $\mathbf{u}$ has a magnitude and direction at every point of a three-dimensional domain, much like a wind or sea current. The quantity $\nabla \cdot \mathbf{u}$ is the amount of $\mathbf{u}$ being created at each point. Integrated over a volume (domain) it represents how much is being created inside this domain. For the sea, where $\mathbf{u}$ is current, or the air, where $\mathbf{u}$ is wind, this is zero because mass can neither be created nor destroyed. We meet this in Chapter 3. Finally we define the "curl" of a vector. The name is not short for anything this time. The definition is

$$\nabla \times \mathbf{u} = \begin{vmatrix} \mathbf{i} & \mathbf{j} & \mathbf{k} \\ \dfrac{\partial}{\partial x} & \dfrac{\partial}{\partial y} & \dfrac{\partial}{\partial z} \\ u & v & w \end{vmatrix}.$$

Written out without using determinants it is

$$\nabla \times \mathbf{u} = \mathbf{i}\left(\frac{\partial w}{\partial y} - \frac{\partial v}{\partial z}\right) + \mathbf{j}\left(\frac{\partial u}{\partial z} - \frac{\partial w}{\partial x}\right) + \mathbf{k}\left(\frac{\partial v}{\partial x} - \frac{\partial u}{\partial y}\right).$$

It is not intuitively obvious what this might represent. However, if $\mathbf{u}$ is the current then it is the fluid equivalent of angular momentum in mechanics and represents vorticity or twisting motion. Perhaps this can best be inferred in a two-dimensional sense by looking at the $\mathbf{k}$ or z component, which is

$$\frac{\partial v}{\partial x} - \frac{\partial u}{\partial y}.$$

If the northerly current v increases as we travel east in the x direction and $u = 0$ then there could be an anticlockwise tendency, which following the right-hand rule, implies a direction upwards (out of the paper in Figure 1.3), that is in the z direction consistent with the

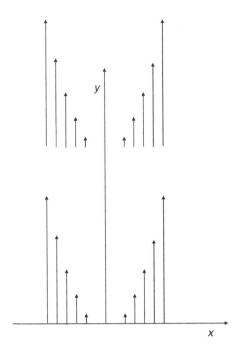

Fig. 1.3   The field $x^2\mathbf{j}$ expressed pictorially.

definition of curl. Maybe an actual example helps: let us calculate the curl of a vector that is given by the expression

$$\mathbf{u} = x^2\mathbf{j}.$$

This corresponds to a vector that is entirely in the y direction (northwards) but increases in magnitude as we travel east provided x is positive. The curl of this vector or vector field, as it is correctly called, is

$$\boldsymbol{\nabla} \times \mathbf{u} = \mathbf{k}\left(\frac{\partial v}{\partial x} - \frac{\partial u}{\partial y}\right) = \mathbf{k}\frac{\partial v}{\partial x} = 2x\mathbf{k}.$$

We see that this is in the z direction, i.e. upwards and also increases with x so long as this is positive. If x is negative then the curl reverses direction and points downwards. This is because although $\mathbf{u}$ increases with x positive it also increases with x negative. Figure 1.3 indicates what is happening here; it displays two manifestations of this vector field, though in reality *every* point of the plane — in fact every point of three-dimensional space — will have an arrow the length of which is x^2 and the direction of which is northwards, but this cannot easily be drawn. Note that along the y (or $\mathbf{j}$) axis the arrows have zero length because $x = 0$ there.

Now there is a result that is important but has nothing to do with calculating these vector derivatives. To get to grips with it we need the definitions of grad and curl. Suppose the current vector $\mathbf{u}$ is of the form of the gradient of some scalar, i.e.

$$\mathbf{u} = \boldsymbol{\nabla}\phi.$$

This means that the components of $\mathbf{u}$ are

$$u = \frac{\partial \phi}{\partial x} \quad v = \frac{\partial \phi}{\partial y} \quad w = \frac{\partial \phi}{\partial z}.$$

If these expressions for u, v and w are substituted into $\boldsymbol{\nabla} \times \mathbf{u}$, each component becomes zero, and we deduce that

$$\boldsymbol{\nabla} \times \mathbf{u} = \mathbf{0}.$$

Fluids that have velocity vectors that have zero curl are called *irrotational* and have an important place in fluid mechanics. Although irrotational fluids are in some sense idealised, their properties do help in the general understanding of how a fluid behaves. It is tempting to believe that the mathematician's concentration on delving into the properties of irrotational fluid mechanics stems from the elegance of the mathematics rather than any practical application. There may be some truth in this, but we will see in Chapter 3 how useful irrotational fluid mechanics can be in studying surface water waves. Advances in aerodynamics could not have occurred without irrotational flow theory, which in this application explains simply and elegantly how flow around an aerofoil generates lift; enough lift for aeroplanes to fly. Its role in coastal engineering and near-shore oceanography is less prominent, but it is there. It turns out that if we have an irrotational flow with its zero curl, then the velocity vector $\mathbf{u}$ must be of the form $\boldsymbol{\nabla}\phi$, but this is more awkward to prove — it is not proved here. Physically, the absence of vorticity in a fluid is closely linked with the neglect of friction. If a current is adjacent to a coast, then frictional forces will act and we expect the current to decrease in magnitude as the coast is approached. This is tantamount to injecting vorticity at the coast. If the current is allowed to slip freely against the coast without reduction — think of a straight coast here with a vertical wall — then the vorticity will be zero. These notions will be picked up again, first in Chapter 3 then again in later applications.

The final topic in this brief run through relevant mathematics is again conceptual rather than technical. In fluid mechanics, we deal with volumes of fluid that are representative of the fluid as a whole. These volumes are arbitrary and we consider them in the light of getting the fluid to obey physical laws such as Newton's second law of motion or the conservation of mass. In order to do this, the properties of the fluid are summed (integrated) over the volume, and this integral can represent a useful quantity such as the momentum or mass of the arbitrary piece of fluid. There are relations between such a volume integral, as it is called, and the integral over the surrounding closed surface. The relationship from which others can be derived is called Gauss' flux theorem, named after perhaps one of the cleverest mathematicians of all time and the founder of the Göttingen School of Mathematics in Germany, Carl Friedrich Gauss (1777–1855). In terms of the current $\mathbf{u}$, it takes the form

$$\int_V \boldsymbol{\nabla} \cdot \mathbf{u} \, dV = \int_S \mathbf{u} \cdot d\mathbf{S}.$$

This has a simple physical interpretation; it states that the amount of fluid being created inside the volume V must be equal to the flux of fluid across the surface S of the same volume. Some may worry that $d\mathbf{S}$ is a vector rather than the scalar that S is, but $d\mathbf{S}$ is an infinitesimally small, therefore flat, bit of area and it is a vector because it has a direction, namely the direction of the normal to it conventionally drawn out of the volume. The dot product between $\mathbf{u}$ and $d\mathbf{S}$ indicates the component of $\mathbf{u}$ out of the surface and integrating this over the entire surface therefore gives the flux.

There is also a relation between the vorticity summed over an open surface S and the flow around its bounding curve. This is called Stokes' theorem, after George Gabriel Stokes (1819–1903) who was born in Ireland and became one of the fathers of fluid mechanics, particularly water wave theory. Stated baldly, this is

$$\int_S \nabla \times \mathbf{u} \cdot d\mathbf{S} = \int_C \mathbf{u} \cdot d\mathbf{s}.$$

This is valid for any surface with a bounding curve, but it is often used for plane surfaces in which case it becomes the non-vectorial Green's theorem in the plane, which is perhaps a bit easier to swallow:

$$\int_S \left(\frac{\partial v}{\partial x} - \frac{\partial u}{\partial y} \right) dS = \int_C (u\,dx + v\,dy).$$

George Green (1793–1841) was a self-educated Nottinghamshire miller who did some very original mathematics, graduated eventually at age 42 and whose remarkable story reads like some kind of fairy tale until his tragically early death from influenza at 47. This is one of several formulae attributed to him and can be derived by directly integrating the left-hand side, but what does it mean? The right-hand side is called the circulation of the fluid around C, the left-hand side is the flux of the vorticity through the surface S. The upshot of Stokes' theorem for a fluid is that vorticity tends to be conserved in a fluid. A theorem called the Kelvin circulation theorem states that the circulation does not change in time if the curve C moves with the flow, so Stokes' theorem leads to the same conclusion for the flux of vorticity. Lord Kelvin was Sir William Thomson (1824–1907) another Irishman and another of the pioneers of fluid mechanics and much else besides, for example electromagnetism; but a terrible teacher by all accounts. This leads on to considerations of vortex shedding and quantifying lift around bodies such as aerofoils, so taking us well into fluid mechanics and beyond the scope of what we can do here. If you are interested in such things, the book by David Acheson (Acheson, 1990) is really good. However, instead of pursuing this, we shall now turn to statistics.

1.3.3 *Linear algebra*

It is beyond doubt that any vector can be expressed as a combination of the three unit vectors $\mathbf{i}, \mathbf{j}$ and $\mathbf{k}$, but why is this? Mathematicians say that these three unit vectors form a *basis* for all the vectors. This means precisely that any vector can be expressed as a linear sum or combination of $\mathbf{i}, \mathbf{j}$ and $\mathbf{k}$. Furthermore, it also states that this combination is unique. Once a linear combination has been found for a vector $\mathbf{q}$ such as

$$\mathbf{q} = q_1 \mathbf{i} + q_2 \mathbf{j} + q_3 \mathbf{k}$$

then the three numbers (components) q_1, q_2, q_3 are uniquely defined. Mathematicians generalise this concept, and we use this generalisation later in the book when finite elements are explained. Instead of vectors, we talk about *elements* or *members* of a vector space. These elements might be vectors, but they could be functions such as $\phi_i(x)$. In fact, it is elements of a vector space that are functions that we use later. In this kind of vector space there is no restriction to just three dimensions. There may be hundreds of dimensions or even, *in extremis*, infinitely many. Our basis here is a sequence of usually simple functions; we will call them

$$\phi_1(x), \phi_2(x), \phi_3(x), \phi_4(x), \phi_5(x), \ldots, \phi_n(x)$$

and we will not have the integer n infinite for the moment. Given that this is a basis for all functions $F(x)$, then we can write

$$F(x) = \sum_{i=1}^{n} a_i \phi_i(x),$$

where the numbers a_i are the components of $F(x)$ in this particular vector space. Usually in coastal and marine modelling the function $F(x)$ is a current in a coastal sea so the domain is a well-defined area and the basis functions $\phi_i(x)$ are defined over the same domain. Applications are normally to functions of two (x, y) variables, but the principles are the same. A simple example would be that any polynomial of degree say five is written as

$$P(x) = a_0 + a_1 x + a_2 x^2 + a_3 x^3 + a_4 x^4 + a_5 x^5.$$

Here, the basis is the set of simple functions $1, x, x^2, x^3, x^4, x^5$ and the components of the function (polynomial in this case) $p(x)$ are $(a_0, a_1, a_2, a_3, a_4, a_5)$. The number of components in the basis is six so the *dimension* of the *vector space* in this case is also six. The number 1 should be written as x^0 in the basis function set for mathematical neatness. Suppose there are n functions $F_n(x)$ with each function expressed in terms of the basis $\phi_1(x), \phi_2(x), \phi_3(x)$, $\phi_4(x), \phi_5(x), \ldots, \phi_n(x)$. The following matrix equation expresses this succinctly

$$
\begin{pmatrix} F_1(x) \\ F_2(x) \\ F_3(x) \\ \vdots \\ F_n(x) \end{pmatrix} = \begin{pmatrix} a_{11} & a_{12} & a_{13} & \cdots & a_{1n} \\ a_{21} & a_{22} & a_{23} & \cdots & a_{2n} \\ a_{31} & a_{32} & a_{33} & \cdots & a_{3n} \\ \vdots & \vdots & \vdots & \ddots & \vdots \\ a_{n1} & a_{n2} & a_{n3} & \cdots & a_{nn} \end{pmatrix} \begin{pmatrix} \phi_1(x) \\ \phi_2(x) \\ \phi_3(x) \\ \vdots \\ \phi_n(x) \end{pmatrix},
$$

as does the notation

$$
\mathbf{F}(x) = \mathbf{A}\phi(x)
$$

even more so. The concept of a scalar product is also generalised into the *inner product*. This is a combination of any two elements of the vector space that results in a scalar. A scalar in this context is a number, not a vector or a function. The quantities a_{ij}, where $i = 1, 2, \ldots, n$; $j = 1, 2, \ldots, n$ are examples of scalars. The scalar product of vectors $a_1\mathbf{i} + a_2\mathbf{j} + a_3\mathbf{k}$ and $b_1\mathbf{i} + b_2\mathbf{j} + b_3\mathbf{k}$ is $a_1 b_1 + a_2 b_2 + a_3 b_3$ and is written

$$
\mathbf{a} \cdot \mathbf{b} = a_1 b_1 + a_2 b_2 + a_3 b_3.
$$

This was introduced in the last subsection. The scalar product of two functions cannot be the same as this of course; it has to be some kind of combination that results in a scalar which is a number. There are choices of inner product, but the most common and the one used later in this book is the following

$$
\langle \phi_i(x), \phi_j(x) \rangle = \int_{x_0}^{x_1} \phi_i(x)\phi_j(x)dx.
$$

The notation $\langle \phi_i(x), \phi_j(x) \rangle$ is standard, and the limits of the integration on the right are fixed numbers. If the scalar product of two vectors is zero then $\mathbf{a}$ and $\mathbf{b}$ are at right angles to each other or orthogonal. This concept of orthogonality is an important one and is generalised to

$$\langle \phi_i(x), \phi_j(x) \rangle = \int_{x_0}^{x_1} \phi_i(x)\phi_j(x)dx = 0 \quad \text{provided } i \neq j.$$

Students who are familiar with Fourier series will also be familiar with this concept with either $\phi_n(x) = \sin(nx)$ or $\phi_n(x) = \cos(nx)$ and the limits of the integral being either 0 and 2π or $-\pi$ and π. The reason for the importance of orthogonality is apparent from considering the inner products of all possible members of a basis as an $n \times n$ matrix

$$\begin{pmatrix} b_{11} & b_{12} & b_{13} & \cdots & b_{1n} \\ b_{21} & b_{22} & b_{23} & \cdots & b_{2n} \\ b_{31} & b_{32} & b_{33} & \cdots & b_{3n} \\ \vdots & \vdots & \vdots & \ddots & \vdots \\ b_{n1} & b_{n2} & b_{n3} & \cdots & b_{nn} \end{pmatrix},$$

where

$$b_{ij} = \langle \phi_i(x), \phi_j(x) \rangle.$$

Each b_{ij} is an integral and would be a non-zero number unless the basis vectors were orthogonal. In the orthogonal case, all off-diagonal entries in this matrix are zero. If in addition all the diagonal entries are unity, then the basis is called *orthonormal*. Orthogonal or orthonormal, such a set of basis functions leads to a vast simplification. In this text, the simplification is to the implementation of the finite element method. A lot more could be said about bases and orthogonality, but it would be out of place in a book such as this. Instead, this topic now waits to be applied in Chapters 4 and 5.

Here, we move on to discuss statistics.

1.4   Statistical Preliminaries

The origins of statistics lie in ancient times, but in the last century and a half "Statistics" has become a scientific study in its own right, separate from mathematics. It owes this status to the pioneers

Florence Nightingale (1820–1910) (yes that one — she was a very able mathematician), Sir Francis Galton (1822–1911) an explorer and meteorologist who became the father of eugenics, selecting parents to "improve" physical and mental abilities of the child, a very dirty word nowadays, and in particular Karl Pearson (1857–1936). Briefly, before these eminent people, statistics was permutations and combinations together with mathematical probability and some fitting of data to predetermined lines, with only the Reverend Thomas Bayes (1702–1761) making a foray into inference in the 18th century. It was really Pearson who invented modern statistics as the science of making sense of large data sets. Later Ronald Fisher (1890–1962) developed the alternative view of being able to deduce and infer from sampling, working directly with the data, and it was the brilliant Russian Andrey Nikolaevich Kolmogorov (1903–1987) who actually showed the rigour of this new statistics and gave it mathematical form. In this text, advanced mathematical structure will be avoided and the basic ideas will be introduced in a very applied way.

There is no doubt that statistics plays an increasingly important role in marine science and coastal engineering; one could even say a pivotal role. The principal difficulty in writing a text such as this is to cater for the wide variety of previous experience amongst the readership. The safest path to take is to assume very little previous knowledge. Those who have managed to get through the last section with a modicum of understanding will already be mathematically quite sophisticated, in which case do pick and choose from what follows. We shall start with the revision of what statisticians call *measures of central tendency*, which means ways of assessing where the middle of a set of data is. The simplest form of data is a list of numbers, although data are also often produced in the form of frequency tables. We shall deal with both.

Example. We wish to find the mode, median and mean of the following list of numbers:

$$5, 3, 6, 5, 4, 5, 2, 8, 6, 5, 4, 8, 3, 4, 5, 4, 8, 2, 5, 4.$$

Solution. First of all, do not worry about the definitions of these words, this will be addressed later; instead, we put the numbers in ascending order as follows:

$$2, 2, 3, 3, 4, 4, 4, 4, 4, 5, 5, 5, 5, 5, 5, 6, 6, 8, 8, 8.$$

The mode is the number that appears the most times. So the mode $= 5$. The median is the number which is in the middle of the distribution, which as the number of numbers is even is a bit tricky, so we'll come back to this. Finally, the mean of the numbers is the sum of the numbers divided by 20 (there are 20 numbers in all), so the mean is computed as $96/20 = 4.8$. In this example, there is a clear mode since there are six 5s, and fewer of each of the other numbers — in general there is often a tie. There is an even number (20) of numbers, therefore the median is the average of the tenth and eleventh numbers. Since both of these are 5, so is the median. The mean is, uniquely, 4.8. Next, let us consider something a little more usual in scientific applications, that is, a situation where the numbers are grouped into classes and we have what is called a frequency distribution. Frequency distributions are usually given in tabular form. Table 1.1 gives the numbers of zooplankton of various lengths as measured by a student marine biologist; this is adapted from research data and considerably simplified. The frequency polygon associated with these data is shown in Figure 1.4.

The median of these data is still the middle number, but this is troublesome to find when the data take this form. There is a formula that we will give later, but it is best to draw the graph. The median is then given by the value taken on the horizontal scale when a vertical line precisely divides the area under the frequency polygon; into two equal halves. The mode is the peak of the frequency polygon; there

Table 1.1   A frequency table.

Length of zooplankton (mm)	No. of zooplankton
0.01–0.50	4
0.51–1.00	10
1.01–1.50	15
1.51–2.00	13
2.01–2.50	7
2.51–3.00	1

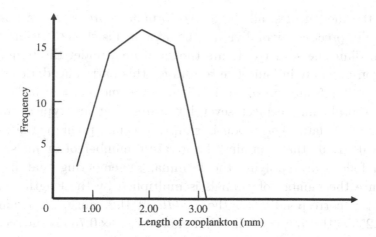

Fig. 1.4   A frequency polygon.

may be more than one. The mean is the quantity μ, which is given by the formula

$$\mu = \frac{\sum f_i n_i}{\sum n_i},$$

where the letters f and n denote the frequency of occurrence of the number and the number, respectively. The subscript is there to designate that there are many numbers — i would run from one to six in our example — and the $\sum$ sign denotes that summation over all i is to occur. The mean and median are always uniquely defined, but the same cannot be said of the mode. The mode is, straightforwardly, the class that contains the largest number but this might not be unique. The median either has to be determined graphically or by a rather messy formula derived from its definition as being the "middle". If the median occurs in a particular class, and the lower boundary of this class is L, then the median itself is determined from the formula

$$\text{Median} = L + \left(\frac{N/2 - (\sum f)}{f_{\text{median}}} \right) c,$$

where N is the total number of items in the data, $\sum f$ is the sum of frequencies of all classes *lower* than the median class, f_{median} is the frequency of the median class and c is the size of the median class interval. Given grouped data, it is easy enough to spot in which class the median lies; all the above formula represents is a mathematically precise way of dividing the area of this class to ensure that the median line so derived cuts the total area under the frequency polygon precisely in half. The results for this particular data set are mode = 1.255 mm, mean = 1.375 mm and median = 1.3969 mm. In this problem, we meet several features that are typical in the handling of data. The mode is simply the mid-point of the interval (1.01–1.50) that contains the greatest number of animals. The mean follows by applying the formula, remembering that in this instance the number of animals is multiplied by the length of zooplankton corresponding to the *middle* of the range, for example, 4×0.255 is the first entry in the numerator, 10×0.755 is the second, etc. Finally, the median is calculated using the given formula with $L = 1.01$, $c = 0.49$, $\sum f = 14$ and $f_{\text{median}} = 15$. This gives the idea of how these measures of central tendency can be calculated. Of course, these days a calculator, laptop or PC takes away the need to do the arithmetic.

The "middle" is not the only parameter that characterises a set of numbers. The two sets of numbers $1, 2, 3, 4, 5, 6$ and $2, 3, 3, 4, 4, 5$ both have six members and a mean of 3.5, but they are not the same. They differ on how the numbers are spread about the mean, and this is a second important characteristic of data. The usual measure of spread is variance or its square root, standard deviation. There are other more subtle measures. In a textbook on marine applications, such as this, it is not possible to go into much detail in the way of statistical theory, nor would it be desirable. The many specialist texts on statistics that start, as we have, by introducing measures of central tendency, go on to discuss topics such as standard deviation, distributions, probability and then to applied topics which include sampling, regression, hypothesis testing and experimental design. All of these have a role to play in marine science and coastal engineering, and we will give them a brief airing in later paragraphs, but it would be over-ambitious to try to cover them in any depth in this book. Perhaps the most important point to make is that the central purpose of statistics is *inference*. The reason why data are analysed is to enable scientists and engineers to establish hypotheses in a statistical sense from the data. It is, however, also necessary to give some ideas about probability as these form a central part in the understanding of the wave spectra that will be met in Chapter 8. Before doing this, having defined and calculated measures of central tendency as statisticians call them, let's discuss measures of spread in more detail.

As mentioned above, the most common of these is called *variance*, together with its square root, *standard deviation*. The variance of a set of numbers measures how spread out they are from their mean. It is defined by the formula

$$\sigma^2 = \frac{\sum (X_i - \overline{X})^2}{N},$$

where the symbols have the following meanings: X_i denotes the data i.e. the numbers themselves, $\overline{X}$ is the arithmetic mean, $\sum$ is the summation sign which means that each number has the mean subtracted from it before it is squared, then the whole is divided by N, the number of numbers in the data set. The reason behind squaring each difference is that this makes all entries under the summation sign positive, hence making sure that the result of this sum is indeed a true representation of the spread of the data from the mean. Statisticians call this a "measure of dispersion", but this is not an appropriate expression to use in a book where dispersion has a physical significance of its own. In order to restore the dimensions, the variance is normally square rooted (hence the square on the left-hand side) and the symbol σ is called the standard deviation.

Again, computers and calculators take away the arithmetical drudgery, but let us go through a simple example:

Example. Find the variance and standard deviation of the numbers:

$$5, 3, 6, 5, 4, 5, 2, 8, 6, 5, 4, 8, 3, 4, 5, 4, 8, 2, 5, 4.$$

Solution. The answers are $\sigma^2 = 3.116$ and $\sigma = 1.765$. If you "cheated" and used a calculator or a computer, this is no problem as long as you are sure of what you have calculated and know what standard deviation and variance actually indicate. The above answers only validate your arithmetic; they do not confirm your understanding. When a frequency table is involved, the definitions are, of course, the same but the method of calculation looks a little different. In fact, there is a very useful formula that can be derived from the definition of variance that proves useful in calculation. This states that the variance is given by the expression:

$$\sigma^2 = \overline{X^2} - \overline{X}^2,$$

which can be read as variance equals the "mean of the squares minus the square of the mean". For grouped data, the following expression is the formula for variance

$$\sigma^2 = \frac{\sum f n_i^2}{N} - \left(\frac{\sum f n_i}{N}\right)^2;$$

the standard deviation is of course the positive square root of the variance. The variance and standard deviation of the data presented in Table 1.1 are 0.408 mm^2 and 0.639 mm, respectively. The calculation of the mean and standard deviation of a set of data is one thing and is easy to do. In a practical context, the scientist — usually a marine biologist — should now go on to discuss the implications of these values. Before leaving means and standard deviation, a word needs to be said about how to compare two (or more) data sets. Everything said above has been about analysing a single data set. However, it is very common to have to compare two or more sets of data. In a marine context there are many examples: sea temperature and/or salinity and the biomass of some phytoplankton, wind speed and direction at different locations, etc. More will be said later on the detail on how such comparisons are done, but the direct measure of comparison of two sets of numbers is a simple extension of variance called *covariance*. Here is the definition:

$$\sigma_{xy}^2 = \frac{\sum (X_i - \overline{X}) \sum (Y_i - \overline{Y})}{N},$$

where $Y_i, i = 1, 2, \ldots, N$ denotes the second set of numbers. In some books the denominator in the definition of both variance and covariance will be $N - 1$ and not N, this is because the definition is for a sampled set and not the entire set of data, and this difference pops out of the mathematics one goes through to define a variance for the sampled set that closely matches the real variance of the entire data set. Do not worry about it as for large N they are very close, and if N is not large any inference one makes will be unreliable anyway. We will return to covariance later when discussing regression and principal component analysis (PCA).

Let us now discuss probability. The notion of probability is tied up with the outcomes of things called events. An example of an event is the tossing of a coin or the drawing of a card from a pack. It can also be the occurrence of a particularly large wave. Defining the *probability p* of an event occurring as

$$p = \frac{\text{number of ways it can occur}}{\text{total number of ways}}$$

is fine for coin tossing or card drawing but is not very useful for practical problems such as wave forecasting. A more workable definition might be to use a large number of trials and define

$$p = \frac{\text{number of successes}}{\text{total number of trials}}.$$

The problem with this definition is that the probability calculated only approximates the actual likelihood of the event occurring. For example, no matter how many trials there are, tossing a fair coin will never result in exactly half being heads and half being tails. One hesitates to say never, but if it occurred one would suspect tomfoolery. In most textbooks, the theory of probability is housed in terms of set theory and Venn diagrams. A set A will contain events and the function $p(A)$, the probability of event A occurring. The set S is the universal set that contains all possible outcomes, and of course $p(S) = 1$ and $A \subseteq S$ and so on. We shall not go this route here. The last chapter of James (2015) is a good introduction if you are keen. This is the fifth and latest edition but there might be another out soon. However, there are some aspects well worth covering here. In this text we shall primarily be concerned with continuous rather than discrete variables. This is because the processes that interest us, such as trying to determine the probability of particularly large waves, stem from continuous processes — there is a continuum of waves in between the smallest and largest — however the following discrete example serves as a useful introduction.

In many practical examples of discrete probability there are only two possible outcomes from an event. Most events can be considered either a success or a failure. This gives the two possible outcomes and two associated probabilities, p and $1 - p$, linked with each and the trial is called a Bernoulli trial. Certainty has the probability 1 and impossibility the probability 0. This follows from either of the definitions given above. For the toss of a fair coin, if p is the probability of getting a head then $p = 1/2$, disallowing the coin landing on its edge. There is a reasonably simple formula that gives the probability of r successes in n trials. If we put $q = 1 - p$ then expand $(p + q)^n$ in a binomial expansion then the rth term gives the probability of r successes in n trials. Mathematically,

$$(p+q)^n = p^n + np^{n-1}q + \frac{n(n-1)}{2!}p^{n-2}q^2 + \cdots + \binom{n}{r}p^r q^{n-r} + \cdots + q^n.$$

The rth term is

$$\binom{n}{r}p^r q^{n-r}.$$

It is the probability of r successes multiplied by the probability of $n - r$ failures multiplied by the number of ways that this particular arrangement can occur in n trials. This arrangement is written using the "n choose r" notation which is defined by

$$\binom{n}{r} = \binom{n}{n-r} = \frac{n!}{(n-r)!r!}.$$

As each arrangement of success or failure is independent, if we add up all possibilities we regenerate the binomial expansion of $(p + q)^n$, which is of course 1 since $q = 1 - p$. It is not difficult to see direct applications of Bernoulli trials in marine biology. Detecting the presence of diseases in phytoplankton or zooplankton, for example, success — no disease, failure — disease is one example that belongs in Chapter 9. Whether or not a toxic level of a contaminant is present in a river, estuary or coastal sea is another example that belongs in Chapter 7. The tossing of coins will be what are called independent events, that is the probability of getting a head or a tail is one half no matter what the previous results are. Drawing cards from a pack without replacement is a different matter. The chances of selecting the ace of spades, for example, will be 1/52 from a full pack but after not drawing it will reduce to 1/51 for the next selection. Once the ace of spades is in your hand, the chances of getting another one drops to zero of course assuming a standard deck; and no card sharps. In many practical examples the independence of events is assured due to the scientific stringency of how the observations or experiments are carried out. However, sometimes events are not independent. For example, suppose the events are A: "the occurrence of waves of height over 10 m" and B: "the occurrence of storms where the winds exceed 50 ms^{-1}" then these are obviously not independent. The probability of A occurring will be smaller than the probability of A occurring given that B has occurred. Sometimes it will be larger; consider if we replaced B by "wind is less than 20 ms^{-1}". In order to distinguish between the two cases, the notation for the former is the straightforward $p(A)$ whereas for the probability of A given B it is $p(A|B)$ and there are rules linking the two; in fact

$$p(A|B) = \frac{p(A \cap B)}{p(B)},$$

where $p(A \cap B)$ is the probability of both A and B occurring. We have no room to explore such subtleties, but hopefully this has made you think about the use of statistics a little more deeply.

For the discrete binomial distribution, the mean is np and the variance $np(1-p)$, but to go any deeper into pure probability theory would not be appropriate here; we do return to the binomial distribution when we discuss maximum likelihood a little later. Meanwhile, note that if n gets very large, it becomes too unwieldy to deal with and instead the Poisson distribution is used, whereby

$$y = \frac{\lambda^n e^{-\lambda}}{n!}$$

with both mean and variance equal to λ. This distribution is useful where the number of events is large but the probability of occurrence of any particular event is small. Mathematically, n gets large, p gets small but $np = \lambda$ remains finite.

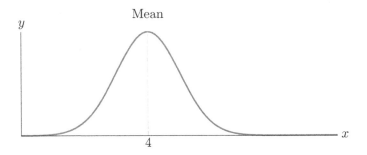

Fig. 1.5   The normal distribution with a mean $\mu = 0$ and a variance of $\frac{1}{2}$ standard deviation of $\sigma = \frac{1}{\sqrt{2}}$.

In other circumstances continuous distributions are required at the outset. In Chapter 8 various distributions will be met that are very useful when considering the prediction of large waves. These distributions are special to the subject. Most of the rest of the world either meet the discrete or continuous normal distribution or the Poisson distribution. The usual formula for the normal distribution is

$$y = \frac{1}{\sigma\sqrt{2\pi}} e^{-(x-\mu)^2/2\sigma^2}, \quad -\infty < x < \infty$$

with mean μ and variance σ^2. The shape is shown in Figure 1.5, although the amplitude has been set to 10 for convenience rather than left at $\frac{1}{\sigma\sqrt{2\pi}}$. Here, y is in what statisticians call the sample space, which means it is wave height or some other quantity that we require to estimate. The random variable x can be time or space, but most commonly in our applications it will be the frequency of the wave. In Chapter 8 the spectrum is indeed defined by a functional form giving its distribution with frequency. Figure 8.4 in Chapter 8 is a good example. Once the mean and standard deviation are fixed, in many diverse applications away from wave prediction it is often acceptable to assume that data presented in the form of a frequency distribution approximate closely to normal. Believe it or not, this assumption tends to be universal. In fact, the normal or Gaussian distribution tends to be assumed even when it is not appropriate; users of statistical routines need to be aware of this. The first thing to remember is that the normal distribution assumes that the variation of frequency $y = f(x)$ with random variable x is the above function. Although the normal distribution is very widely used, the particular mean and standard deviation of each problem is going to be different. It is obviously desirable to have only a single standard normal distribution to cater for predictions, and this is easily accomplished by adjusting the variables to a mean of zero and a standard deviation of one. Remember, therefore, to transform your data X into the normal variable z, sometimes called the z-statistic, through the simple transformation

$$z = \frac{X - \mu}{\sigma},$$

before doing any statistical testing, and testing is what it's all about. The reason for proposing a normal distribution is to test whether your particular data fit this distribution. The commonest of tests to use is the χ^2 test, which can be used to test whether or not a particular set of data fits a given hypothesis. In order to get the idea of testing a hypothesis, let us return to tossing a coin. If a given coin is tossed 1,000 times, say, and the outcomes are recorded, then this test can be used to decide whether the coin is biased or fair. Similarly, the χ^2 test can be used to decide whether or not data fit the conclusions drawn from a particular model. As hinted at above, however, one never gets *the* answer, and the criterion for acceptance or rejection of a hypothesis, to the applied marine scientist or coastal engineer, is not God-given but is in fact dependent on assumptions involving the normal distribution.

Before we can do examples, we need to introduce the subject of hypothesis testing a bit more formally. This is the traditional first step on the road to *inference*, the main purpose behind most of statistics. Suppose we have some data, perhaps from observations taken on a field trip. These data form what statisticians call a population. It is a collection of numbers arranged in a table or represented graphically. There will be certain statistics associated with the data — we have calculated the mean and standard deviation, but there are others. Now suppose further that we suspect that these data obey the form dictated by, say, the normal distribution. That is, we suspect that the mean and standard deviation conform to a certain normal, bell-shaped curve. We can use the χ^2 test to ascertain the truth of this hypothesis. This hypothesis is called the *null hypothesis* and is given the symbol H_0. If H_0 is rejected when in fact it is true, we say that a type I error has occurred. If we accept H_0 when it is actually false, we say that a type II error has occurred. Unfortunately, it is all too easy to make both sorts of errors, and it is always best to take a cynical look at the data, looking for oddities — *outliers* as statisticians call them — which may be due to human error in observing or instrumental failure, and which could distort the data and be the underlying cause of the type I or type II error. Finally, statisticians give the symbol H_1 to an alternative to the null hypothesis. Hopefully some of this will come alive through the next two examples.

The first of these examples is an introductory one involving, as said above, that old standby, the tossing of coins; the second is a more practical example involving real marine data.

Example. Suppose a coin is tossed 1,000 times, and the outcome is 530 heads and 470 tails. We might expect the outcome to be 500 heads and 500 tails, but then again it is the nature of chance that most of us would actually be surprised at such a precise obedience of the laws of probability. The pertinent question to ask is: is the coin fair? In other words, can the deviation from the ideal answer be attributed to chance, or is there a bias in the coin? In this case, the null hypothesis might be:

H_0: heads and tails occur with equal frequency.

Solution. We shall use the χ^2 test. In order to do this, we need an appropriate distribution. The χ^2 distribution can be found on the web: I found a .pdf at http://sites.stat.psu.edu/~mga/401/tables/ Chi-square-table.pdf for example, but type "chi-square table" into Google. In the χ^2 table, the top row, labelled χ^2 which denotes the *levels of significance*, gives a choice of 13 numbers. These numbers represent significance levels so that, respectively, the columns that they head are appropriate to testing at the 99.5%, 99%, 97.5%, 95%, 90%, 75%, 50%, 25%, 10%, 5%, 2.5%, 1% and 0.5% levels. Let us choose the value 0.01, so that we are testing at the 99% significance level. Coin tossing is a process that has two possible outcomes (heads or tails); therefore the first row is chosen. The number in this row is $\chi^2 = 6.635$. Now we calculate the value of χ^2 according to the formula

$$\chi^2 = \sum \frac{(\text{Observed} - \text{Expected})^2}{\text{Expected}}.$$

Remember, the summation sign is not a sum over 1,000 trials, but a sum over all possible outcomes. The calculated value is

$$\chi^2 = \frac{(530 - 500)^2}{500} + \frac{(470 - 500)^2}{500},$$

so that $\chi^2 = 3.6$. This value is less than the value in the table, so we accept the null hypothesis H_0 and conclude that, at the 99% significance level, the coin is not biased. This is probably the correct conclusion, but if on examining the data we found 200 consecutive heads, we would want to research further into how the coin was tossed, etc. This latter point may seem a little silly here, but if we were dealing with real data, it is analogous to "eye-balling" the figures and spotting if anything suspicious is present in the data. Mind you, one is much more likely to look if the hypothesis is rejected.

To the relief of most of you, the next part comprises a marine-related example.

Here, is an example involving fish. Table 1.2 gives the actual and expected values for catches of five species of fish.

First, the null hypothesis for this problem H_0 states that: "the expected catch and the actual catch are the same". Using a χ^2 test with parameter 0.01, do we reject H_0? To answer this we calculate χ^2 from the formula and get the appropriate value of χ^2 from a χ^2 table; these are

$\chi^2(\text{calculated}) = 17$,
$\chi^2(\text{table}) = 13.3$.

Table 1.2 Actual and expected catches of five species of fish.

	Species A	Species B	Species C	Species D	Species E
Expected catch	25	5	7	31	35
Actual catch	20	4	17	26	30

On the face of it, these results indicate that we should reject the null hypothesis. However, if we glance at the table of data, there is a very large discrepancy between expected and actual catch for species C. Without species C data the calculated value of χ^2 would have been well below 13.3 and H_0 would have been accepted. The correct conclusion to draw, therefore, is that the figures for species C need to be re-examined and the reason for the glut of fish or the serious underestimation of the catch ascertained. In passing, note that for an n-variable problem ($n = 2$ for the coin, and $n = 5$ for the fish) we look at the line $n - 1$ rather than line n in the χ^2 table. The reasons for this are rather technical and have to do with the (statistical) degrees of freedom of the system.

When using hypothesis testing, it is always necessary to put data in the form of frequency. The χ^2 test simply does not work for dimensional data in the form of lengths or masses. If your data are in such a form, classify them in some way; batch them up to rid the numbers of dimension. Once the data are put into these m classes, then the degrees of freedom, the row along which to look up the value in the χ^2 table, is $\nu = m - k - 1$, where k is the number of parameters to be estimated (usually zero for us). Let us go no further here; interested readers are directed to statistics books — there are plenty to choose from. The statistics chapters in James (2015) are particularly accessible.

1.4.1   *Maximum likelihood estimation*

At the beginning of this section about statistics, certain names were mentioned as pioneers of the subject. Maximum likelihood estimation (MLE) belongs securely in the Ronald Fisher (1890–1962) realm whereby sampling methods are used. He first used it in the context of agriculture around the time of the First World War, but it remains a useful technique. The idea is to be able to estimate the value of some parameter given the outcome of a number of trials and assuming that these trials fit a particular distribution. In keeping with this section, general theory is avoided it — looks very daunting — and instead we do some examples to get the general idea.

Example. Suppose a die is thrown 12 times with the result: $5, 6, 6, 4, 2, 6, 6, 6, 3, 1, 6, 6$. It is suspected that there is a bias. What is the maximum value of the probability of getting this result?

Solution. Assuming p is the probability of throwing a 6. Then the probability of getting seven 6s and five other numbers in 12 throws will be

$$\binom{12}{7} p^7 (1-p)^5.$$

In this example, the order that the seven successes come is not relevant. To maximise this for p we need to differentiate and put the derivative equal to zero. We could differentiate $p^7(1-p)^5$ as it stands, but the trick is to take logarithms and differentiate these as follows

$$y = p^7(1-p)^5$$

$$\text{so} \quad \ln y = 7 \ln p + 5 \ln(1-p)$$

$$\text{differentiating, we get} \quad \frac{1}{y}\frac{dy}{dp} = \frac{7}{p} - \frac{5}{1-p}$$

which has to equal zero in order for p to be a maximum. Hence

$$p = \frac{7}{12} = 0.5833.$$

Now if the dice was a fair one, the probability of throwing a 6 or any other number would be $1/6$. The value 0.5833 seems to indicate that, with this die, there is a better than even chance of throwing a 6. Note the phraseology here; in particular the use of "seems" and the presence of "likelihood" in MLE. Nothing is certain, and it just might be the case that the 12 throws were a lucky sequence as far as throwing sixes were concerned.

Here is a different kind of example. Suppose that samples are taken from a population that has a normal distribution. Normal distributions have two free parameters: the mean usually denoted by μ and a variance usually denoted by σ^2. The square root of variance, σ, is the standard deviation. For this example, suppose that the variance is known but that N samples are taken. Let the sizes of all these samples be given by the numbers $X_1, X_2, \ldots, X_N$. By definition, the probability density function of each of these samples will be

$$f(X_i|\mu) = \frac{1}{\sigma\sqrt{2\pi}}e^{\frac{(X_i-\mu)^2}{2\sigma^2}},$$

where each sample has the same mean and variance as the population from which the samples are taken. However we do not know the value of μ and we can use the MLE to estimate this. If all the samples are independently taken, then the likelihood function is simply the product of the density functions as follows

$$L(X_1, X_2, \ldots, X_N) = \frac{1}{\sigma\sqrt{2\pi}}e^{-\frac{(X_1-\mu)^2}{2\sigma^2}} \times \frac{1}{\sigma\sqrt{2\pi}}e^{-\frac{(X_2-\mu)^2}{2\sigma^2}}$$
$$\times \cdots \times \frac{1}{\sigma\sqrt{2\pi}}e^{-\frac{(X_N-\mu)^2}{2\sigma^2}}.$$

The right-hand side can be tidied up to give

$$L(X_1, X_2, \ldots, X_N) = \left(\frac{1}{\sigma\sqrt{2\pi}}\right)^N e^{-\sum_{i=1}^{N}\frac{(X_i-\mu)^2}{2\sigma^2}}.$$

As before, logarithms are taken before the right-hand side is differentiated with respect to μ to estimate its likelihood. So,

$$\ln(L) = N\ln\left(\frac{1}{\sigma\sqrt{2\pi}}\right) - \frac{\sum_{i=1}^{N}(X_i - \mu)^2}{2\sigma^2}$$

and differentiating gives

$$\frac{d}{d\mu}(\ln(L)) = \frac{\sum_{i=1}^{N}(X_i - \mu)}{\sigma^2},$$

noting that the first term does not contain μ so differentiates to zero, and differentiating the second term gives rise to a series of terms in which the 2s cancel and the minus sign disappears. Setting this quantity to zero gives

$$\sum_{i=1}^{N}(X_i - \mu) = 0$$

or

$$\sum_{i=1}^{N}X_i - \sum_{i=1}^{N}\mu = 0.$$

The second of these terms is simply $N\mu$, thus we have

$$\mu = \frac{\sum_{i=1}^{N} X_i}{N}$$

as the MLE for the mean. Thus, for populations having a normal distribution, the MLE of the mean is simply the mean of the sample means of the population.

1.4.2   *Regression*

The next topic to cover in this briefest of excursions into statistics is fitting lines to data. The most common example of this is the regression line, which is a line of best fit through a set of data points. Some time will be spent going through the principles of regression. After this, there will be a discussion of more sophisticated techniques for fitting data; these include empirical orthogonal functions (EOFs), which are really part of PCA, which itself is in fact a special case of factor analysis (which one hesitates to abbreviate to FA).

Let us start with a simple scatter plot as shown in Figure 1.6, for which there is a quite straightforward procedure for drawing a line of best fit through the data. An arbitrary line is drawn, then the square of the perpendicular distance of each point from this line is calculated. These are all added together, and the minimum value of this is

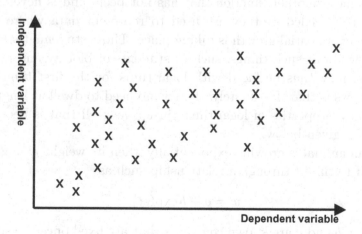

Fig. 1.6   A scatter diagram.

found. The parameters of the line that correspond to this minimum value give the line of best fit. Difficulties arise only when the data are so scattered that there is virtually zero correlation, in which case the line of best fit has no meaning. In fact, there are always *two* regression lines. If x and y denote the standard axes, these regression lines are called "y on x" and "x on y", and if there is no correlation then these two regression lines are at right angles to each other. The presence of these two lines is one reason for using the more complex PCA and factor analysis. There will be more about PCA later. Recall that the term correlation refers to the measure of agreement between two sets of data. A correlation of one denotes perfect agreement, a correlation of minus one denotes perfect disagreement. As an example of the latter, the rainfall at one point of an estuary, and the salinity of the water at the same point; as the rainfall increases, the salinity decreases and vice versa and a correlation of zero denotes no relationship at all. Other complications occur when there is obviously a relationship between two variables, but this relationship is not a linear one. This takes us into log–linear and log–log plots. These days, most schools seem to have abandoned logarithms because they are no longer of any practical use as a calculating tool. Teachers think that they have gone the way of the *ready reckoner* and the *slide rule*. The news for such teachers is that they have not gone completely. Logarithms have another function that has not been, and is never likely to be, superseded — they are used to represent data where some kind of exponential growth is taking place. Those students that need to know about such things, such as students of biology and marine science, may thus be faced with logarithms for the first time. The good news is that, fortunately, there is no need to dwell at length on the many properties of logarithms, just a few; all that is necessary in fact is given below.

If an animal is growing exponentially, then its weight w might be related to time t through a relationship such as

$$w = a + b\exp(ct),$$

where a, b and c are known constants that are fixed once the species and its environment are also fixed. In reality, of course, this growth will stop, and these more sophisticated models are introduced in Chapter 9, but this is a simple illustration only. If we wanted to make t the subject of this formula, then we would subtract a from both sides before taking logs to obtain

$$t = \frac{1}{c} \ln \left(\frac{w - a}{b} \right),$$

where the symbol "ln" denotes the natural or Naperian logarithm. This particular logarithm function is the inverse of the exponential function, and is the "log" referred to in the phrase "log–linear", as in graph paper. We have still not given the reasons for needing to know about such graph paper. To do so, consider the expression just derived,

$$t = \frac{1}{c} \ln \left(\frac{w - a}{b} \right).$$

If data corresponding to $(w - a)/b$ were to be plotted on one axis of log–linear paper, and data corresponding to t be plotted on the other, then provided w and t were related in the way dictated by the above equation, the plot would be a straight line with slope c. Once a scatter plot can be assumed to contain within it an implied linear relationship, then all the regression methods developed for straight lines can be brought to bear on the data. Table 1.3 gives some examples of relationships and the correct page of a spreadsheet, for example the graphics facility of Excel that should be used to display them as a straight line. In what follows, X and Y are the independent and dependent variables, respectively.

There is specialist software that renders variables that are related logistically, the last entry in Table 1.3, as a straight line. However, a log–linear relationship can be used provided the equation is transformed into exponential type by treating $(1/Y) - c$ as a variable.

One important question we have not yet addressed is how to assess whether or not a particular law is suitable for a given set of data; we cannot always rely on simply "eye-balling" it. If we wish to compare two sets of figures in a quantitative manner, then we calculate a correlation coefficient. There are several such coefficients to choose from,

Table 1.3   Some common relationships.

Equation	Straight line	Description
$Y = \frac{1}{a+bX}$	$\frac{1}{Y} = a + bX$	A hyperbola: use ordinary linear spreadsheet
$Y = ab^X$	$\ln Y = \ln a + X \ln b$	An exponential curve: use a log–linear relation
$Y = aX^b$	$\ln Y = \ln a + b \ln X$	Geometric curve: use a log–log relationship
$Y = \frac{1}{ab^X + c}$	$\frac{1}{Y} = ab^X + c$	Logistic curve: use a log–linear relationship (with care)

but the one most commonly used is the Pearson correlation coefficient, which is 1 for perfect agreement, -1 for perfect disagreement, and 0 for no relationship at all. To calculate the Pearson correlation coefficient, r_{XY}, the formula

$$r_{XY} = \frac{(1/N)\sum[(X_i - \overline{X})(Y_i - \overline{Y})]}{S_X S_Y},$$

where

$$S_X^2 = \frac{1}{N-1}\sum(X_i - \overline{X})^2 \quad \text{and} \quad S_Y^2 = \frac{1}{N-1}\sum(Y_i - \overline{Y})^2$$

is used. All summations are over all the data points. As mentioned on page 42, the presence of $N - 1$ rather than N in some of these expressions may perplex some readers, but as stated there, when N is large this difference is unimportant, and when N is small any correlations will have large uncertainty anyway. Although the above formula gives the definition of r_{XY}, we give below the most widely used practical formulae for calculating not only r_{XY} but also the regression line of Y on X in the form $Y = AX + B$. Purists will also notice a missing factor of $(N - 1)^2/N^2$ in the formula for r_{XY}, but again this quantity is very close to one in most practical examples. In fact, if it is not, then any straight line drawn through such sparse

Table 1.4   A table of river discharge data.

Discharge	1972	1973	1974	1975	1976	
N (t yr^{-1})	13600	8900	13500	12700	7000	
P (t yr^{-1})	310	210	250	220	120	
Q (m^3 s^{-1})	150	365	505	515	240	
Discharge	1977	1978	1979	1980	1981	1982
N (t yr^{-1})	16700	14900	13600	18700	18000	16400
P (t yr^{-1})	350	290	310	390	330	270
Q (m^3 s^{-1})	535	535	435	645	620	535

data has only scant value.

$$r_{XY}^2 = \frac{(N\sum XY - \sum X \sum Y)^2}{|N\sum X^2 - (\sum X)^2||N\sum Y^2 - (\sum Y)^2|},$$

$$B = \frac{N\sum XY - \sum X \sum Y}{N\sum X^2 - (\sum X)^2}, \quad A = \frac{\sum Y - B\sum X}{N}.$$

Let us now do an example.

Example. Table 1.4 gives the discharges of nitrogen (N) and total phosphorus (P) through the River Göta in tonnes per year, as measured in the years 1972–1982 (inclusive). The quantity Q denotes the river discharge in m^3 s^{-1}.

Solution. First, we need to plot the two scatter diagrams of the discharges of nitrogen and phosphorus. These are shown in Figures 1.7 and 1.8, respectively. The variable Q is the independent variable, and it is seen that the data are suitable for a linear regression line to be appropriate. Calculate the two correlation coefficients r_N and r_P using the formula to obtain $r_N = 0.73$, $r_P = 0.53$.

Although both correlations are positive, they are not particularly high, so it is not obvious that linear regression is the best way to obtain reliable predictions. One may find a better nonlinear relationship, but looking at scatter plots does not immediately suggest any obvious alternative candidates. We therefore still press ahead and

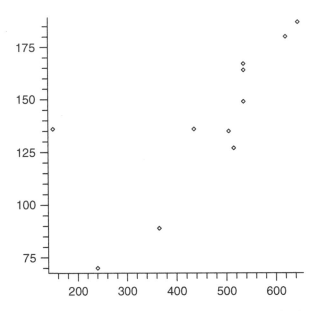

Fig. 1.7   Total nitrogen against river flow: the x-axis is Q m^{3}s^{-1}, the y-axis is nitrogen N in tonnes per year.

calculate the linear regression lines, but bearing in mind that predictions need to be treated with some caution. It is possible in fact to place error bars on the values of r_{XY}, but such refinements are considered outside the scope of this introductory text. The regression lines for nitrogen and phosphorus are

$$N = 17.06Q + 6121.06 \quad \text{and} \quad P = 0.258Q + 158.22.$$

Note that no attention has been paid to the units here. The data are given in a mixture of units, as is quite typical (one might even say prevalent) in marine science with its long nautical traditions and no conversions to, say, standard SI units have been made. This may annoy the purists, but in the calculation of lines of best fit, the geometric distance between the data points and the regression line has been minimised and this process is independent of units. Only if we wish for sensible units for these constants A_N, A_P, B_N and B_P does it become necessary to standardise. Finally in this example, let us do some predicting. We use the regression lines to predict the values of nitrogen and phosphorus in the River Göta when the

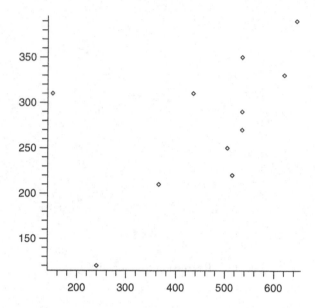

Fig. 1.8 Total phosphorus against river flow: the x-axis is as before and the y-axis is phosphorus, P, in tonnes per year.

river discharge is 800 m^3s^{-1}. Either from drawing these lines on the graphs shown in Figures 1.9 and 1.8 or, more accurately, from inserting $Q = 800$ into the formula for each line in turn we get

$$\text{N} = 19769 \quad \text{and} \quad \text{P} = 364.$$

The first, albeit less accurate, method is acceptable, particularly because it keeps you in touch with the data, reminding you how scattered the points are, and hence how low the correlation is. Most importantly, it indicates how much (how little?) reliance can be put on these predictions. A correlation of 0.9 would be considered a reasonable figure, and the data fall well below this.

1.4.3 *Principal component analysis*

There are several more advanced techniques that are now readily used by not just marine scientists and coastal engineers but all kinds of researchers. Many of them are not particularly new, but the advance in software has meant that they are much more available and easy to

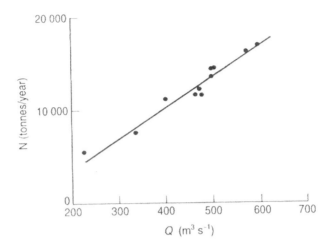

Fig. 1.9   Regression line for nitrogen.

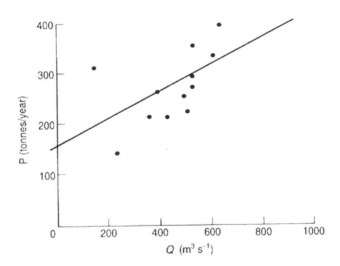

Fig. 1.10   Regression line for phosphorus.

use than used to be the case. Let us start with Factor Analysis. The development of Factor Analysis occurred well away from marine science and coastal engineering; indeed well away from science. It finds its greatest use in marketing, psychology and social science. In these subjects, there are a plethora of ill-defined variables and to make sense of any large data sets is bewilderingly difficult. Factor Analysis is a technique for reducing the number of variables to a salient few. In these less exact fields, the variables are usually assumed to be linear combinations of factors. When it comes to more precise sciences, we home in on the method PCA, which is a very useful method for analysing the variability, spatial or temporal, of physical fields. Not only scalars such as temperature or salinity but vectors such as currents too. We have already discussed regression in the previous section, and in some ways PCA is a generalisation of regression. In other ways it is far more sophisticated. Briefly, PCA is a method of identifying patterns in data. The identification of the line of best fit or regression line is one. The line that is perpendicular to this, also a regression line of course, but hardly a line of best fit is another. The point about PCA is that the data can be n dimensional and it can pick out the main trend, second best, third best, etc. For those that know about Fourier series or harmonic analysis (of tides), PCA is very like retaining the first two or three terms as a picture of the main ingredients. Neglecting all but the first few principal components does mean the loss of information, but in practice it is these first two or three that contain most of the information and with PCA it is in a far more digestible form. So having sold it, how do we do it?

Many of the building blocks are already here, and as with useful methods, to do the general theory will look far too daunting. Let us go through it in two dimensions. This is useful in two respects. First of all, being two dimensional means data can be displayed on paper or screen, and secondly, the relationships with regression are easier to see. The first thing that is needed is a reasonably full data set. Suppose that this is in the form of pairs (x_i, y_i) with $i = 1, 2, \ldots, n$ displayed on a scatter diagram. It is necessary for these to be standardised in some way before any analysis takes place, in much the same way as data are "normalised" for use with the normal distribution. This is important and needs to be remembered when doing real data analysis, as in Chapter 8 when an example involving coastal erosion is done. Here, suppose that this has been done and the scatter diagram is ready. Figure 1.11 shows the data. One can see that

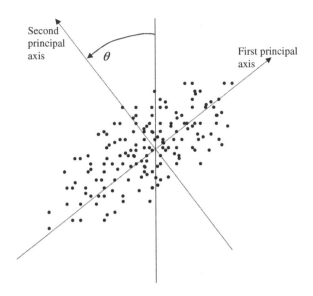

Fig. 1.11   A bivariate scatter plot showing the first and second principal axes and the angle θ through which the graph needs to rotate.

the data in some sense follow a linear trend, and two lines are drawn that are the first and second principal axes. For those that have a mechanics background, these are analogous to the principal axes of a rigid body. In order to calculate them here, recall the definition of covariance given earlier in this chapter: if the points are (x_i, y_i) with $i = 1, 2, \ldots, n$ then the covariance is

$$\sigma^2_{xy} = \frac{\sum (x_i - \overline{x}) \sum (y_i - \overline{y})}{n}.$$

However, it is best if the mean is subtracted from the data first, in effect translating the origin of the scatter plot to the middle of the points. The covariance is then

$$\sigma^2_{xy} = \frac{\sum (X_i) \sum (Y_i)}{n},$$

where $X_i = x_i - \overline{x}$ and $Y_i = y_i - \overline{y}$. Of course σ_{yx} will be the same as σ_{xy} by symmetry. The variances of the data are

$$\sigma^2_x = \frac{\sum (X_i) \sum (X_i)}{n}$$

and

$$\sigma_y^2 = \frac{\sum(Y_i)\sum(Y_i)}{n}.$$

In order to carry out PCA, the *covariance matrix* needs to be found. For the bivariate case it is the following 2×2 matrix:

$$\begin{pmatrix} \sigma_x^2 & \sigma_{xy} \\ \sigma_{yx} & \sigma_y^2 \end{pmatrix}.$$

The most natural way to proceed would be to rotate the axes such that one axis went through the "axis" of the data and resembled the regression line, then the other axis would be perpendicular to this. The way to do this is to define new axes ξ, η that are rotated through an angle θ (shown in Figure 1.11). These new co-ordinates are related to (X, Y) through

$$\xi = X \cos\theta + Y \sin\theta \quad \text{and} \quad \eta = -X \sin\theta + Y \cos\theta,$$

where there should be no confusion using η here as its normal meaning as surface elevation does not feature until Chapter 3. The angle θ is for the moment arbitrary. The variance of the data along the direction dictated by θ is easily calculated as $\sigma^2(\theta)$, where

$$\sigma^2 = \sum_{i=1}^{N}[X_i \cos\theta + Y_i \sin\theta]^2 = \sigma_{xx}^2 \cos^2\theta + 2\sigma_{xy}^2 \sin\theta \cos\theta + \sigma_{yy}^2 \sin^2\theta.$$

Using double angle formulae, this can be written

$$\sigma^2(\theta) = \frac{1}{2}(\sigma_{xx}^2 + \sigma_{yy}^2) + \sigma_{xy}^2 \sin 2\theta + \frac{1}{2}(\sigma_{xx}^2 - \sigma_{yy}^2)\cos 2\theta.$$

The angle θ should be chosen so that the line goes through the data in such a way that $\sigma^2(\theta)$ is an extremum. This is found by choosing the variance to be an extremum — usually maximum or minimum with respect to θ, so that using calculus for a maximum or minimum the derivative is set to zero and we get

$$\frac{d}{d\theta}(\sigma^2(\theta)) = (\sigma_{yy}^2 - \sigma_{xx}^2)\sin 2\theta + 2\sigma_{xy}^2 \cos 2\theta = 0$$

so that the values of θ are given by

$$\tan 2\theta = \frac{2\sigma_{xy}^2}{\sigma_{xx}^2 - \sigma_{yy}^2}.$$

This formula gives two angles, 90 degrees apart. Using the second derivative as a check reveals that the two principal variances are given by

$$\frac{1}{2}\left[(\sigma_{xx}^2 + \sigma_{yy}^2) \pm \sqrt{(\sigma_{xx}^2 - \sigma_{yy}^2)^2 + 4\sigma_{xy}^2} \right],$$

where the plus sign gives the maximum variance, the principal axis, and the minus sign the axis perpendicular to this. It is found that for these special values of θ the covariance is zero, so the covariance matrix is diagonal. Anyone who knows about diagonalising matrices will perhaps have heard of eigenvalues and eigenvectors, of which more in a moment. The two directions in the plane of the data are sometimes called EOFs; "use of EOFs" is often preferred to PCA in oceanography but they are the same thing. EOFs are often used in association with the use of data assimilation and what amounts to a smoothing routine called Kalman filtering, but discussion of all this is postponed until the next chapter where more specific details are given.

In two dimensions this looks reasonably straightforward, but if the data had a larger number of dimensions instead of (x_i, y_i) we would have to write $(x_{1i}, x_{2i}, \ldots, x_{Ki})$ where perhaps $K = 20$; then the covariance matrix would be $K \times K$ and using calculus and geometry, as we did above, is not an option. Instead one needs some matrix algebra. Some of you will know enough about eigenvalues and eigenvectors of a matrix to follow what comes next, but others will not. For those in the latter category, here is a quick diversion.

If A is a square matrix, then a number λ such that

$$|A - \lambda I| = 0$$

is called an *eigenvalue* of the matrix A. Here I is the unit matrix that has ones along the diagonal but zeros everywhere else. The vertical bars denote the determinate of the matrix which when multiplied out leads to an equation of the same order in λ as the size of the matrix A. That is, a 2×2 matrix leads to a quadratic in λ, a 3×3 matrix a cubic and so on. Solving this equation, called the characteristic equation, for λ will therefore give as many different values for λ as there are rows or columns of the matrix A. This is not always so: some quadratics can have only one root, $\lambda^2 - 2\lambda + 1 = 0$ for example, but we will gloss over these special cases. Here is an example; consider the matrix:

$$\begin{pmatrix} 2 & 3 \\ 1 & 4 \end{pmatrix}.$$

This has eigenvalues given by the roots of the equation

$$\begin{vmatrix} 2 - \lambda & 3 \\ 1 & 4 - \lambda \end{vmatrix} = 0.$$

This is expanded to

$$(2 - \lambda)(4 - \lambda) - 3 = 0$$

or

$$\lambda^2 - 6\lambda + 5 = 0.$$

This has the two solutions

$$\lambda = 1 \quad \text{and} \quad \lambda = 5$$

and these are the two eigenvalues of the matrix

$$\begin{pmatrix} 2 & 3 \\ 1 & 4 \end{pmatrix}.$$

Having found the eigenvalues, there are eigenvectors associated with each of them. The theory says that if there is a scalar λ such that for the matrix A we have

$$|A - \lambda I| = 0$$

then associated with each of these λ there is at least one vector $\underline{x}$ such that

$$A\underline{x} = \lambda\underline{x}$$

and these vectors $\underline{x}$ are the eigenvectors of the matrix A. In order to find these for our 2×2 matrix we solve the two sets of simultaneous equations

$$\begin{pmatrix} 2 & 3 \\ 1 & 4 \end{pmatrix} \begin{pmatrix} x \\ y \end{pmatrix} = 1 \begin{pmatrix} x \\ y \end{pmatrix} \quad \text{and} \quad \begin{pmatrix} 2 & 3 \\ 1 & 4 \end{pmatrix} \begin{pmatrix} x \\ y \end{pmatrix} = 5 \begin{pmatrix} x \\ y \end{pmatrix}.$$

These two sets of equations can be written

$$\begin{pmatrix} 1 & 3 \\ 1 & 3 \end{pmatrix} \begin{pmatrix} x \\ y \end{pmatrix} = 0 \quad \text{and} \quad \begin{pmatrix} -3 & 3 \\ 1 & -1 \end{pmatrix} \begin{pmatrix} x \\ y \end{pmatrix} = 0.$$

You may have noticed that neither of these are true simultaneous equations. The first is the one equation $x + 3y = 0$ written twice and the second is the one equation $-x + y = 0$ twice over. This is correct because the eigenvalues λ are precisely those values that render the determinant of the matrix $(A - \lambda I)$ singular (having zero determinant), and a zero determinant means redundancy, which means for two equations that they are the same. To solve $x + 3y = 0$ we can choose $x = 3$ and $y = -1$, and to solve $-x + y = 0$ choose $x = y = 1$. Therefore, we have eigenvalues 1 and 5 with associated eigenvectors $\begin{pmatrix} 3 \\ -1 \end{pmatrix}$ and $\begin{pmatrix} 1 \\ 1 \end{pmatrix}$ respectively. These vectors can be multiplied by any number (scalar) and they would still remain eigenvectors, that is because it is the *direction* that is important not the length of the vector. In PCA, it turns out that it is the *normalised* eigenvectors that are required, that is the eigenvectors of unit length. The length of a vector is the square root of the sum of the squares of the components, so for the example here, the normalised eigenvectors are, uniquely

$$\begin{pmatrix} 3/\sqrt{10} \\ -1/\sqrt{10} \end{pmatrix} \quad \text{and} \quad \begin{pmatrix} 1/\sqrt{2} \\ 1/\sqrt{2} \end{pmatrix}.$$

The method generalises to any square matrix, though of course the arithmetic gets more involved, as do the number of special cases to consider. Never mind about the technical stuff, what does this all mean. We will come to the context of PCA in a bit, but if the matrix A represented a geometrical description of a quadric surface — an ellipsoid perhaps — then the eigenvectors would represent the main axis; there is only one and the axes of symmetry perpendicular to this. If the cross-section of the ellipsoid is circular then there would be an infinite choice of the two perpendicular axes. It is precisely this case that arises when eigenvalues are a multiple root of the characteristic equation. Turning to mechanics, if A represented the matrix of moments of inertia, then the eigenvectors would represent what are the principal moments of inertia, if you know about such things. In PCA, the eigenvectors represent the direction through the data where there is greatest variation. A regression line will be one such, but for multivariate data these will be more than one. In Figure 1.11, one eigenvector or principal axis is obvious and drawn, a second is less so but is also displayed. If there were 20 variables, then the matrix of covariances would be 20×20; there would be 20 eigenvalues and associated with each of these an eigenvector. There will always be enough eigenvectors, but sometimes the eigenvalues will be less; the coincident roots case of the characteristic equation gives rise to a whole plane of eigenvectors from which one can choose any two at right angles; this is the case alluded to above. The role of finding eigenvalues and eigenvectors is to pick out directions through data where there is most variation. Once the direction of maximum variation is found, we then seek the second largest, then the third largest, etc. We have not said anything about how to order these yet, but it turns out that the first two or three eigenvectors or principal components carry most of the variation and so analysing the data is much simplified. A 20-dimensional problem has been reduced to two or three dimensions. Also, it is the largest eigenvalue that is associated with the greatest variation, so although there will be as many eigenvectors as dimensions, as said above, only a few will be required to capture most of the variation. In Chapter 8 there will be an example, but this is as far as we will go into the theory of using statistics for modelling. The next step would be to look at nonlinear regression and to include placing confidence intervals on predictions. Those interested in these topics need to consult more specialist statistics texts.

1.5 Exercises

(1) For the following phenomena, list the effects that need to be included in a *first* model:

  (a) Gulf stream.
  (b) A pollution accident in an estuary.
  (c) A surface oil slick 5 km, offshore of Aberdeen, Scotland.

(2) Explain how you think modelling can help in the clean-up of Eastern European rivers.
(3) Explain why calculus is important to know for the modeller of coastal and marine processes.

(4) Say which of the following variables are scalars and which are vectors: pressure; wind; salinity; temperature; force; current; stress.

(5) Use Green's theorem in the plane to show that if the current is in the form $\mathbf{u} = \nabla\phi$, then the circulation around any closed curve in the fluid must be zero.

(6) The following table gives two data sets for measurements of inputs into the North Sea; units are 10^3 km^3:

Location	A	B	C	D	E	F
Expt. 1	0	9.5	34	3.4	0.4	0.4
Expt. 2	10	11	40	5.2	0.2	0.3

A is the Pentland Firth, *B* between Orkney and Shetland, *C* between Shetland and Norway, *D* through the straits of Dover, *E* river discharges and *F* represents input from precipitation. Use the χ^2 test to decide whether or not the two sets of data are in agreement, and comment accordingly.

(7) Suppose a die is thrown six times and the results are: $3, 6, 6, 1, 6, 2$. The probability of throwing a six should be 1/6 but these results indicate a bias. Use MLE to estimate the maximum value of the probability of throwing a six.

(8) It is suspected that the relationship between the probability of an extreme wave occurring p and time t is of the form

$$p = \exp\{-\exp\{(t - a)/k\}\},$$

where a and k are constants. Find a function of p that renders this relationship linear.

Chapter 2

Modelling Tools and Techniques

2.1 Introduction

In the last chapter, we examined the motivation for modelling and some of the basic mathematical and statistical tools required. Here, we introduce some other tools that will prove useful in carrying out the modelling procedure. The mathematical formulation behind the balances and the assumptions are left until the next chapter, but here some of the basic physics and modelling notions are covered. We also examine data assimilation and signal some new modelling techniques that, as yet, are not widely used but are expected to make an impact on the modelling community shortly. First we discuss notions of *typical* lengths, *typical* times, *typical* speeds etc. It is by no means obvious what is meant by this, therefore a few words by way of introduction are appropriate. Many, not to say most, variables can be expressed in terms of time (measured in seconds), length (measured in metres) and mass (measured in kilograms). Examples are speed (length divided by time), density (mass divided by volume, hence mass divided by length cubed) and pressure (mass divided by (length times time squared)). The quantities mass, length and time are called the *fundamental* quantities. In a given situation, most users of models, as well as the builders of the models themselves, wish to concentrate on phenomena which have a specific limited range of any of the fundamental quantities, or of variables derived from them. For example, a tidal modeller would not be interested in small time scales of a few seconds or large time scales of many centuries, but something in between. An estuarine modeller would not be concerned with length scales of hundreds or thousands of kilometres but something much less. Most phenomena in coastal and marine science are scale specific, and the art of successful modelling is more often than not linked to the ability to screen out the unwanted in order to focus on what is desired. Dimensional analysis forms an essential element of this simplification. Once this simplification has taken place, the terms that remain constitute a simplified dynamic balance, and these balances are useful in describing the fundamental movements of the sea.

Oceans have a general structure which is dictated by their geographical origin. There are mid-oceanic ridges, trenches and, most important for us here, continental shelves and coasts. These shelves border the margins of the oceans, the seas over them are only about 200 m deep, and the transition from this 200 m to the oceanic 3000–4000 m is achieved through a relatively small region, the continental slope — which is aptly named, since the slope is commonly quite mild, only 4°. Simple trigonometry shows that the depth can sink from 200 m to 3000 m in a horizontal distance of 40 km with the slope. Since the width of the ocean is measured in thousands of kilometres, and most continental shelves are hundreds of kilometres wide, this slope region is of insignificant horizontal extent. It is, of course, only the dimension that is insignificant. The continental slope contains important currents and is a significant source and sink of energy for many different types of flow. It also marks the boundary between significantly different biology. However, importantly for continental shelf modellers, it serves to mark the border between deep ocean and continental shelf and is the site of the open boundary condition. The dimensions of the continental shelf render models of continental shelf seas quite distinct from ocean models. In the next chapter and Chapter 8 we shall be looking at modelling the waves and currents that occur on the world's continental shelves and also the waves that exist and may be trapped on the continental slope. There are also the better known surface waves; some discussion of these is appropriate here.

Surface water waves are a very important part, one might say essential property, of the sea. Waves are what we see first when we look at it, and they certainly have a great influence on man's interaction with it. The length scale associated with visible surface waves is normally a few metres at most, but waves associated with tides have a typical wavelength of a thousand kilometres and tsunamis associated with submarine seismic activity have wavelengths that are even longer. These different kinds of waves that occur in coastal engineering and coastal oceanography are all important to be able to model and all will receive attention in this text. So, we see that the surface water waves that finish their existence by crashing on to the beach are but one type of wave. Waves do not have to be associated with an up and down movement of the sea surface, for meanders in longshore current are waves too.

Currents can also obviously vary on a number of widely differing length scales. The variation has the same scope as that of waves. Given this, perhaps the first useful task is to make a list of relevant quantities and their dimensions, including the breakdown in term of the fundamental quantities mass, length and time, M, L and T, respectively.

2.2 Dimensional Analysis

In this section we shall do some calculations based on a knowledge of the dimensions of the quantities and not much else. Dimensional analysis can be usefully employed to determine likely relationships between variables or parameters, but first we have to know enough about the situation to state what the parameters or variables are, which can be ignored and which must be retained. If we get this wrong, dimensional analysis can be misleading, which can be worse than useless. A useful tool in dimensional analysis goes by the rather grandeur name of the Buckingham Pi theorem. It was developed by Lord Buckingham in 1915 and he used the symbol Π to denote products of quantities, so the name Buckingham Pi theorem has stuck, which is a bit unfortunate as the technique is given a gravitas it scarcely deserves. So, rather than emphasising its theoretical status, let us first do an example that utilises this law, which we rename the law of indices. It is also called the Rayleigh method.

Table 2.1   A table of parameters and their dimensions.

Quantity	Unit	Dimensions
Mass	Kilogram (kg)	M
Length	Metre (m)	L
Time	Second (s)	T
Temperature	Kelvin (K)	Dimensionless
Salinity	ppt	Dimensionless
Velocity	Metre per second $(\mathrm{m\,s^{-1}})$	LT^{-1}
Acceleration	Metre per second per second $(\mathrm{m\,s^{-2}})$	LT^{-2}
Area	Square metre $(\mathrm{m^2})$	L^2
Volume	Cubic metre $(\mathrm{m^3})$	L^3
Discharge	Cubic metre per second $(\mathrm{m^3\,s^{-1}})$	L^3T^{-1}
Force	Newton (N)	MLT^{-2}
Pressure	Pascal (Pa)	$ML^{-1}T^{-2}$
Pressure gradient	Pascals per metre $(\mathrm{Pa\,m^{-1}})$	$ML^{-2}T^{-2}$
Density	Kilograms per cubic metre $(\mathrm{kg\,m^{-3}})$	ML^{-3}
Dynamic viscosity	Newton second per square metre $(\mathrm{Ns\,m^{-2}})$	$ML^{-1}T^{-1}$
Kinematic viscosity	Square metre per second $(\mathrm{m^2\,s^{-1}})$	L^2T^{-1}
Surface tension	Newtons per metre $(\mathrm{N\,m^{-1}})$	MT^{-2}
Weight (same as force)	Newton (N)	MLT^{-2}
Angular velocity	Radians per second $(\mathrm{rad\,s^{-1}})$	T^{-1}
Angular acceleration	Radians per second square $(\mathrm{rad\,s^{-2}})$	T^{-2}
Vorticity	Radians per second $(\mathrm{rad\,s^{-1}})$	T^{-1}
Circulation	Square metre per second $(\mathrm{m^2\,s^{-1}})$	L^2T^{-1}
Energy	Joule (J)	ML^2T^{-2}
Work (same as energy)	Joule (J)	ML^2T^{-2}
Power	Watt (W)	ML^2T^{-3}
Temperature gradient	Degrees per metre $(\mathrm{K\,m^{-1}})$	L^{-1}

Consider first the reasonably straightforward problem of finding the period of swing of a simple pendulum. The pendulum consists of a string of length l which is fixed at one end. To the other end is attached a bob of mass m_0 and this bob is displaced a small distance from its vertical equilibrium position and swings freely. Now, we have to reason which parameters are important in determining the period of oscillation. Certainly the mass m_0 is one candidate, as is the length of the string l. One other parameter is the acceleration due to gravity g; this is because it is the component of the weight of the bob $m_0 g$ perpendicular to the string which is the restoring force. It is in this part of the process that errors are prone to be made; some insight into the working of the simple pendulum is mandatory to get the right parameters. First of all, there are four parameters, P, l, m_0, g (period, length of string, mass of bob and acceleration due to gravity, respectively), and three fundamental quantities, mass M, length L and time T. The assumption that is not provable is that there is a relationship

$$Pg^B m_0^C l^D = \text{const.}$$

This stems from each variable having units that are powers of M, L and T, so it is reasonable to assume that such a relationship as this exists. As the combination on the left is assumed dimensionless, it remains dimensionless if raised to a power. This means that one of the powers, in our case the power to which P is raised, can be assumed to be unity. The dimensions of the four parameters are

$$P \sim T, \quad g \sim LT^{-2}, \quad m_0 \sim M \quad \text{and} \quad l \sim L,$$

from which

$$T(LT^{-2})^B M^C L^D = \text{const.}$$

Equating powers of T, L and M to zero gives the equations

$$1 - 2B = 0, \quad B + D = 0, \quad C = 0,$$

from which

$$B = 1/2, \quad C = 0, \quad D = -1/2.$$

This means that the quantity

$$Pg^{1/2} l^{-1/2} = \text{const.}$$

Making the period P the subject of this formula gives

$$P = \text{const} \sqrt{\frac{l}{g}}.$$

The correct formula is obtained by setting the constant equal to 2π. There is no way dimensional analysis can give the value of this constant. It has, however, given the correct dependence of P on g and the length of the pendulum l, and it correctly predicts no dependence on the mass of the bob m_0. Thus, had we forgotten about the mass of the bob, we would have still obtained the correct answer. This would have been simply luck; if we had forgotten g, we would have obtained nonsense (try it!) So, in order not to get nonsense we must *not* trust to luck but be sure of our knowledge of the physics that governs a particular situation before rushing into dimensional analysis and the method of indices. Sometimes there are too many variables to get a unique solution. Here is the problem of determining a drag law in a viscous liquid.

The variables that drag will depend on are size L, density ρ with dimensions ML^{-3}, dynamic viscosity μ with dimensions $ML^{-1}T^{-1}$, speed U with dimensions LT^{-1} and of course gravity g with dimension LT^{-2}. Drag is a force with dimensions MLT^{-2}, so the method of indices (the Rayleigh method) gives the equation

$$MLT^{-2} = L^a(ML^{-3})^b(ML^{-1}T^{-1})^c(LT^{-1})^d(LT^{-2})^e.$$

This time all the quantities on the right are raised to a power labelled a to e. This is because the left-hand side has dimensions and is not a dimensionless constant as it was in the pendulum example. Equating the powers of M, L and T gives three equations, but there are five unknowns this time:

$$M : 1 = b + c,$$
$$L : 1 = a - 3b - c + d + e,$$
$$T : -2 = -c - d - 2e.$$

To make progress, we need to choose two of the five unknowns and express the others in terms of them. The M equation gives

$$b = 1 - c,$$

the T equation gives

$$d = 2 - c - 2e$$

and finally, eliminating b between the M and L equations by writing $b = 1 - c$ results in

$$a = 2 - c + e.$$

Now we put these values into the right-hand side of the drag dimensional equation giving

$$L^a(ML^{-3})^b(ML^{-1}T^{-1})^c(LT^{-1})^d(LT^{-2})^e,$$
$$= L^{2-c+e}(ML^{-3})^{1-c}(ML^{-1}T^{-1})^c(LT^{-1})^{2-c-2e}(LT^{-2})^e,$$
$$= L^{2-c+e}\rho^{1-c}\mu^c U^{2-c-2e}g^e,$$
$$= \rho L^2 U^2 \left(\frac{Lg}{U^2}\right)^e \left(\frac{\mu}{L\rho U}\right)^c.$$

This is the best that can be done with this method, but take a look at the right-hand side. By construction, the two expressions in the two parentheses are dimensionless numbers and they are in fact reciprocals of well known dimensionless quantities in fluid mechanics. The group $L\rho U/\mu$ is the Reynolds number and U^2/Lg is the Froude number. In this text, the Reynolds number takes a back seat as motion is dominated by the rotation of the Earth, which is absent from this example. The Froude number makes an appearance when considering the dynamics of estuaries and there is more about it below. Given that in a regime both the Reynolds and Froude numbers will be constant, the drag is thus proportional to the square of the fluid speed and the square of the size dimension of the body. We will return to this later when considering friction in the next chapter. For completeness here is the Buckingham Pi Theorem stated in general:

Buckingham Pi Theorem

Given n parameters such as length, speed, density, viscosity, force etc. and given that these parameters are composed of a set of m quantities (in our case three; mass, length and time), then it is possible to express the relation between the parameters in terms of $n - m$ dimensionless products, formed from any n parameters regarded as primary.

If there is but one solution to the equations obtained for the indices, then we solve as in the pendulum example. If, on the other hand, there are not enough equations; in terms of the Buckingham Pi Theorem $n - m$ $(= n - 3)$ is non-zero — the fluid drag example above it was 2 — then we get exactly this number of dimensionless groups of numbers.

Let us now pay more attention to dimensionless groups of numbers. The use of *dimensionless quantities* outside the method of indices is very important. Back to the Froude number; in the sea there are two quantities that have the dimension of length divided by time, L/T. One is the current or water velocity, the other is the wave speed or celerity; we shall be meeting both in the next and subsequent chapters. The current needs little explanation as it is reasonably obvious what is meant by it. Wave speed is a slightly more tricky concept. Waves are an important part of our study, and a great deal of time will be devoted to the study of various types of water waves in particular. The speed that the crest of a wave takes as it travels is called its celerity — it is definitely not the same as the current; in fact waves seldom carry any fluid along, they pass through the water as a vibration. A different kind of vibration (due to pressure variations) is what allows sound to travel in air and allows us to hear, but the principle is the same. Waves transmit signals without transmitting mass. We shall see in Chapter 3 that the letter c is used to denote wave speed or celerity, and it too has dimensions L/T. Let us form the quantity U/c, that is current speed divided by wave speed. It is a dimensionless number, and its value will tell us something about the regime in which both quantities have been measured. If it is large, then U, the current speed, will dominate over waves; if it is small, the reverse is true. It is this dimensionless ratio that is called the Froude number and plays an important role in civil engineering. In the drag example above the "wave speed" did not exist, but in deep water it is $\sqrt{gH}$, where H is the water depth, so the ratio U^2/gH is dimensionless and when compared to the dimensionless group in the fluid drag example above can be seen to be the Froude number. In a river, where U might indicate the flow of the river towards the sea for example, when the Froude number is large, waves cannot propagate against the current. In some rivers there is an interesting case of a wave due to a tide trying to propagate against a river, and because the wave due to the tide has a celerity that is large, the Froude number changes from greater than one (no tide) to less than one (tide) and at the boundary we have what is called a tidal bore. Tidal bores can be walls of water travelling up rivers and can be explained reasonably simply using knowledge of the local Froude number. Upstream of the bore, the river flow dominates and any waves that might be present in the river are unable to propagate upstream because the local value of the Froude number is too large (greater than one). The incoming tide might be funnelled up into the river through local geometry (as happens in the Bay of Fundy (US) or the River Severn (UK)). When this happens there is enough energy in the tide to halt the river flow and cause the wave to "break through". This can be explained by calculating the new local Froude number. Because the river flow has slowed, U will be smaller, and for the tide, the celerity (approximated by $\sqrt{gH}$) is large enough for the Froude number to be less than one. At the bore itself is the transition where the Froude number actually equals one. The bore then propagates as a shock wave up stream. Thus, the basic physics is explained in terms of dimensional analysis. In reality, there are different types of bores, and some of them are modelled using solitons (solitary waves with a single crest and no troughs, and a theoretically infinite wavelength). Dimensional analysis cannot model this kind of detail. This is typical of dimensional analysis; it is a crude tool for getting a first grip on the dynamics. Let us now utilise dimensional analysis in a slightly different way for dynamic balances. This is a preliminary look at dynamic balances that will be derived more fully in the next chapter.

2.3   Dynamic Balances

In order to derive dynamic balances in terms of dimensions all that is required is some knowledge of what the balances represent and the magnitudes of fundamental quantities.

The balance we choose to derive is Newton's second law, which states that force is mass times acceleration. In fluids this is called the Navier–Stokes equation. In terms of dimensions the Navier–Stokes equation holds true since acceleration has dimension LT^{-2} and force per unit mass has dimension $MLT^{-2} \times M^{-1}$, which is also LT^{-2}. Let us now dissect what is meant by acceleration. There are three different kinds of acceleration that occur in marine science (oceanography or meteorology) and to some extent coastal engineering and coastal dynamics of a large enough scale. The first is straightforward and is the fluid acceleration; the exact counterpart of acceleration of particles in mechanics. To distinguish this from the others it is sometimes called point acceleration. There is no doubt that this has dimension LT^{-2}. The second kind of acceleration occurs because the Earth is rotating. This is not normally important in most of coastal engineering, but we cannot ignore it completely as it most certainly features in tidal dynamics, large current systems and tsunamis. It can also feature in quite small scale stratified flows. When Newton formulated his second law, it had to be in what he called an *inertial* frame of reference. This means that the origin of co-ordinates and the axes are fixed or at most travel with constant speed in a straight line. On the Earth this is not so, and so Newton's second law has to be modified. The upshot is that we have to define the Coriolis parameter as $2\Omega \sin(latitude)$ where Ω is the angular velocity of the Earth $(= 7.29 \times 10^{-5} \text{ s}^{-1})$ and is usually denoted by the letter f (or γ if f has to be used elsewhere). The acceleration caused by the rotation of the Earth affecting the speed of an ocean current turns out to be $f \times$ current. This has dimensions $T^{-1} \times LT^{-1}$, which is once again LT^{-2}, i.e. acceleration. *Coriolis* acceleration, as this is called, turns out to be the most important acceleration for oceanographers and meteorologists and can also be important for coastal engineers, as has been said above. The third kind of acceleration is perhaps the most difficult to grasp. In mathematical terms it will be derived in the next chapter and has the form $(\mathbf{u} \cdot \nabla)\mathbf{u}$ with dimension $U^2 L^{-1}$ where U is speed. Acceleration formed this way does not contain time explicitly, it is called *advective* acceleration and can be present even when the current is steady. A river with a steady flow rounding a bend is quite a good illustration of advective acceleration. Let us examine this scenario. As the water rounds the bend, adjacent water particles — although travelling with constant speed and therefore not possessing point acceleration — change position relative to one another. Therefore, there is a change with time in some respect, even though the flow is a steady one. This acceleration is in fact exactly the same as the acceleration experienced by someone going around a corner at constant speed. It is erroneously called centrifugal force and more properly called centripetal acceleration. Centripetal acceleration is u^2/r, where u is speed and r is the radius of curvature of the bend. This is obviously of dimension $U^2 L^{-1}$, consistent with the form mentioned earlier. We have thus talked our way through the three kinds of acceleration. This will be useful when the more traditional mathematical treatment is given in Chapter 3.

The right-hand side of the Navier–Stokes equation of motion consists of the forces. In fact, it is the force per unit mass as, for fluids, the Navier–Stokes equation takes the form "acceleration equals force per unit mass" rather than "force equals mass times acceleration". One usually thinks of a small piece of sea and considers what acts on this in terms of accelerations and forces. The forces that act are pressure gradient forces, the gravitational force and friction. First of all consider gravity. This is the force that pulls the small piece of fluid towards the centre of the Earth. It is constant; but is it? First of all, the Earth is not exactly a sphere but is flatter at the poles. Secondly, it is not homogeneous and certain rocks can change the value of gravity locally, and thirdly, the Earth is spinning around on its axis once every day — this means that there is a centrifugal force (more correctly centripetal acceleration) acting at each point of the Earth (except the poles), directed towards the axis. However, it is easy to show that each of these variations is small. For example, the centripetal acceleration is much smaller than gravity. To do the sums, true gravity is around 10 m s^{-2} whereas the largest magnitude of centripetal acceleration, at the Equator, is $\Omega^2 R$, where R is the radius of the Earth. This is $(7.29 \times 10^{-5})^2 \times 6.36 \times 10^6$ in SI units. This is of order 4.6×10^{-3}; certainly a lot less than g, by a factor of over 1000. The other variations of gravity, although slightly larger, are also small enough to be ignored. Let us consider the "pressure force" dimensionally. The force that acts on our small piece of sea that is due to pressure will actually be due to the *difference* in pressure on either side. Think of a pipe that contains water; the reason water runs along a pipe is due to the pressure difference not the absolute value of pressure. The force is thus a pressure gradient force. Pressure gradient divided by density is $ML^{-2}T^{-2}/ML^{-3}$ which is LT^{-2}

as required. The dimension of the frictional force due to turbulence is also LT^{-2}; much more is said about friction in the next chapter. Newton's second law is often more accurately called the equation of conservation of linear momentum; let us now turn briefly to another conservation law, the conservation of matter.

In our small piece of sea, mass can neither be created nor destroyed. This is the conservation of mass and receives mathematical attention in the next chapter; here let us discuss some of its properties. In most marine science there is usually an important difference between horizontal and vertical length scales. The horizontal scale is indicated by L and the vertical scale is indicated by D. The ratio D/L is typically 10^{-3}. This is even true in a lot of coastal engineering, although it must be said there are examples where it is not the case. This difference is only absent in very small scale dynamics — sea surface waves and some aspects of river flow are these exceptions. The vertical velocity can be deduced in order of magnitude terms by looking at mass conservation, derived in the next chapter and expressed mathematically by the divergence of u being zero (see page 25)

$$\frac{\partial u}{\partial x} + \frac{\partial v}{\partial y} + \frac{\partial w}{\partial z} = 0.$$

The dimensions of each term are the same $1/T$ and for the first two terms they are U/L, where U is a typical horizontal current. For the third term it is W/D, where W is a typical vertical current. If all three terms are of similar magnitude we must have

$$\frac{U}{L} = \frac{W}{D},$$

whence

$$W = \frac{UD}{L}.$$

The dimensionless ratio D/L is called the *aspect ratio*. A similar term is used in aeronautical engineering, and its oceanic value 10^{-3} gives a good indication of the magnitude of typical vertical currents once the horizontal current can be estimated. In an upwelling region, where there are vertical currents that bring nutrient rich upwelled water to the surface which in turn fosters biological activity important to many an economy, a typical value of W might be 5×10^{-5} m s^{-1}, which is very small indeed. There are many cases when the third term $\partial w/\partial z$ is much smaller than the horizontal terms, so that continuity is well approximated by

$$\frac{\partial u}{\partial x} + \frac{\partial v}{\partial y} = 0.$$

In this case, no deduction can be made about the magnitude of the vertical current, except to say that it is less than (by at least an order of magnitude) the value of UD/L. Such are the limitations of dimensional analysis.

2.4   Measurement and Empirical Orthogonal Functions

Although this is a book about modelling, there are some aspects of numerical values that are not about numerical modelling. Aspects such as errors and stability concern the outcome of numerical approximation and belong firmly in Chapters 4 and 5; here measurement is discussed with special reference to its impact on modelling. When models are discussed in detail, the need to compare the results of models with data are always emphasised, but just what is meant by data? Data can be characterised in various ways; first of all there are general classifications similar to dimensional analysis. Some data are quite crude and take the form of collections; this is common in collections of biological data where a trawl through the sea collects numbers of different species. The result of this trawl might be classified by biological type, such as plant (phytoplankton) and animal (zooplankton), and the numbers of each might be counted. Perhaps the weight of each too. There might be a further subdivision into different plants and animals. Weights are measured to a certain degree of accuracy, and it is important that this accuracy is known by the modeller. Some measurements can be repeated; lots of trawls might be taken and some kind of average taken. It might be assumed that this improves the precision of the measurements. It might, but it might not for various reasons, the commonest of which is not being aware of a bias such as time of year, month or day, state of the tide, size of net, etc. Experimental scientists preserve the word *precision* for the repeatability of the collection of data whereas *accuracy* is how well (accurate) the measurements are made. The point about units of measurement has already been made in relation to dimensional analysis, so it is re-emphasised here that units must be correct and the scale of unit (SI when possible) correctly chosen. Compare this with current measurement, which is usually repeatable and can be made precise within the error of the device. Modern devices that are non-invasive are very accurate too. The remainder of this section is about the implementation of EOFs, which is the oceanographic counterpart of Factor Analysis introduced briefly in Chapter 1. It is tempting to believe that Factor Analysis is only used for biological data as the origins of Factor Analysis lie in the traditionally non-quantitative area of the social sciences with which biological science shares the features of being data dominated and less precise than, say, physics. This is incorrect these days as, for one, satellite data have made Factor Analysis and its more quantitative child PCAs (or EOFs) very useful for physical data too.

A bit was said about EOFs in the first chapter. The simple example there was two dimensional, which was given for presentational purposes. Let us now consider a data set that is more complex. Satellites routinely measure sea surface temperature as they follow their path around the globe. So, each time a reading is taken the time and temperature are recorded. One might expect this to be a single column of figures but that is not a useful way to present the data. More usefully, the temperature is taken over an oceanic region timed (say daily) over several years. The data are then arranged as a matrix to represent the track and the time, typically monthly mean temperatures over a whole year. For those familiar with MATLAB such an example can be found in Chapter 4 of Glover, Jenkins and Doney (2011). If the eigenvalues of the matrix are found and arranged in order of magnitude, it is found that almost all of the variation is in the first one. This means that the data are, in fact, largely one dimensional despite looking complex. So, although there are literally hundreds of variables each corresponding to an eigenvalue, looking like a bewildering complexity, apparently this is an illusion and it is a simple one-dimensional data set. This dominant variation however is the seasonal one; it is cold in the winter and warm in the summer. Just looking at the first eigenvalue would tell us this and not a lot more, so the next few are examined for more subtle variations and it is these that inform us about the detailed structure of the temperature, a structure not at all obvious from eyeballing the data. After the first few eigenvalues, the rest are less important as they could be artificially generated ("noise" to use signal processing terminology) by inaccuracies in the data. Clever things can be done with these data though. Filtering techniques can screen out the obvious seasonal data and these help even more detailed structure to appear. Similar techniques are used to enhance photographs; they are all over our TV screens these days in series such as CSI and in the UK the late lamented Spooks. However, back to the present example. So, processing such data and carefully scrutinising them after having found the first few EOFs reveals underlying structure, cycles, etc. This is modelling for understanding the processes, not predictive modelling. Let's now discuss EOFs in relation to Kalman filtering, which helps us to do this optimally.

2.5 The Kalman Filter and Data Assimilation

In the last 20 or so years, the Kalman filter has become a very useful tool in oceanography. There are entire texts devoted to its application so all we can hope to do here is introduce it and the part it plays in modelling. Briefly, the Kalman filter is an estimator; in oceanography and allied subjects it is used in conjunction with data assimilation and the computation of EOFs, as outlined above, to improve predictions. In order to read about its use, and perhaps get to use it yourself, there is a lot of technical material to wade through that make this not possible to do in this text; instead an explanation will be attempted using very little mathematics, but even so we will start gently. In later chapters, numerical methods are dealt with explicitly. In these chapters there are specific examples that involve interpolation and data assimilation that are now termed *kriging*. Kriging is used more and more in oceanography and coastal engineering since the availability of satellite data, but we postpone detailed exposition until after the introduction of numerical methods in Chapters 3 and 4. In Chapter 1 the idea of systems modelling was introduced, albeit only verbally. However, there it was proposed that modelling an environment like shallow sea dynamics could be viewed as a hard system governed by precise laws as opposed to social systems that could be termed soft governed, if they are governed at all, by statistical trends deduced by data usually obtained via questionnaires. These soft systems are difficult to simulate. For hard systems, more precise simulation is possible. If the output of a model is treated as a signal then the driving mechanisms such as wind and tide can be taken as inputs and the current so produced is an output. Other outputs could be a predicted sea level, the temperature or the salinity. This way of looking at modelling is allied to that of control theory and is completely natural for engineering devices, but perhaps less so for systems with environmental input. One of the most useful concepts to introduce when dealing with signals is that of "white noise". In order to define this we need to back track a bit. In the language of control theory, an input $\underline{y}$ acts upon some system represented by an operator L and this produces the output $\underline{x}$. An everyday example would be a child on a swing. Your push is the known input, the motion of the child the output and the swing would be governed by the operator. The problem is to find the output $\underline{x}$. In symbols

$$L\underline{x} = \underline{y}.$$

The operator L is commonly differential, as in the swing example, so that this equation is an ordinary differential equation that has to be solved for $\underline{x}$ given $\underline{y}$. However, L could be a matrix in which case a system of linear equations has to be solved. Often L represents more than one differential equation called a system of differential equations, but almost always the derivative is a time derivative, and that is key to the applications to oceanography. The output is usually in the form of a time series (these are met in Chapter 8 when real sea waves are analysed, but have a look at Figure 8.3 to see a typical time series). The tools for analysing time series are autocorrelation and cross correlation. The strict mathematical definitions of these can be set aside, but the autocorrelation measures the agreement that a time series has with itself some time later. Cross correlation measures the agreement between two separate signals. Control theory is a mature subject with its own terminology and vast literature. However, it tends to focus on deterministic problems and there are all sorts of techniques for solving these deterministic control theory problems. Probably the best known is the Laplace transform that conveniently converts ordinary differential equations into linear algebraic equations, see Dyke (2014). For problems with a stochastic input, analysis based around Fourier transforms is preferred. This is all with the proviso that the original problem was linear to start with of course. Nonlinear control is the next level and not discussed at all in this short section. It is the operator L that interests us here, as it is L that represents the system. Trying to find out about L can be straightforward for simple problems, but can also be like solving a murder mystery. One way to proceed is to look at the response of the system to particular inputs. The most fruitful input to use is the impulse or Dirac-δ input. This is what engineers call hitting it with a hammer; mathematicians refer to Green's functions. It is the input that excites only the basic natural frequencies. This has to be so as there is no frequency inherent in an impulse, so any in the output must have been produced internally, that is via the properties of the operator L. Any other input would also (probably) excite these frequencies but also "contaminate" the output with a frequency only present in the input. Another useful input is *white noise*. White noise is a signal where all frequencies are present with equal amplitude; it is literally at the other end of the spectrum to an impulse. A white noise signal has an autocorrelation function that is an impulse because there is no relationship at all between one point of the signal and any other. All time series are represented as wriggly lines, but white noise is the most wriggly it can get so that on the page it is simply a grey band.

In ocean science these days there are a lot of available data, often so much that there are too many to use in any particular model.

Then there are data that are in some sense better than the original and should be incorporated into the model in order to improve it. If this can be done as the model runs, it is called data assimilation and the use of data assimilation is now done in association with Kalman filters. The key feature of the Kalman filter is that it is sequential, so that it only needs the values of variables at a single (the previous) time step in order to operate. It is therefore very suited to prediction problems. Details about numerical methods are the subject of Chapter 4, but briefly we write

$$\frac{dx}{dt} \approx \frac{x(t + \Delta t) - x(t)}{\Delta t},$$

so that given $x(t)$ and Δt if there is a differential equation for dx/dt then it is possible to approximate the value that x has at this later time Δt. It is Δt that is the time step. It is the model that comes up with the value of $x(t)$ based on knowledge of the processes. If there are many alternatives for finding $x(t+\Delta t)$ because of uncertainties in data or whatever, then the Kalman filter can look at all the possible versions of $x(t + \Delta t)$ and, using statistical estimation, get the best one. Traditional Kalman filtering is based securely on least squares estimation. In the language of control theory, there is a proposed relationship between the state of the system and the measurement of this same state. The least squares method selects the parameters that can be adjusted in the description of the state in such a way that the square of the distance between the state and the measurements is a minimum. In this way, the state and measurements are as close as possible. Having achieved this, the linear Kalman filter is cleverer. It is tempting just to give the mathematics, but this will not be done. There have to be a lot of observations for the method to bear fruit; and the Kalman filter is a method that minimises the variance of the estimate of the difference between the model variables and the observations. This is done by using analysis similar to PCA met in Chapter 1. Kalman filtering depends on various assumptions such as the "noise" in both the variation of the model predictions and observations being white, and both being independent of initial conditions. The consequence of a successful application of the Kalman filtering technique would be the best possible predictions given the observations. This idealised state is not reached, however. Practical problems include data whose variance do not obey the white noise rule and have a bias, in which case the Kalman filter is not optimal. Nonlinearities will always be a problem if they are significant, although there are nonlinear modifications to the Kalman filter. Some do not agree with using Kalman filter to estimate parameters by fixing the observations and getting the best model fit. How the process converges is also not well enough understood; however, it does seem to work.

With the advent of remote sensing and much more sophisticated data, traditional Kalman filtering techniques are limited due to computational constraints associated with such a rich multidimensional data set. Different techniques based on systematic reduction of the dimensions to render the estimation problem tractable are now being developed. Some of you might have spotted a link to the last part of Chapter 1 here. The use of PCA is the selection of just the right data to represent a more complex set and so would seem to solve precisely this kind of reduction problem. Forward referencing Chapter 9, techniques based upon the Kalman filter will be very useful in ecosystem modelling as the number of variables get large and the data set even larger. However, let us leave this here and wind up this chapter by looking at other different modelling techniques.

2.6 New Ideas on Modelling

In the 1960s the scientific world was agog with the new power of the computer, and in particular all sorts of possibilities were being sold as being "just around the corner" for the new technology. Amongst the most intriguing was the possibility of artificial intelligence (abbreviated to AI). Science fiction writers had been writing about robots for years, and they now seemed not only possible but probable. The idea of robots programmed to do all those awful jobs that humans do not like doing, and even jobs that humans are unable to do, was extremely enticing. It was quickly learned that computers could do very quickly what humans took a long time to do, for example, any routine computation, searching through lots of possibilities quickly, that sort of thing. What was not realised was that tasks that humans found very simple, such as recognising faces, or the correct use of idioms in language, were not just hard for a computer but virtually impossible. A two-year old can crawl across a toy ridden floor with ease; a sophisticated robot cannot, even today. This slowed the development of robotics and other forms of AI so much so that it fell out of the public eye. Viewing 40-year old television programmes about the dawn of the AI era became a source of amusement. However, AI has not stood still and in the last 10 or so years there have been two developments that will impact upon modelling. These are the use of genetic algorithms (GAs) and the use of neural networks (NN). They are sometimes confused, so a section will be devoted to each here. It is fair to say that neither has impacted enough in coastal engineering or nearshore oceanography for an example to be done. When this does occur, this section will move to follow Chapter 4 because numerical methods would certainly be required.

Without turning this text too much into one on computer science, there are a few basics to know about. The first is about the speed and architecture of a computer. Many will be aware of Moore's Law that states that the speed of computation will double every two years. The original version was expressed in terms of hardware, but essentially it states that growth of computer power is exponential. Only in recent times has this slowed down, principally due to size limitations; the innards of a computer chip have dimensions approaching that of the atoms and molecules that make up the substance of the chip. Recently, the architecture of graphics is influencing computational schemes as well as speed. In graphics, scenes have to be produced and this is now done in a parallel way, so the Central Processing Unit (CPU) of a computer is being superseded by a new Graphics Processing Unit (GPU) that works in parallel. Parallel processors have been with us for 20 years or more, but the architecture of recent GPUs is different. So much so that some of the numerical algorithms that are introduced in Chapters 4 and 5 will have to be modified. A (forgive the pun) parallel development is quantum computing, another new development that takes account of the quantum nature of very small atomic scale. In theory, quantum computers are as parallel as possible, with all possible outcomes of a computation calculated residing in a kind of probabilistic mist until the final results drop out. Quantum computers are now tantalisingly close to being a reality. Their presence will not only change modelling but also revolutionise internet security. So these new developments will come; let us return to the computing basics required in order to understand the operation of GA and NN.

The first of these to consider is a decision tree. Perhaps a simple example will explain the basics. Twenty or so years ago you might have been faced with a machine that asks you questions of a medical nature, then this machine operates using decision trees. It asks you questions to which the answer is yes or no. "Do you have a history of heart problems?" (Answer yes or no.) "Chest pain less than 24 hours ago?" (Again answer yes or no.) This would go on until the machine — called an expert system back then, now probably termed a decision support system — could effect a diagnosis, usually linked to a probability. There are many topics that are essential to making a computer more flexible than just a number cruncher, amongst which are predicate logic, other subjects to do with the processing of language such as syntactic analysis, then there are vision systems to enable computers to recognise objects (fashionable now for security devices alternative to PIN numbers) and similar voice recognition systems. Combining these can give rise to sophisticated search techniques. However, these cannot be directly applied to modelling. GAs can.

2.6.1 *Genetic algorithms*

A genetic algorithm (GA) is really just an extremely sophisticated search technique, it builds upon all the techniques mentioned above but is also fundamentally distinct. GAs use tools that were introduced in Chapter 1, in particular many use PCA to help with their search. A simple use of a GA would be timetabling in a school or university. There is a database of classrooms including location, size and various facilities and of course all students, their modules or classes, and the same for all the courses. Constraints can be specified such as the Dean requires certain people to be available at certain times, or for health and safety reasons certain facilities cannot be used for more than three hours without a break, and so on. Classes are then asked to be scheduled such that neither students nor staff are asked to be in two places at once. Preferences can be built in but they need not be hard and fast. For example, try to avoid (university) lectures at 16.00 on a Friday, but if there is no other choice allow them. Avoid students being without a lunch break (one hour in the 12.00–14.00 slot). The name genetic algorithm is inspired by biology, and GAs are often called evolutionary algorithms. As in nature, it is the survival of the fittest, and it is this Darwinesque natural selection process that distinguishes it from simple search techniques. If the computer has a set of rules that conflict in some way, then by looking at all the outcomes in a trial the computer "realises" that there is a preferred combination (the child as it were) and gets a solution that works. The way that the computer achieves this is via sophisticated programming and is called either crossover or mutation in the computing literature. Crossover swaps over two elements in a operation to produce a new operation, whilst mutation, as the name implies, changes an operation by a small adjustment. This kind of breeding is done under the umbrella of an evaluation function in which there is a measure that is used to ensure improvement. Inefficient characteristics that do not improve the algorithm are discarded whereas efficient characteristics are retained. In this way, biologically almost, the algorithm is improved in efficiency until some kind of optimum is reached. So, for the timetabling problem the best solution is arrived at. GAs are beginning to be used in oceanography; for example in time series and in estimating other oceanographic parameters. In Chapter 8, wave forecasting will be introduced alongside the concept of a time series. Briefly, a time series is the kind of wriggly line one might meet when looking at the Dow Jones or FTSE financial index. Success in forecasting how such wriggly lines will behave in the future is what financial forecasting is all about. Similarly, the prediction of flooding, forecasting sea level changes and the like is the life blood of environmental forecasting. Traditional techniques are detailed in Chapter 8, however, if one has lots of different possibilities for choosing values of critical parameters, then the situation is similar to the timetabling problem outlined above. GAs can then be employed to select optimal values by using crossover and mutation techniques. There is evidence that this is being done successfully in recent publications (Google "evolutionary algorithms for time series forecasting" and see what appears under Google Scholar). Another use of GAs is in the use of unstructured grids, which are mentioned in Chapter 4. Selection of exactly how the grid morphs will depend on the weight attached to each factor that changes the dimensions and other geometry of the grid. Precisely how this is done is the subject of Chapter 4, however, if there are many contributory factors then a GA could be used to optimise the selection in terms of the minimisation of truncation error, for example. Most papers on this at the moment tend to be theoretical, but the future certainly points to the greater uses of GAs.

2.6.2 *Neural networks*

An NN is another, and again very different, kind of approach to modelling. As the name implies, it is based on how the brain works: by connecting together in a sophisticated way a number of very simple processing units. The ways the neurons are connected have weights associated with them. These weights determine how the network operates and can be changed by subjecting the network to data. The secret is to subject the network to trial data of the type you need to analyse: this is called *training* the NN. When the network is fully trained it is in some sense optimal and can be used to solve the problem. Of course, there is no real way of determining when this optimality is reached as it is in the connections not the neurons themselves that the developments occur. In this sense, this is a "black box" model. The capacity of current computers means that the complexity of a computer-based NN can be large. Therefore, problems that are best solved on an NN are those with masses of appropriate data upon which the NN can be trained. Good examples of problems particularly suited to NN modelling are computer speech, whereby a computer can learn to mimic human speech, by learning, and in the computer recognition of handwriting. These and similar pattern recognition tasks are very suited to analysis using NNs. The medical expert system mentioned on page 80 can be improved markedly by the injection of lots of (anonymous of course) medical data upon which to train the model. Are there any oceanographic or coastal engineering applications? The answer is not yet, but it is not difficult to see that as more and more data are acquired there could be. Allowing a finite difference model to be designed as an NN, one could design each neuron as a specific module containing a specific way to model a particular process. The weights could then be changed depending on the environment or type of process being modelled. For now though, this has not yet happened.

Let us now return to tried and tested modelling, and in the next chapter the basic relationships vital to the modelling of shallow seas and continental shelf sea dynamics are derived — hopefully in a user-friendly way.

2.7   Exercises

(1) Using dimensional analysis, propose a formula for the flow rate Q of a viscous fluid along a pipe of radius a; the pressure of the fluid is p and the dynamic viscosity is μ. Assume that the flow rate is inversely proportional to the length of the pipe l.

(2) A sphere is moving through a viscous liquid. Assuming the drag F depends on its diameter D, speed U, density ρ and dynamic viscosity μ, determine a power law formula for F given the Reynolds number $\frac{\mu}{\rho U D}$ is constant.

(3) Calculate the dimensions of the following terms

$$\nu \frac{\partial^2 u}{\partial z^2} \quad \text{and} \quad fu,$$

where ν is eddy viscosity, u is the current and f is the Coriolis frequency. If the ratio of these two terms is the Ekman number E_V, find E_V in terms of ν, f and a typical depth D.

(4) If two distinct signals are both white noise, what form does the cross correlation take?

(5) There are two signals that are perfect sine waves. What is the cross correlation if the signals are (i) in phase, (ii) 90 degrees out of phase, (iii) 180° out of phase? There is no need to use any mathematics, but if you want to, take one signal as $\sin kt$ and the others as, respectively, $\sin kt$, $\cos kt$ and $-\sin kt$. What is the autocorrelation of any of them? (They are all the same.)

(6) Consider GAs and NNs. Give one characeristic that is common to both, and give a characteristic of a GA that is not part of an NN and vice versa.

Chapter 3

Mathematical Foundations

3.1 Introduction

In the last chapter, the ideas of modelling in general were discussed. In particular, the very important technique of dimensional analysis and other notions of modelling were introduced. However, dimensional analysis techniques, useful as they are, can only give information about which terms to omit from a model. They do not give any information about the precise form important terms might take, except perhaps power law possibilities. They do not, for example, quantify rates of change due to mass conservation or the laws of fluid motion. This chapter begins to address these shortcomings. By its very nature, there is some mathematics to wade through here. It is possible to leave out the mathematics if the vector calculus of Chapter 1 was all too much, but some effort to follow the calculations through is rewarding. There is very little actual mathematical calculation (except in the theory of linear water waves), it is more understanding what the symbols mean and one or two relations that connect them. It has very little to do with the usual reasons for the dislike of mathematics — not being able to remember integration techniques or get your head around proofs and the like — so the hope is that most will find the derivations of these basic laws useful.

Once the equations are there they can be used to build simple models of linear water waves, tides, wind-driven flow, etc.

3.2   Fluid Mechanics

It has now been over 300 years since Sir Isaac Newton formulated his famous view of mechanics. Despite the recent advances in physics which have led to new theories of the very large and fast (general relativity, *circa* 1915) and the very small and also fast (quantum mechanics, *circa* 1926), the motion of virtually all known objects closely follows the laws set down by Newton all those years ago. Relativity and quantum mechanics are in fact not only incompatible with Newtonian mechanics, they are at odds with each other. Fortunately, neither are relevant to ocean science which deals with less extreme mass, length and time scales. In marine science and coastal engineering, Newton's laws hold. The adaptation of Newton's laws, notably his second, to fluids was done during the 18th century by Euler and (Daniel) Bernoulli. In words, this law is "force = mass × acceleration", which in fluids reads "force per unit mass = acceleration". Bernoulli's contribution was to convert the equation of fluid motion into the form of an equation which expresses the conservation of mechanical energy.

The next logical step is to derive these equations in their mathematical glory. In order to do this, some assumptions have to made about the knowledge of you, the reader. We have already assumed that you know about differentiating and integrating, and maybe after reading Chapter 1 you also know something about vectors up to and including div., grad and curl ($\nabla.$, ∇ and $\nabla\times$), so we can proceed with some ease. If this is not the case and even after working through Chapter 1 you still get a headache trying to understand these symbols, simply skip the derivation and accept the result. Or, if this sticks in the throat, you need a more substantial introduction to vector calculus (the book by Marsden and Tromba (1988) is a good one, but there are plenty of others). The easiest equation to derive is the conservation of mass. Consider an arbitrary volume of the sea, depicted schematically in Figure 3.1. The total amount of mass leaving this volume does so through the surface surrounding the volume. This has to be zero because there are no sources or sinks in the sea. Hence, if **u** is the current (fluid velocity) and $d\mathbf{S}$ is the

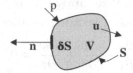

Fig. 3.1 An arbitrary volume of fluid.

(vector) element of the surrounding closed surface of our arbitrary volume, then

$$\int_S \mathbf{u} \cdot d\mathbf{S} = 0,$$

where the direction of $d\mathbf{S}$ is the unit outward drawn normal. There is a well known result in vector calculus for transforming a surface integral, which is what this is into a volume integral. This was outlined verbally in Section 1.3.2 and is called Gauss' flux theorem and takes the form

$$\int_S \mathbf{F} \cdot d\mathbf{S} = \int_V \boldsymbol{\nabla} \cdot \mathbf{F} dV.$$

Using this we see that

$$\int_V \boldsymbol{\nabla} \cdot \mathbf{u} dV = 0,$$

where V is our arbitrary volume. As it *is* arbitrary, this can only be true if

$$\boldsymbol{\nabla} \cdot \mathbf{u} = 0$$

throughout the sea. This is the equation of conservation of mass, often called the continuity equation. This is valid as long as the density of the fluid remains constant. If the density is time dependent, then its rate of change causes a change in mass of our volume and appears as an extra term. Fortunately, this is only worth considering if one is interested in acoustics where wave speeds similar in magnitude to the speed of sound occur. This is not the case in coastal engineering. Density can change, for example in estuaries and rivers, but this change does not interact with the dynamics and we use what is called the Boussinesq approximation. Under this approximation the continuity equation remains as above. It is a single scalar equation, and not using the symbols of vector calculus can be written

$$\frac{\partial u}{\partial x} + \frac{\partial v}{\partial y} + \frac{\partial w}{\partial z} = 0,$$

where $\mathbf{u} = (u, v, w)$ and we have adopted Cartesian co-ordinates (x, y, z). The double use of the symbol u is less confusing than alternatives. This equation was stated in the last chapter. There is a version of this equation that is particularly useful for coastal engineers and shallow sea oceanographers and we will return to it when we consider shallow water approximations. For now, let us move on to the conservation of momentum, which is Newton's second law. We still consider a small, not infinitesimal but arbitrary, volume of the sea that does not touch any boundary, refer back to Figure 3.1 again. The acceleration of the current within this volume is

$$\frac{d\mathbf{u}}{dt}.$$

This may seem a straightforward extension to acceleration of particles moving with velocity $\mathbf{u}$, but there is a subtlety. The volume itself moves around, so technically we have

$$\mathbf{u}(x, y, z, t) = \mathbf{u}(x(t), y(t), z(t), t).$$

This has the consequence that when we calculate the rate of change of $\mathbf{u}$ we really need to take account of the dependencies of x, y and z on t to do a proper job. The chain rule comes to our aid here

$$\frac{d\mathbf{u}}{dt} = \frac{\partial \mathbf{u}}{\partial x}\frac{dx}{dt} + \frac{\partial \mathbf{u}}{\partial y}\frac{dy}{dt} + \frac{\partial \mathbf{u}}{\partial z}\frac{dz}{dt} + \frac{\partial \mathbf{u}}{\partial t}$$

or using vector notation and that

$$\frac{dx}{dt} = u, \quad \frac{dy}{dt} = v \quad \text{and} \quad \frac{dz}{dt} = w,$$

we can write

$$\frac{d\mathbf{u}}{dt} = (\mathbf{u} \cdot \nabla)\mathbf{u} + \frac{\partial \mathbf{u}}{\partial t}.$$

You may question why it is we have go through all this. The problem is that Newton's second law has to have a fixed origin and fixed axes. We have an arbitrary volume of a fluid which does not have either fixed origin or fixed axes, hence the need to reference x, y and z back to t, and to state that the motion starts when $t = 0$. This is known in fluid mechanics as "differentiation following the fluid" or the Lagrangian point of view. To make doubly sure that the distinction from "differentiation at a point" or the Eulerian point of view is drawn, the notation with a capital D is often used, so that we write

$$\frac{D\mathbf{u}}{Dt} = \frac{\partial \mathbf{u}}{\partial t} + (\mathbf{u} \cdot \boldsymbol{\nabla})\mathbf{u}.$$

The partial derivative with respect to time is also often written first as shown. One immediate consequence is that for flows that do not explicitly involve time, called steady flows, there is still an acceleration. This will be familiar to those who know about rigid body motion — even if a body rotates steadily there is still an acceleration v^2/a directed towards the centre of rotation (v is the speed and a is the radius); we talked through this in Chapter 2. So the acceleration of our small mass of sea is, using our new "big D" notation

$$\frac{D}{Dt}\left(\int_V \rho\mathbf{u}dV\right) = \int_V \rho\frac{D\mathbf{u}}{Dt}dV = \int_V \rho\left(\frac{\partial \mathbf{u}}{\partial t} + (\mathbf{u} \cdot \boldsymbol{\nabla})\mathbf{u}\right)dV,$$

where we have adopted the Boussinesq approximation and assumed that the density ρ remains constant in this context. Marine scientists also need to consider the Earth's rotation, but we shall leave this for later. Instead, let us turn to what is here the right-hand side of Newton's second law. This contains the forces that act on our small mass of sea. In a fluid, there is always pressure. This is a force per unit area that acts on the surface of our small volume, but points inwards (in a direction $-\mathbf{n}$). Gravity also acts, but throughout the volume. Frictional forces (viscosity in laminar flow, but turbulence in general) also act throughout. (Forward referencing, it turns out that the mathematical form of friction, like pressure, stems from considering the action on the surface of our volume. For now though, we are simply using the word "friction", so it acts throughout.) Therefore we have

$$-\int_S p\mathbf{n}dS + \int_V \rho\mathbf{g}dV + \int_V \rho(\text{friction})dV$$

as the total force. Before equating force to mass times rate of change of velocity in this volume, we need to convert the one surface integral into a volume integral. A corollary to Gauss' flux theorem, rather than the theorem, itself, tells us that

$$\int_S p\mathbf{n}dS = \int_V \boldsymbol{\nabla}pdV.$$

Newton's second law is therefore

$$\int_V \rho\left(\frac{\partial \mathbf{u}}{\partial t} + (\mathbf{u}\cdot\boldsymbol{\nabla})\mathbf{u}\right)dV = -\int_V \boldsymbol{\nabla}pdV + \int_V \rho\mathbf{g}dV + \int_V \rho(\text{friction})dV,$$

and since the volume is arbitrary, and dividing by ρ, we have

$$\frac{\partial \mathbf{u}}{\partial t} + (\mathbf{u}\cdot\boldsymbol{\nabla})\mathbf{u} = -\frac{1}{\rho}\boldsymbol{\nabla}p + \mathbf{g} + (\text{friction}).$$

This is Newton's second law for fluids and is called the Navier–Stokes equation. As already mentioned in Section 2.3, the question of how to model friction will exercise us, but later. Also keep in mind that for some applications the rotation of the Earth needs to be considered.

3.3   The Shallow Water Equations

In coastal seas, there is often a direct relationship between the fluid pressure at a point and the height of water above this point. This is called hydrostatic pressure and is valid for most situations. The only exception occurs if one is looking at the dynamics of surface waves, although with smaller and smaller resolution in numerical models non-hydrostatic effects are assuming some importance in local area models. Pressure that is hydrostatic is given by:

$$p(x, y, z, t) = p_A + \rho g(\eta(x, y, t) - z),$$

where p_A is atmospheric pressure and $\eta(x, y, t)$ is the sea surface elevation. A schematic diagram is shown as Figure 3.2.

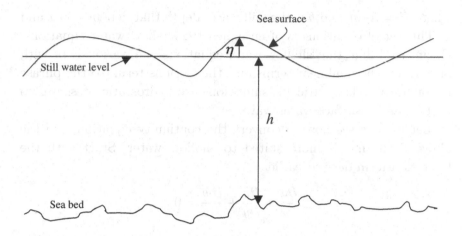

Fig. 3.2   A schematic diagram of a free surface η in water of depth h.

This means that the pressure gradient terms in the equations of motion can be written in terms of gradients in the sea surface as follows

$$\frac{\partial p}{\partial x} = \rho g \frac{\partial \eta}{\partial x} \quad \text{and} \quad \frac{\partial p}{\partial y} = \rho g \frac{\partial \eta}{\partial y},$$

provided the density does not depend on x or y. It can be argued that p_A, the atmospheric pressure, *does* depend on the horizontal co-ordinates. However, rates of change of p_A are much smaller than others as the atmosphere operates on a much larger length scale than the sea, especially the coastal sea. So the two horizontal (x and y) Navier–Stokes equations become, under these circumstances

$$\frac{\partial \mathbf{u}}{\partial t} + (\mathbf{u} \cdot \boldsymbol{\nabla})\mathbf{u} + 2\boldsymbol{\Omega} \times \mathbf{u} = -g\boldsymbol{\nabla}\eta,$$

where friction has been ignored but Coriolis acceleration has been included. The z equation is

$$\frac{\partial w}{\partial t} + (\mathbf{u} \cdot \boldsymbol{\nabla})w = 0$$

and the continuity equation takes the form

$$\frac{\partial H}{\partial t} + \boldsymbol{\nabla} \cdot (H\mathbf{u}) = 0,$$

where $H = h + \eta$, and h is the still water depth that depends on x and y. This set of equations is often called the shallow water equations. Having said that, the "shallow water equations" can variously be with or without the nonlinear terms and the Coriolis term, so the phrase is not precise. The critical assumptions are hydrostatic pressure, no friction and a surface water wave.

Let us now see how to convert the continuity equation into the above form that is more suited to shallow water. Start with the Cartesian form derived earlier

$$\frac{\partial u}{\partial x} + \frac{\partial v}{\partial y} + \frac{\partial w}{\partial z} = 0$$

and integrate this between the sea bed $z = -h(x, y)$ and the sea surface $z = \eta(x, y, t)$. The third term can actually be integrated so we obtain

$$\int_{-h(x,y)}^{\eta(x,y,t)} \left(\frac{\partial u}{\partial x} + \frac{\partial v}{\partial y} \right) dz + [w]_{-h(x,y)}^{\eta(x,y)} = 0.$$

We now perform some mathematics on this equation that involves use of what is sometimes known as the Leibniz rule but is more helpfully termed differentiation under the integral sign. The rule in its generality, but for only one variable, is

$$\frac{d}{d\alpha} \left\{ \int_{a(\alpha)}^{b(\alpha)} f(x, \alpha) dx \right\} = \int_{a(\alpha)}^{b(\alpha)} \frac{\partial f}{\partial \alpha} dx + \frac{db}{d\alpha} f(b(\alpha), \alpha) - \frac{da}{d\alpha} f(a(\alpha), \alpha).$$

It concerns the derivative of a function of one variable (α) but this variable is present in the limits of an integral, hence the two extra terms on the right-hand side. For us, we have x and y occurring in both the upper and lower limits of the integral, and so extra terms need to be added when changing the integration and differentiation operators. Additionally, we need to use the rule backwards, as it were. Here are the steps: first the two results

$$\int_{-h(x,y)}^{\eta(x,y,t)} \left(\frac{\partial u}{\partial x} \right) dz = \frac{\partial}{\partial x} \int_{-h(x,y)}^{\eta(x,y,t)} u(x, y, z, t) dz - u \frac{\partial \eta}{\partial x} + u \frac{\partial h}{\partial x}$$

and

$$\int_{-h(x,y)}^{\eta(x,y,t)} \left(\frac{\partial v}{\partial y}\right) dz = \frac{\partial}{\partial y} \int_{-h(x,y)}^{\eta(x,y,t)} v(x,y,z,t)dz - v\frac{\partial \eta}{\partial y} + v\frac{\partial h}{\partial y},$$

where we note that both u and v can depend on z. At the surface of the sea, the vertical velocity is the same as the rate of change of the sea surface elevation with respect to time. This rate of change is the total derivative which incorporates the assumption that surface water particles stay at the surface. This is called the kinematic surface condition. So

$$[w]^{\eta(x,y,t)} = \frac{D\eta}{Dt} = \frac{\partial \eta}{\partial t} + u\frac{\partial \eta}{\partial x} + v\frac{\partial \eta}{\partial y}$$

and also at the sea bed

$$[w]_{-h(x,y)} = -\frac{Dh}{Dt} = -u\frac{\partial h}{\partial x} - v\frac{\partial h}{\partial y}.$$

Inserting these expressions into the integrated continuity equations, after the cancellation of terms, gives

$$\frac{\partial}{\partial x}\left(\int_{-h(x,y)}^{\eta(x,y,t)} u\,dz\right) + \frac{\partial}{\partial y}\left(\int_{-h(x,y)}^{\eta(x,y,t)} v\,dz\right) + \frac{\partial \eta}{\partial t} = 0. \qquad (3.1)$$

This is true even when u and v are dependent on the vertical coordinate z. However, if they are independent of z then the integration can be performed explicitly and we get

$$\frac{\partial}{\partial x}(Hu) + \frac{\partial}{\partial y}(Hv) + \frac{\partial \eta}{\partial t} = 0, \qquad (3.2)$$

where we have written

$$H(x,y,t) = h(x,y) + \eta(x,y,t),$$

known as the total height.

Perhaps the best known form of the shallow water equations in coastal sea dynamics is the linear version but including Coriolis acceleration. This is defined and derived in section 3.6.1, so take the extra terms on trust for now. These are also called the Laplace tidal equations and take the form

$$\frac{\partial u}{\partial t} - fv = -g\frac{\partial \eta}{\partial x}, \qquad (3.3)$$

$$\frac{\partial v}{\partial t} + fu = -g\frac{\partial \eta}{\partial y}, \qquad (3.4)$$

$$\frac{\partial \eta}{\partial t} + \frac{\partial}{\partial x}(hu) + \frac{\partial}{\partial y}(hv) = 0. \qquad (3.5)$$

Here, $h = h(x, y)$ but non-linearities with the sea surface elevation η and velocity components u, v and w have been neglected. We return to the Laplace tidal equations for Kelvin waves later in Chapter 6.

3.3.1   *Bernoulli equation*

Let us once more consider the Navier–Stokes equation but this time ignore friction. This version is often called Euler's equation for fluid flow. This distinguishes it from all the other equations attributed to that most prolific of mathematicians Leonhard Euler (1707–1783). Here it is:

$$\frac{\partial \mathbf{u}}{\partial t} + (\mathbf{u} \cdot \boldsymbol{\nabla})\mathbf{u} = -\frac{1}{\rho}\boldsymbol{\nabla}p + \mathbf{g}.$$

When Newton's second law is considered in mechanics, there is often mileage in solving problems using an energy equation. In fluid mechanics the same is true. The energy equation for fluids is called the Bernoulli equation (after Daniel Bernoulli (1700–1782), one of the prolific and talented Bernoulli family and a contemporary of Euler). Let us now derive it, or more accurately them, since there are two versions valid under different circumstances. Let us start with perhaps the most straightforward case, that of an irrotational fluid. A fluid is called irrotational if, at all points of the fluid, the vorticity (defined as $\boldsymbol{\zeta} = \boldsymbol{\nabla} \times \mathbf{u}$, see section 3.4) is zero. This seems a strange quantity to define, but the vorticity is a quantity that tends to be preserved in a fluid. In fact, it plays an analogous role in fluids to angular momentum in mechanics and gets a section to itself next. In order to use the fact that the fluid is irrotational, we need to utilise one of the worst looking vector identities; here's the one we need

$$\nabla(\mathbf{F} \cdot \mathbf{G}) = \mathbf{F} \times (\nabla \times \mathbf{G}) + \mathbf{G} \times (\nabla \times \mathbf{F}) + (\mathbf{F} \cdot \nabla)\mathbf{G} + (\mathbf{G} \cdot \nabla)\mathbf{F}.$$

Putting $\mathbf{F} = \mathbf{G} = \mathbf{u}$ then rearranging gives the slightly simpler looking formula

$$(\mathbf{u} \cdot \nabla)\mathbf{u} = \mathbf{u} \times (\nabla \times \mathbf{u}) + \nabla \left(\frac{1}{2}\mathbf{u}^2\right).$$

Under the assumption of irrotationality, the first term has to be zero, hence

$$(\mathbf{u}.\nabla)\mathbf{u} = \nabla \left(\frac{1}{2}\mathbf{u}^2\right).$$

The left-hand side is the advective acceleration, and the right-hand side is grad of a quantity. This is useful to us and is why we needed to raid the larder of vector identities. If $\nabla \times \mathbf{u} = \mathbf{0}$ then there exists a scalar ϕ, called a potential, such that $\mathbf{u} = \nabla\phi$. Note that it is *a* potential rather than *the* potential, but potentials of a given current only differ at most by constants. On the other side of the Navier–Stokes equation there is the pressure term. This is already a gradient, so maybe that's fine. Not quite. As we have said before, sometimes the density varies and although this variation can be ignored with respect to dynamics (the Boussinesq approximation) it still varies, so can we take the $1/\rho$ inside the gradient operator? If the sea is stratified (the density varies with depth) then density is a function of pressure alone and we can write

$$\frac{1}{\rho}\frac{\partial p}{\partial x} = \frac{\partial}{\partial x}\int \frac{dp}{\rho}$$

so that

$$\frac{1}{\rho}\nabla p = \nabla \int \frac{dp}{\rho}.$$

The equation of motion — Navier–Stokes or here without friction called the Euler equation — now looks like

$$\frac{\partial}{\partial t}(\nabla\phi) + \nabla \left(\frac{1}{2}\mathbf{u}^2\right) = -\nabla \int \frac{dp}{\rho} + \nabla(-gz),$$

where we have used that

$$\nabla(-gz) = \mathbf{g},$$

that is $-gz$ is the (gravitational) potential. $\mathbf{g}$ is assumed to point directly downwards, which is true (ignoring variations due to the oblate nature of the Earth, local geology, and the centripetal acceleration of the Earth — these are all very small and all very negligible). Hence,

$$\nabla\left(\frac{\partial\phi}{\partial t} + \frac{1}{2}\mathbf{u}^2 + \int\frac{dp}{\rho} + gz\right) = \mathbf{0}$$

and thus

$$\frac{\partial\phi}{\partial t} + \frac{1}{2}\mathbf{u}^2 + \int\frac{dp}{\rho} + gz = F(t),$$

where $F(t)$ is an arbitrary function of time. Maybe the integration was a bit quick. What actually happens is that we use a bit more vector calculus. Since

$$\nabla\Phi = \mathbf{0}$$

for some scalar function Φ then it is also true that

$$\mathbf{dr}\cdot\nabla\Phi = d\Phi = 0,$$

where $\mathbf{dr}$ is, for us, the infinitesimal arc of a streamline in the sea. Integration then proceeds at once. With the time dependence absent from the integration process, the arbitrary "constant" of integration has to be allowed to be dependent on time.

Suppose now that there is no time dependence at all; so called steady flow. Sure, the time derivative term disappears and the constant on the right really is constant, but there is something more subtle. If we relax the irrotationality condition, then it turns out that Bernoulli's equation can still be derived. Recall that vector identity

$$(\mathbf{u}\cdot\nabla)\mathbf{u} = \mathbf{u}\times(\nabla\times\mathbf{u}) + \nabla\left(\frac{1}{2}\mathbf{u}^2\right),$$

but this time do not assume that $\nabla\times\mathbf{u} = \mathbf{0}$. Provided $\mathbf{u}\times(\nabla\times\mathbf{u}) = \mathbf{0}$ then the derivation of the Bernoulli equation proceeds as before because the time derivative term has gone. This means that, for steady flow, the Bernoulli equation of the form

$$\frac{1}{2}u^2 + \int \frac{dp}{\rho} + gz = \text{constant}$$

is valid on surfaces where

$$\mathbf{u} \times (\nabla \times \mathbf{u}) = \mathbf{u} \times \zeta = 0,$$

where ζ is the vorticity. These are surfaces that contain the streamlines. (This is not shown here, but stems from taking the curl of Euler's equation. Those interested in the details need to see more advanced texts on fluid mechanics.) The sea surface is such a surface, and hence, on the sea surface the Bernoulli equation will hold. This is useful as it relates pressure to current, albeit nonlinearly.

One of the main applications of the Bernoulli equation for coastal engineers comes through the modelling of surface waves. If waves are objects never before modelled, after a section on vorticity the following sections will provide a useful introduction and are certainly necessary before tackling linear water wave theory.

3.4 Vorticity

Vorticity is a very important fundamental quantity and has a special place in all fluid mechanics; geophysical fluid dynamics in particular. As was mentioned in the last section, it is the fluid counterpart of angular momentum that is the vector cross product of radial distance and momentum. Physically, the angular momentum of a rigid body is the perpendicular distance of a rotating body from the centre of rotation multiplied by the linear momentum possessed by that body, namely, using vector notation

$$\mathbf{r} \times m\mathbf{v},$$

where the direction is defined at right angles to the plane of the radius vector and the velocity of the centre of mass of the rigid body. The magnitude is the mass times the speed times the perpendicular distance of the centre of mass from the origin. It is the quantity that is preserved by an ice skater pirouetting; as the skater draws her arms in, the distribution of the mass of the skater about the centre of mass is closer to the axis (that is interpreted as a decreasing $|\mathbf{r}|$, clear to those who know about moments of inertia and in particular the radius of gyration), so the speed of rotation has to increase in order to preserve this angular momentum. The increase in $|\mathbf{v}|$ exactly balancing the decrease in $|\mathbf{r}|$. In mechanics, the angular momentum of a body is a quantity that tends not to vary. This fact lies behind the behaviour of the planets and satellites outside the Earth and tops and gyroscopes closer to home. In *fluid* mechanics, the vorticity has an analogous role. We have already defined vorticity as the curl of the velocity vector, but this does not do it any justice. Its stark definition in terms of mathematically well-defined quantities is indeed singularly unilluminating. Let us instead begin by first discussing some properties of a fluid. If a river or stream flows at a constant rate, then near to the bank there is generation of vorticity. If a neutrally buoyant float with a line drawn on its topmost surface was placed in the stream next to the bank, it would drift with the stream, but because the side nearest the bank was always travelling slower than the opposite side (due to the friction of the bank) the line would rotate. This rotation is a measure of vorticity. Paradoxically the same float put near the vortex of a draining bath would not rotate at all (until it hit the vortex itself); the vorticity of a fluid near a vortex is zero. Now the vorticity of a fluid tends to be preserved by motion not influenced by friction, and many properties of quite complex looking flows can be deduced from following the vorticity and assuming it is constant. As we shall see later in this chapter, in meteorology and oceanography, an important source of vorticity arises from the fact that the Earth rotates. The details can wait, but because of this rotation there is no fixed origin, so to use Newton's second law we need to introduce a force called the Coriolis force that compensates for having to refer distances to an origin that moves in a circle. We shall see that only the vertical component of this Coriolis force (strictly Coriolis acceleration) matters and we call twice this the Coriolis parameter. It depends on the sine of the latitude and is given the symbol f and is zero at the Equator and negative in the southern hemisphere. It is this change in the Coriolis parameter with latitude (it is zero at the Equator but $\pm 2\Omega$, where Ω is the angular speed of the Earth at each pole) that contributes vorticity to the flow. The vorticity arising from the Coriolis acceleration only depends on the position of a current on the Earth's surface and so is distinctive. The total vorticity of a large scale ocean current has three influences: the local changes ("shear") in the current; the local value of the Coriolis parameter; and the ocean depth. The whole is given the distinctive name *planetary vorticity*, and its conservation is crucial to the full understanding of ocean scale physics. It is less crucial to those who are more concerned with smaller scale modelling appropriate to coasts, but it is still well worth discussing for the insight it gives. There is more on large scale ocean physics in Chapter 10.

Before deriving relationships that are very useful, but only apply to the rotating Earth, let us start with the equation of motion of a fluid due to Euler written in the form

$$\frac{\partial \mathbf{u}}{\partial t} + \boldsymbol{\nabla}\left(\frac{1}{2}\mathbf{u}^2\right) - \mathbf{u} \times (\boldsymbol{\nabla} \times \mathbf{u}) = -\boldsymbol{\nabla}\int \frac{dp}{\rho} + \boldsymbol{\nabla}(-gz),$$

where the advection term $(\mathbf{u} \cdot \boldsymbol{\nabla})\mathbf{u}$ has been expanded as before. Take the curl of this equation, and remembering that "curl of grad" is identically zero, we get the equation

$$\frac{\partial \boldsymbol{\zeta}}{\partial t} - \boldsymbol{\nabla} \times (\mathbf{u} \times \boldsymbol{\zeta}) = 0,$$

where $\boldsymbol{\zeta} = \boldsymbol{\nabla} \times \mathbf{u}$ is the vorticity vector. Another of those vector identities is now used to expand the curl term

$$\boldsymbol{\nabla} \times (\mathbf{u} \times \boldsymbol{\zeta}) = (\boldsymbol{\zeta} \cdot \boldsymbol{\nabla})\mathbf{u} - (\mathbf{u} \cdot \boldsymbol{\nabla})\boldsymbol{\zeta} + \boldsymbol{\zeta}(\boldsymbol{\nabla} \cdot \mathbf{u}) - \mathbf{u}(\boldsymbol{\nabla} \cdot \boldsymbol{\zeta}).$$

The divergence terms (the last two) are both zero. $\boldsymbol{\nabla} \cdot \mathbf{u} = 0$ is the continuity equation and $\boldsymbol{\nabla} \cdot \boldsymbol{\zeta} = 0$ as div of curl is always identically zero. Further, remembering the definition of total (Lagrangian) derivative

$$\frac{D\boldsymbol{\zeta}}{Dt} = \frac{\partial \boldsymbol{\zeta}}{\partial t} + (\mathbf{u} \cdot \boldsymbol{\nabla})\boldsymbol{\zeta},$$

leads to the equation

$$\frac{D\boldsymbol{\zeta}}{Dt} = (\boldsymbol{\zeta} \cdot \boldsymbol{\nabla})\mathbf{u}.$$

This is called the vorticity equation (for inviscid flow). The directional derivative is defined by

$$\frac{d\mathbf{q}}{d\alpha} = (\boldsymbol{\alpha} \cdot \boldsymbol{\nabla})\mathbf{q}.$$

This is the spatial rate of change of the vector quantity $\mathbf{q}$ in the direction of $\boldsymbol{\alpha}$. The vorticity equation thus expresses that the total (Lagrangian) derivative of the vorticity vector $\boldsymbol{\zeta}$ is given by the rate of change of the local velocity vector $\mathbf{u}$ in the direction of the vorticity. For two dimensional non-divergent flow this has to be zero for the following reason. The velocity vector $\mathbf{u}$ lies in a plane, say the (x, y) plane, and the vorticity vector $\boldsymbol{\zeta}$, defined as $\boldsymbol{\nabla} \times \mathbf{u}$, is in the z direction which is therefore everywhere perpendicular to the plane of $\mathbf{u}$. There can be no spatial rate of change of $\mathbf{u}$ in this direction as $\mathbf{u}$ has no component in this direction. Therefore, for two dimensional flow the vorticity equation is

$$\frac{D\boldsymbol{\zeta}}{Dt} = \mathbf{0},$$

which in turn means that the vorticity vector does not change following the fluid. This is also true for rotating flows, but the form is slightly different as there is the Coriolis term. Also, on the Earth, flows are seldom non-divergent. In three dimensional flow, the vorticity is not conserved following the flow, instead the current is changed (sometimes referred to as *stretched*) in the direction of the local vorticity vector. So in an inviscid environment, the vorticity is conserved following the fluid by this local stretching of the fluid streamlines. In this text, the main thrust of vorticity dynamics is felt through the rotation of the Earth, however, let us pursue vorticity in a non-rotating environment a little further. The equation

$$\frac{D\boldsymbol{\zeta}}{Dt} = (\boldsymbol{\zeta} \cdot \boldsymbol{\nabla})\mathbf{u}$$

is a differential equation in a single unknown $\boldsymbol{\zeta}$. Using the theory of differential equations, if the solution of such an equation is initially zero then, as long as the gradients of **u** remain continuous, the single time derivative can (theoretically at least) be integrated and the value of the vorticity must remain zero for all time. This is sometimes referred to as Helmholtz' first law or the persistence of vorticity. In the case quoted, more the persistence of the lack of vorticity. Vorticity is thus a conserved quantity in inviscid fluid dynamics, and if the properties of a complicated but largely inviscid flow are sought, one can do a lot worse than calculate the vorticity and trace its behaviour in the domain of the flow.

3.5   Modelling Waves

The essential ingredient common to all waves is the to-and-fro motion, without there being any overall movement in any direction. Offshore engineers and oceanographers need to understand many things that oscillate; a few examples are ships, offshore structures subject to high winds and the sea itself. At first sight, all of these seem quite different and very difficult to model. However, they all exhibit a to-and-fro motion, and it is this that we will attempt to describe. We will initially restrict our attention to motion in one dimension, or more strictly, to what engineers and applied scientists call a single degree of freedom system. Later, this restriction will be lifted as ocean scientists sometimes have to deal with two dimensional waves. Consider a mass attached to a spring on a smooth horizontal table, as shown in Figure 3.3. If the spring is neither stretched nor compressed, the mass will not be subject to any force, and therefore

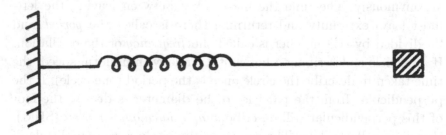

Fig. 3.3   A mass connected to a spring.

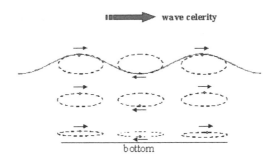

Fig. 3.4   A surface water wave.

it will be in equilibrium. If the mass is pulled (or pushed) in the line of the spring and then released, it will vibrate back and forth. The spring will always try to restore the mass to its equilibrium position, but it will overshoot and, in the absence of any damping or friction, never come to rest other than instantaneously. Perfect springs have no damping, so the simple spring system shown in Figure 3.3, once set in motion, will vibrate forever. Compare this with the water wave shown as Figure 3.4. This too exhibits the to-and-fro motion and it should come as no surprise that it is modelled in an identical way, at least initially.

In order to explain the terms used in describing waves, let us look at the mass spring system of Figure 3.3. The distance between the position of equilibrium, where the mass has no forces acting on it and thus does not move, to the furthest point reached by the mass in either direction of its motion is called the *amplitude* of the vibration or oscillation. The words *vibration* and *oscillation* are used synonymously. The time the mass takes between leaving the leftmost (say) extremity and returning there is called the *period*, and 2π divided by this number is called the *frequency* of the oscillation. If a particle is describing a horizontal circle with uniform speed, the time taken to describe the circle once is the period (one cycle). If the perpendicular from the particle to the diameter is drawn, the foot of this perpendicular will describe *simple harmonic motion* (SHM), that is it will oscillate. The radius of the circle is the amplitude of the vibration. This provides a useful alternative view of vibration.

3.6 Simple Harmonic Motion

SHM is ideal in the sense that in order to describe the motion of springs the springs have to be well enough behaved to obey Hooke's Law, whereby extension is proportional to the applied force. To derive an equation to describe it, we must use the second of Newton's laws of motion, which is not really appropriate here.

Figure 3.5 is the same as Figure 3.3 but with the addition of an origin, axis, force and some labels. The equation obeyed by this mass spring system, writing k for the stiffness of the linear spring, is

$$m\frac{d^2x}{dt^2} = -kx. \tag{3.6}$$

Dividing by m and writing $\omega^2 = k/m$ yields

$$\frac{d^2x}{dt^2} = -\omega^2 x, \tag{3.7}$$

where ω is a number with dimensions of $(\text{time})^{-1}$ called the *natural frequency* of the mass spring system. This name arises from the fact that it is the frequency at which the mass oscillates when pulled to one side and released. We can see this because the solution to equation (3.7), a second order differential equation with constant coefficients, is

$$x = A\sin\omega t + B\cos\omega t, \tag{3.8}$$

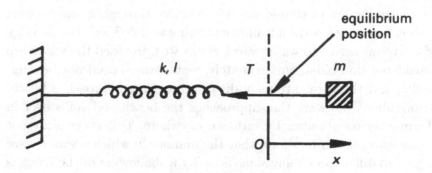

Fig. 3.5 Defining some terms.

where A and B are constants obtained by using given conditions on x or dx/dt at specific times (usually at $t = 0$). An example should make the determination of A and B clear.

Example. Calculate the displacement x in terms of ω and t if, at time $t = 0$, $x = 3$ and the velocity is zero.

Solution. The velocity of a particle that is displaced a distance x is its rate of change with respect to time, dx/dt, so if x is given by equation (3.8), then differentiating with respect to t gives

$$\frac{dx}{dt} = \omega A \cos \omega t - \omega B \sin \omega t. \tag{3.9}$$

At time $t = 0$, we obtain

$$\left.\frac{dx}{dt}\right|_{t=0} = \omega A. \tag{3.10}$$

If this is to be zero, we must have $A = 0$.

Setting $A = 0$ results in the simplification of equation (3.8) to

$$x = B \cos \omega t \tag{3.11}$$

and if $x = 3$ at $t = 0$

$$B = 3. \tag{3.12}$$

We have thus determined the particular values of A and B that satisfy $x = 3$ and $dx/dt = 0$ at time $t = 0$. The solution is

$$x = 3 \cos \omega t. \tag{3.13}$$

The introduction of simple sinusoidal waves through a mass spring system certainly has the advantage of being well-defined. The analogy of the terms across to water waves works well provided the waves are considered sinusoidal. Unfortunately, water waves need not be sinusoidal, and they are extremely distorted as they approach a beach. Think about the wave that approaches the beach and spills over in the manner sought after by surfers everywhere. This is certainly not a sine wave. It can be shown that the manner in which a water wave begins to differ from being sinusoid is by a shallowing of the troughs and a steepening of the crests, but there are plenty of examples of waves in the ocean which are virtually sinusoidal (see Leblond and

Mysak, 1978). Tides are one of these, and we consider tides a little later in this chapter and more thoroughly in Chapter 6.

3.6.1 *Waves and Coriolis acceleration*

In coastal oceanography, waves can be quite complex and some more terms come in useful. For example, for waves that are not perhaps cleanly sinusoidal the *total height*, H, is a good concept. This is the vertical distance between the lowest trough and topmost crest of the waves, but the letter H has already been designated to mean the total height from sea bed to surface wave, so to avoid confusion this wave definition of H is not used very much beyond this paragraph. In fact, it is only used in the section on real waves when its use is unavoidable and the depth is not of prime concern. The *wave steepness* is the ratio H/L where L remains the *wavelength*, which is the horizontal distance between successive crests (or troughs). The quantity $2\pi/L = k$ is called the *wave number*. The natural frequency (remember this is $2\pi/$period) of a wave in oceanography often tells us about the mechanism(s) being modelled. For tides, this would be a single tidal frequency or perhaps a sum of frequencies of several different tides. Adding up different waves in this way forms the basis of the harmonic theory of tides, which still is a useful model. Harmonic analysis is a realisation of Fourier decomposition, which might be familiar to some of you. It does not matter if it is not. Another natural frequency for larger scale ocean dynamics at a horizontal scale of several kilometres is associated with the vertical component of the angular velocity of the Earth and arises out of the Coriolis acceleration. The Coriolis acceleration is an important acceleration for offshore and coastal engineering modellers as well as ocean scientists, so it will now be derived in some mathematical detail. The derivation that follows only demands elementary geometry and some knowledge of vector algebra. So it's back to section 1.3 if you need to brush up on the mathematical background. The arbitrary point on the Earth's surface is labelled O and is the origin of the local (x, y, z) co-ordinates. x points east, y points north and z points up. In this co-ordinate system, shown in Figure 3.6, the velocity of the fluid is **u**, but of course the origin is moving with respect to the centre of the Earth. (The Earth's centre

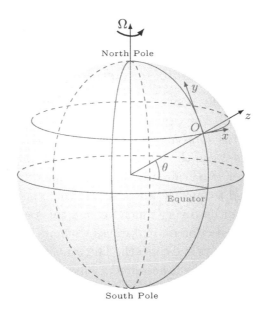

Fig. 3.6   Motion relative to a rotating Earth: x points East, y points North and z points upwards.

will be assumed fixed as its motion around the Sun has such a large radius of curvature that it is straight to a very good approximation.) The motion of the point O is by virtue of the angular velocity of the Earth. In the plane of the line of latitude that contains O, the x and y axes rotate with angular velocity $\Omega \sin\theta$, where θ is latitude. This is the vertical (z) component of Ω. The vector Ω has direction south pole to north pole. Thus, when the derivative (rate of change) with respect to time of any vector is calculated, and this calculation is performed relative to the rotating axes, the changes of the unit vectors that point along Ox and Oy, call them $\mathbf{i}$ and $\mathbf{j}$, need to be taken into account. So, take any vector $\mathbf{A}(t)$ which is a function of time then the calculation of its rate of change proceeds as follows

$$\frac{d\mathbf{A}(t)}{dt} = \frac{d(A_1\mathbf{i} + A_2\mathbf{j})}{dt} = \frac{dA_1}{dt}\mathbf{i} + \frac{dA_2}{dt}\mathbf{j} + A_1\frac{d\mathbf{i}}{dt} + A_2\frac{d\mathbf{j}}{dt},$$

where A_1 and A_2 are the components of $\mathbf{A}(t)$ in the x and y directions, respectively. By examining how the axes change in an infinitesimally small time, it can be easily shown that

$$\frac{d\mathbf{i}}{dt} = \Omega\mathbf{j}, \quad \text{and} \quad \frac{d\mathbf{j}}{dt} = -\Omega\mathbf{i}$$

and hence

$$\frac{d\mathbf{A}(t)}{dt} = \frac{dA_1}{dt}\mathbf{i} + \frac{dA_2}{dt}\mathbf{j} + \Omega A_1\mathbf{j} - \Omega A_2\mathbf{i}.$$

The first two terms are the rate of change of $\mathbf{A}(t)$ with respect to time as if the axes were not rotating. So the extra two terms represent the effects of the rotation. These can be put succinctly in terms of vector quantities by using the cross product. Thus, we can write

$$\left[\frac{d\mathbf{A}(t)}{dt}\right]_{\text{fixed}} = \left[\frac{d\mathbf{A}(t)}{dt}\right]_{\text{rotating}} + \Omega \times \mathbf{A}(t).$$

The derivation of this has been two dimensional and may not look very general. However, it is mathematically rigorous as the above vector equation is co-ordinate independent, and a mathematical result that is co-ordinate independent but derived using specific co-ordinates (in our case two-dimensional Cartesian co-ordinates) remains true in all orthogonal curvilinear co-ordinate systems. This is a useful theorem and illustrates the usefulness of pure mathematics.

The Coriolis acceleration is now easily derived by a double application of the formula. If $\mathbf{r}$ denotes the position vector relative to an inertial frame of reference (the centre of the Earth, for example) then we have

$$\frac{d^2\mathbf{r}}{dt^2} = \frac{d}{dt}\frac{d\mathbf{r}}{dt},$$

where all derivatives are relative to fixed axes. The left-hand side is the true acceleration. In terms of actually measurable quantities, which are of course measured relative to a rotating frame of reference, we thus have,

$$\frac{d}{dt}\frac{d\mathbf{r}(t)}{dt} = \left[\frac{d}{dt} + \mathbf{\Omega}\times\right]\left[\frac{d}{dt} + \mathbf{\Omega}\times\right]_{\text{rotating}}\mathbf{r}(t)$$

$$= \left[\frac{d^2\mathbf{r}(t)}{dt^2}\right]_{\text{rotating}} + 2\mathbf{\Omega}\times\left[\frac{d\mathbf{r}(t)}{dt}\right]_{\text{rotating}} + \mathbf{\Omega}\times(\mathbf{\Omega}\times\mathbf{r}(t)).$$

The Coriolis acceleration is the term

$$2\mathbf{\Omega}\times\left[\frac{d\mathbf{r}(t)}{dt}\right]_{\text{rotating}} ;$$

the term

$$\mathbf{\Omega}\times(\mathbf{\Omega}\times\mathbf{r}(t))$$

is the centripetal acceleration, which is directed towards the axis of rotation of the Earth but is so small compared to gravity that it already has been discarded. We have also assumed that the angular velocity of the Earth $\mathbf{\Omega}$ does not vary with time. (Otherwise there would be yet another term to consider.)

With respect to the chosen co-ordinate system (see Figure 3.6), the Coriolis acceleration is

$$2\mathbf{\Omega}\times\mathbf{u} = 2(0, \Omega\cos\theta, \Omega\sin\theta)\times(u, v, w),$$

where $(u, v, w) = \mathbf{u}$ is the fluid velocity, u is the easterly current, v the northerly current and w the upward current. This latter is of course very small and is neglected except for very specialist modelling. w is always neglected when compared with u or v, and it is this approximation that gives the Coriolis acceleration the components $(-fv, fu, 0)$, where $f = 2\Omega\sin\theta$ and is called the Coriolis parameter. f is twice the vertical component of the Earth's angular velocity, and it is only this component that plays an important part in the dynamics of the ocean and atmosphere.

Thus, we have derived the *Coriolis frequency*, denoted by f where $f = 2\Omega\sin(latitude)$. This frequency occurs all over the place in oceanography; here is an idealised example. If an iceberg is floating in a frictionless boundless sea and is given a push and allowed to move freely it will travel in a circle of radius U/f, where U is the initial speed imparted by the push. This circle is called an *inertial circle*; the velocity of the iceberg (u, v) is given by

$$u = U \cos ft, \quad v = U \sin ft$$

and the circle itself by the Cartesian equation

$$x^2 + y^2 = \left(\frac{U}{f}\right)^2,$$

provided the direction of push is along the x axis and the co-ordinate system is appropriately chosen (which means that the iceberg has co-ordinates $(0, -U/f)$ at time $t = 0$). This can be formally derived from the dynamical equations

$$\frac{\partial u}{\partial t} - fv = 0$$

$$\frac{\partial v}{\partial t} + fu = 0,$$

from which the SHM equations

$$\left(\frac{\partial^2}{\partial t^2} + f^2\right)(u, v) = 0$$

are easily obtained, and from which the above solution immediately follows. This SHM is called an *inertial oscillation*. The distance U/f is termed the Rossby radius of deformation and is a naturally occurring length scale for the width of coastal currents and upwelling, provided the Coriolis parameter is assumed constant. What we have here is clearly a two dimensional application of oscillations. Truly two dimensional sinusoidal waves are called *plane waves* and have the general form

$$a_0 \cos(kx + ly - \omega t + \phi) \quad \text{or} \quad ae^{i(kx+ly-\omega t)},$$

where the symbols k, l are wave numbers and the constant ϕ is a phase. In the second expression, only the real part has physical significance (a is complex, $a = a_0 e^{i\phi}$, a_0 is real, the *amplitude*; ϕ is real, the *phase*). It is convenient sometimes to adopt the powerful vector notation and write the plane wave in the form

$$ae^{i(\mathbf{r}\cdot\boldsymbol{\kappa}-\omega t)},$$

where $\mathbf{r}$ is the two-dimensional position vector ($\mathbf{r} = x\mathbf{i} + y\mathbf{j}$) and $\boldsymbol{\kappa}$ is the two dimensional wave number vector defined by $\boldsymbol{\kappa} = k\mathbf{i} + l\mathbf{j}$. The trouble some of you will have to go through to master this vector description will be well worth it if the need is to be able to deal with waves that are bending (due to refraction over topography perhaps) or being reflected obliquely. It opens up the use of wave ray theory. This notation also extends naturally to three dimensional "plane" waves which are not considered here. In the context of tides and waves on this kind of geophysical scale, these two dimensional plane waves are sometimes called Poincaré waves after the brilliant French mathematician Jules Henri Poincaré (1854–1912).

The speed at which a wave travels is labelled c and is called its wave speed or *celerity*, and for purely, progressive waves of the type $a_0 \sin(kx - \omega t)$ it is the ratio ω/k, which is wave frequency divided by wave number. However, for waves that are not purely progressive the wave speed needs to be found by other means. One property of waves yet to be alluded to but nevertheless important is *dispersion*. This is the tendency for a wave to decrease in amplitude as it travels. For all waves a relationship can be derived that connects the allowable wavelengths and wave frequencies. For us in the field of coastal engineering or coastal science this relationship arises from considering the dynamics of water waves in a sea of constant depth, details of which are given later. The depth is h and the dispersion relation takes the form

$$c^2 = \frac{g}{k} \tanh(kh),$$

where tanh is a mathematical function called the hyperbolic tangent, found on scientific calculators and defined by

$$\tanh x = \frac{\sinh x}{\cosh x} = \frac{e^x - e^{-x}}{e^x + e^{-x}}.$$

For waves over very deep water, the depth h plays no part and very nearly

$$c^2 = \frac{g}{k}.$$

On the other hand, for shallow water the hyperbolic tangent is more or less a straight line so $\tanh x \approx x$ and this time the wave speed does not depend on the wavelength, instead

$$c^2 = gh.$$

These dispersion relationships can tell us a great deal about the behaviour of waves in various sea environments and are particularly useful for coastal and environmental engineers. In the real sea, one does not just have isolated sine waves, but a complicated surface of crests and troughs. Assuming that these can be approximated by a wave in one dimension, then in turn they can be thought of as a sum of sinusoidal waves (a Fourier series in fact). Each component of this combination of sine waves will, according to the dispersion relation, travel at different speeds. In order to demonstrate this, let us look at the simplest case; a wave train that consists solely of waves of two slightly differing wavelengths. Consider two waves of the same amplitude but different wavelengths

$$\eta = a(\cos(k_1 x - \omega_1 t) + \cos(k_2 x - \omega_2 t)).$$

Using elementary trigonometry (the formula that converts the sum of two cosines into a product of two cosines) we get

$$\eta = 2a \cos\left(\frac{(k_1 - k_2)}{2}x - \frac{(\omega_1 - \omega_2)}{2}t\right)$$

$$\times \cos\left(\frac{(k_1 + k_2)}{2}x - \frac{(\omega_1 + \omega_2)}{2}t\right). \tag{3.14}$$

Now, if we suppose that k_1 and k_2 are very close, then so of course are ω_1 and ω_2, and the first factor with $(k_1 - k_2)$ in equation (3.14) has very different properties from the one with $(k_1 + k_2)$. In fact, this second term behaves very similarly to either of the two waves that make up the original pair. On the other hand, the first term has a very long wavelength and low frequency and so exhibits very different behaviour. Thus, superimposing two waves with wavelengths and frequencies that are very close in value will give "beats" of the type met when tuning stringed musical instruments. Figure 3.7 shows this graphically. This graph is related to equation (3.14) straightforwardly. In order to draw the graph, the values $k_1 = 1.0$ and $k_2 = 0.9$ have been chosen; the basic wave with 20 wavelengths shown is the wave of wave number 0.95, the mean of these values, and the "carrier" wave (the envelope wave) has wave number 0.05, being half of the difference. This means that the basic wave is in fact modulated

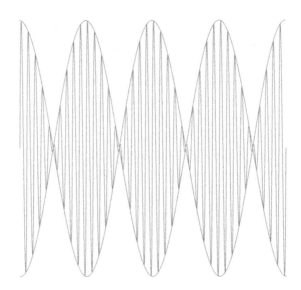

Fig. 3.7   Two waves showing 20 wavelengths of the basic wave but only two beat wavelengths.

by a slowly varying wave and it is this that is demonstrated mathematically in equation (3.14), made clearer perhaps by writing

$$A(x,t) = 2a \cos\left(\frac{(k_1 - k_2)}{2}x - \frac{(\omega_1 - \omega_2)}{2}t\right)$$

so that the surface elevation has a slowly varying amplitude

$$\eta = A(x,t) \cos\left(\frac{(k_1 + k_2)}{2}x - \frac{(\omega_1 + \omega_2)}{2}t\right).$$

From equation (3.14) it can be seen that these beats progress at a speed that is slower than the speed of either primary wave. If we label the speeds of the waves c_1, c_2, then the speed of the "beat" is given by

$$c_g = \frac{\omega_1 - \omega_2}{k_1 - k_2},$$

which in turn can be written

$$c_g = \frac{\Delta\omega}{\Delta k}.$$

In the limit as the two speeds c_1 and c_2 get closer and closer

$$c_g = \frac{d\omega}{dk}.$$

The speed c_g is called the *group velocity* and in general is the speed at which the energy of a group of waves travels. As said earlier, the group velocity of a wave train is always less than the celerity of any of the sinusoidal waves of which it is composed. In some cases, remarkably, the group velocity can actually oppose the direction of the wave train. This is counterintuitive, but is true for Rossby waves (planetary waves) which have a group velocity that always has a westerly component (see Exercise 10.2). These waves are particularly interesting to study as they have unusual properties, however they are only found in the ocean as meanders in the Gulf Stream and currents of similar scale and therefore traditionally play a small part in a book on coastal sea modelling. However, the jet stream and its part in climate change are now important to understand, and it is meanders on the jet stream that are also Rossby waves. In the ocean, it is the persistent westerly migration of energy that lies behind the western intensification of ocean currents (Gulf Stream and Kuroshio). We shall return to consider the group velocity of a wave spectrum in Chapter 8.

3.6.2 *Energy of water waves*

It turns out to be useful to get an expression for the energy of water waves, then to do some calculations on the rate of change of energy with respect to time. It is also useful to find a reasonably general expression for the rate of change of energy. For those who followed the vector calculus in section 1.3 and earlier in this chapter, this runs along similar lines. It is straightforward for those in the know, but the advice is the same as it was there — skip it if you have to. Suppose the wave is described by a velocity potential ϕ so that

$$u = \frac{\partial \phi}{\partial x}, \quad \text{and} \quad w = \frac{\partial \phi}{\partial z}.$$

The total energy is then the quantity

$$\frac{1}{2}\rho \left[\left(\frac{\partial \phi}{\partial x} \right)^2 + \left(\frac{\partial \phi}{\partial z} \right)^2 \right] + \rho gz,$$

summed (integrated) over the volume of the wave. The total energy is for the moment conveniently written in terms of vector notation as

$$\rho \int_V \left\{ \frac{1}{2} (\boldsymbol{\nabla} \phi)^2 + gz \right\} dV,$$

assuming that the density ρ is a constant. In order to find the rate of change of this with respect to time, it is simply differentiated. In general, this is not straightforward as the volume will be changing with time and we would have to resort to using a more general form of Leibniz' Rule that came to our aid when deriving a convenient form of the continuity equation see the formula on page 91. Fortunately, this does not trouble us as our volume will not change with time — we will follow a progressive wave. Thus, the rate of change of energy is

$$\rho \int_V \frac{\partial}{\partial t} \left\{ \frac{1}{2} (\boldsymbol{\nabla} \phi)^2 + gz \right\} dV = \rho \int_V \boldsymbol{\nabla} \phi_t \cdot \boldsymbol{\nabla} \phi \, dV$$

(where the suffix t denotes the derivative with respect to t) and this is

$$\rho \int_S \frac{\partial \phi}{\partial t} \frac{\partial \phi}{\partial n} dS$$

because

$$\int_V \boldsymbol{\nabla} (\phi_t \boldsymbol{\nabla} \phi) dV = \int_V \{ \phi_t \nabla^2 \phi + \boldsymbol{\nabla} \phi_t . \boldsymbol{\nabla} \phi \} dV.$$

The first term is zero because $\nabla^2 \phi = 0$ for these linear waves, and the left-hand side is the surface integral

$$\int_S \frac{\partial \phi}{\partial t} \frac{\partial \phi}{\partial n} dS$$

because of Gauss' flux theorem. Thus, the rate of change of energy of the wave is a surface integral

$$\frac{DE}{Dt} = \int_S \rho \phi_t \phi_n dS,$$

where n is the normal to the surface, which for us is the same as x. Note that the potential energy term has differentiated out and so does not contribute to the overall change of energy with time. Now we are interested in progressive waves as it is these that generalise to real sea waves and help us to understand their behaviour. Progressive waves of any form, not just sinusoids, have the factor "$x - ct$" built in, so that

$$\phi(x, z, t) = \phi(x - ct, z)$$

and in particular

$$\frac{\partial \phi}{\partial t} = -c\frac{\partial \phi}{\partial x},$$

no matter what the function is. Thus we have, for progressive waves,

$$\frac{DE}{Dt} = -\int_{\text{wavelength}} \int_{-h}^{0} \rho c \left\{ \frac{\partial \phi}{\partial x} \right\}^2 dz dx, \qquad (3.15)$$

where the big "D" notation has been used to emphasise that this is the total derivative and not just a partial derivative. To the lowest order of approximation, which is what we are doing, they are the same of course. The surface over which the integral is evaluated is the total depth (linearised so that the top limit is 0 and not η) and one wavelength. It is this equation that will be used in Chapter 8.

3.6.3 *Linear surface water waves*

It is now time to study water waves more mathematically. Historically, the first application of fluid mechanics to water waves was done nearly 200 years ago and forms what today is called linear water wave theory. It is still useful, although mainly only as background. First of all, ignore the rotation of the Earth and all nonlinear effects; also assume that motion is in two dimensions, the horizontal x direction and the vertical z direction. The two equations of motion and continuity equation are then

$$\frac{\partial u}{\partial t} = -\frac{1}{\rho}\frac{\partial p}{\partial x},$$

$$\frac{\partial w}{\partial t} = -\frac{1}{\rho}\frac{\partial p}{\partial z} - g,$$

$$\frac{\partial u}{\partial x} + \frac{\partial w}{\partial z} = 0.$$

The hydrostatic assumption has not been made here as the waves are short and there is significant vertical motion. The last (continuity) equation is solved by setting

$$u = \frac{\partial \psi}{\partial z}, \qquad w = -\frac{\partial \psi}{\partial x},$$

where the function $\psi(x, z)$ is called the streamfunction. If the pressure is eliminated from the first two equations, then integrating with respect to t gives

$$\frac{\partial u}{\partial z} - \frac{\partial w}{\partial x} = 0.$$

There are mathematical reasons for taking the constant of integration to be zero, but they need not concern us here. What this does mean is that the flow is indeed irrotational, and so a potential can be defined

$$u = \frac{\partial \phi}{\partial x}, \qquad w = \frac{\partial \phi}{\partial z},$$

giving the continuity equation as

$$\nabla^2 \phi = 0.$$

Cross differentiating (that is the z derivative of the first (x) equation minus the x derivative of the second (z) equation) also shows that

$$\nabla^2 \psi = 0,$$

which incidentally is the same as the flow being irrotational. Thus, we have simple equations for the streamfunction ψ and the potential ϕ. Both Laplace's equation. ψ and ϕ are called conjugate functions and a whole raft of mathematics can be brought to bear to analyse them. This was a favourite hobby before the computer age, but it is a little passé these days. Nevertheless, the classic text, Stoker (1958) is largely based upon these assumptions, as indeed are some of the design codes used by coastal engineers. The boundary conditions at the sea surface are worthy of note. There are two of them. First of all a kinematic condition which says that a water particle once at the surface remains there. This means that the sea surface elevation is related to the vertical velocity at the sea surface by

$$w = \frac{D\eta}{Dt} = \frac{\partial \eta}{\partial t} + u\frac{\partial \eta}{\partial x} + w\frac{\partial \eta}{\partial z}.$$

Using our assumptions, this reduces to

$$\frac{\partial \phi}{\partial z} = \frac{\partial \eta}{\partial t} \quad \text{at } z = \eta.$$

Since, also, the pressure is constant on the free surface, we can use the Bernoulli equation to give the linearised dynamic condition

$$\frac{\partial \phi}{\partial t} + g\eta = -\frac{p_A}{\rho} \quad \text{at } z = \eta,$$

where p_A is the atmospheric pressure, often taken as zero. These two surface boundary conditions can be combined conveniently by eliminating η to give

$$\frac{\partial^2 \phi}{\partial t^2} + g\frac{\partial \phi}{\partial z} = -\frac{1}{\rho}\frac{\partial p_A}{\partial t} \quad \text{at } z = \eta.$$

At the sea bed, the condition can be as simple as stating that all quantities must decay for large negative values of z, but the more precise

$$\frac{\partial \phi}{\partial n} = 0 \quad \text{at } z = -h(x,y),$$

where **n** is a direction normal to the sea bed, is better. To calculate this derivative, the relationship

$$\frac{\partial \phi}{\partial n} = \hat{\mathbf{n}} \cdot \boldsymbol{\nabla}\phi = \frac{\boldsymbol{\nabla}h}{|\boldsymbol{\nabla}h|} \cdot \boldsymbol{\nabla}\phi$$

(where $\hat{\mathbf{n}}$ is a unit vector in the direction of **n**) can be used.

Therefore, we have a theory of linear water waves that can be solved in various domains in the x, y plane with a free surface $z = \eta(x, y, t)$ and a sea bed at $z = -h(x, y)$.

One simple example would be to let the surface wave be sinusoidal; we can then assume a velocity potential of the form

$$\phi(x, z, t) = F(z)\sin(kx - \omega t),$$

so that

$$\nabla^2 \phi = 0$$

gives

$$F''(z) - k^2 F(z) = 0,$$

whence

$$F(z) = Ae^{kz} - Be^{-kz},$$

where A and B are constants. Differentiating gives

$$w = \frac{\partial \phi}{\partial z} = F'(z)\sin(kx - \omega t) = k(Ae^{kz} - Be^{-kz})\sin(kx - \omega t),$$

so the boundary condition at the (flat) sea bed $z = -h$ gives

$$A = Be^{2kh},$$

which means that

$$w = C\sinh k(z + h)\sin(kx - \omega t),$$

where C is a constant. From the continuity equation, by differentiation with respect to z then integration with respect to x we can derive that

$$u = C\cosh k(z + h)\cos(kx - \omega t),$$

so finally, integrating again gives

$$\phi = \frac{C}{k}\cosh k(z + h)\sin(kx - \omega t).$$

The condition

$$\frac{\partial^2 \phi}{\partial t^2} + g\frac{\partial \phi}{\partial z} = 0 \quad \text{at } z = 0$$

then gives

$$-\omega^2 \cosh kh + gk\sinh kh = 0$$

or

$$\omega^2 = gk\tanh kh.$$

This is called the dispersion relation for linear surface waves and tells us the allowable wavelengths $(2\pi/k)$ given the period $(2\pi/\omega)$, or indeed vice versa. More often it is written in terms of wave celerity or wave speed, $c = \omega/k$, as

$$c^2 = \frac{g}{k} \tanh kh,$$

an expression that came out of the blue earlier — now you know where it comes from. We reiterate the important special cases: If the waves are short or the water deep, or both, then $kh \gg 1$, $\tanh x \approx 1$ and, approximately,

$$c^2 = \frac{g}{k}.$$

If the waves are long or the water is shallow, or both, then $kh \ll 1$, $\tanh x \approx x$ and approximately,

$$c^2 = gh.$$

This last relation is valid for tsunamis and gives a good indication of their speed of travel. Many applications of linear water wave theory are to waves on a scale where the Coriolis acceleration is important, in which case the above theory needs modification. There are also applications to the real sea. Both of these will be addressed later in Chapters 6, 8 and 10.

3.7   Overall Energy Balance

Although the energy of linear water waves has been calculated, the overall energy of the sea has not been addressed. Those familiar with mechanics might think that this should be the first integral of the momentum equation. This is not the case when dealing with a geo-physical fluid such as the air or sea, for in both there are consider-ations of heat. The equation that governs heat is termed the energy equation and this will now be derived. To get there takes some knowl-edge of physics — one might almost say thermodynamics — but it does not involve much more than college (or 'A' level) physics. The internal energy, U_E, is given by

$$U_E = c_V T,$$

where c_V is the specific heat or heat capacity at constant volume and T is temperature. The time rate of change of this quantity will be the difference between the rate of heat gain, Q, and the work done by the pressure on the unit mass of sea, W. Denote this balance by

$$\frac{d}{dt}(c_V T) = Q - W.$$

The next part presupposes the knowledge of a subject that in this text we have yet to explore, but is to be covered in Chapter 7. So take a glance at that chapter or for the moment take it on trust that the rate of heat gain Q obeys the equation

$$\bullet \qquad Q = \frac{\kappa}{\rho}\nabla^2 T.$$

This equation arises from modelling the diffusion of heat, and the whole of Chapter 7 is devoted to the subject of diffusion, which is what heat does in the sea. In particular, the derivation of the factor $\frac{\kappa}{\rho}\nabla^2 T$ is done in section 7.2.1. For an incompressible unit mass of sea, there is no change in the volume due to pressure, so it does no work. Hence $W = 0$. This is not always the case in the atmosphere where thermodynamics can assume greater importance, but here in the sea $W = 0$; always. With c_V being a constant, the above equation is thus

$$c_V \frac{dT}{dt} = \frac{\kappa}{\rho}\nabla^2 T.$$

The left-hand side is a total derivative, so this equation is

$$\frac{\partial T}{\partial t} + (\mathbf{u} \cdot \boldsymbol{\nabla})T = \frac{\kappa}{c_V \rho}\nabla^2 T.$$

There is a similar equation for salt balance,

$$\frac{\partial S}{\partial t} + (\mathbf{u} \cdot \boldsymbol{\nabla})S = \kappa_S \nabla^2 S,$$

where all the constants are absorbed into the salt diffusion coefficient κ_S. Not much more will be said about these equations as they only figure in large scale ocean models where heat and salt budgets assume importance. In coastal seas they are not so important to consider; both temperature and salinity do vary but wind and tide assume far greater importance. Only at the mouth of estuaries does salinity feature in models and this is considered in Chapter 11. However, what both equations state is that T and S are conserved apart from diffusion, as dictated by the coefficients $\frac{\kappa}{\rho c_V}$ for T and κ_S for S, and Chapter 7 takes this to the next level. The next section looks at how to model friction in a fluid, a topic that has been put aside so far in this text but is very important to the coastal engineer and scientist, and it has been given a great deal of attention in the last 50 years or so.

3.8 Modelling Turbulence

If we have a current adjacent to the sea bed, then at the sea bed the water sticks to the bed itself and so the current must be zero due to the frictional forces there. However, it is maximum somewhere in the mid depths or possibly at the sea surface, therefore there has to be a shear just above the bed itself that brings the current to zero (see Figure 3.8). To model this, we first duplicate the thoughts of Sir Isaac Newton and propose that the magnitude of the shear must be proportional to the stress in the fluid caused by the friction, mathematically

$$\tau_{zx} \propto \frac{\partial U}{\partial z}$$

or

$$\tau_{zx} = \rho \nu \frac{\partial U}{\partial z},$$

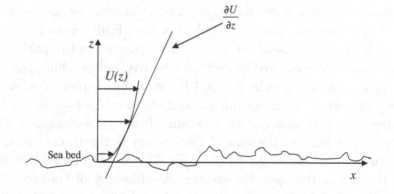

Fig. 3.8   The shear due to friction.

where ν is a constant called the kinematic viscosity of the sea and τ is stress. The zx suffix denotes that the surface along which the stress acts has a normal in the direction of z but that the stress itself points in the x direction. The quantity $\rho\nu$ is sometimes written μ and is called the dynamic viscosity. This density dependent viscosity is not used in the context of oceanography and so will not be met again in this text. The kinematic viscosity is due to molecular friction and its magnitude is therefore constant for particular liquids at particular temperatures. Now the physics behind viscosity involves quantifying Brownian Motion which can be seen by placing fine powder (e.g. lycopodium or fine talc) in a beaker of water and examining a particle of the powder under a microscope. The particle will be seen to vibrate due to molecular bombardment. This molecular action can be analysed by kinetic theory, which in turn leads to a precise definition of viscosity. In the sea, there is a much larger mechanism that can be thought of as doing something similar to molecularly driven friction. This is called turbulence. Turbulence is the chaotic motion of a fluid that consists of whirling eddy motion. Turbulence will certainly slow down a current, whether or not it is near a boundary, and slow it down much faster than molecular friction. Scientists borrowed Newton's formulation of molecular viscosity and wrote

$$\tau_{zx} = \rho\nu_V \frac{\partial \bar{U}}{\partial z}.$$

Here, τ is still stress but, this time, stress due to turbulence in the sea. The bar over $\bar{U}$ denotes that a time average has taken place to eliminate turbulent fluctuations, and ν_V is the much larger viscosity, called *eddy* viscosity, due to turbulent action. Eddy viscosity has to be much larger because of the length scale. The "mean free path" of a molecule can be measured in microns whereas that of a fluid particle in a turbulent eddy can be 10^4 or 10^5 times this. If mean free paths leave you cold, then simply think about the size of a molecular eddy compared to a turbulent eddy. The same disparity of size exists. Size is important here as the crucial mechanism is the transmission of momentum at right angles to the flow. This is done by such eddies and the bigger the eddy the greater the efficiency of transfer. The molecular viscosity is a measure of this transfer for a molecular eddy

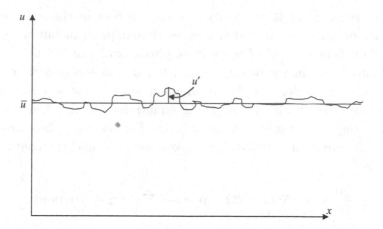

Fig. 3.9   Diagrammatic representation of $u = \bar{u} + u'$ with $\bar{u} \gg u'$.

and the eddy viscosity the measure for a turbulent eddy. Hence, eddy viscosity is 10^4 or 10^5 times molecular viscosity, which is why molecular viscosity is largely ignored in marine modelling. One problem with eddy viscosity is that it is certainly not a constant for a given fluid; it will vary from place to place in the sea and not just with temperature — in fact temperature has little influence here. There will be more turbulence where currents meet, for example, than in the Sargasso Sea. Therefore, is the eddy viscosity model useful? The answer is yes it is, but there are others that work better, especially given sophisticated modern models. However, let us work with eddy viscosity for now. So, we are faced with a turbulent sea.

Consider Figure 3.9, which displays a typical current: One typical feature is that the current has a non turbulent part that is displayed as steady in Figure 3.9 and a fluctuating part that is entirely due to the turbulence. Of course, the "steady" part will usually fluctuate too as it might be a tide or other oscillatory current, it is just that the period of any such non turbulent fluctuations will be much longer than periods associated with turbulence, so on the scale of Figure 3.9, $\bar{u}$ will appear constant, the overbar averaging out the turbulence but nothing else. $\bar{u}$ is thus still a function of time.

We have introduced the basic idea of applying turbulence ideas to oceanography. This is now taken further using vector notation; after all, the sea current is a velocity. So, any current in the sea can be thought of as a mean flow that may be time dependent but is steady on a short time scale of a few seconds, plus a small random deviation. Mathematically in vectors this is written $\mathbf{u} = \bar{\mathbf{u}} + \mathbf{u}'$ where $\bar{\mathbf{u}}' = \mathbf{0}$ and the overbar denotes the mean taken over a few seconds. This is now inserted in all the governing equations and the overall time average (again over a few seconds) taken. For reference, here are the two main equations, the Navier–Stokes equation and the continuity equation

$$\frac{\partial \mathbf{u}}{\partial t} + (\mathbf{u} \cdot \nabla)\mathbf{u} + 2\mathbf{\Omega} \times \mathbf{u} = -\frac{1}{\rho}\nabla p + \mathbf{g} + (\text{friction})$$

and

$$\nabla \cdot \mathbf{u} = 0 = \frac{\partial u}{\partial x} + \frac{\partial v}{\partial y} + \frac{\partial w}{\partial z} = 0.$$

Inserting $\mathbf{u} = \bar{\mathbf{u}} + \mathbf{u}'$ then taking the mean of both equations gives the same two equations with $\mathbf{u}$ replaced by $\bar{\mathbf{u}}$ everywhere, except for the nonlinear advective acceleration, which produces the extra non-zero term

$$\overline{(\mathbf{u}'.\nabla)\mathbf{u}'}$$

on the left-hand side of the Navier–Stokes equations. So, averaging the two equations over a turbulent time scale gives

$$\frac{\partial \bar{\mathbf{u}}}{\partial t} + (\bar{\mathbf{u}} \cdot \nabla)\bar{\mathbf{u}} + \overline{(\mathbf{u}'.\nabla)\mathbf{u}'} + 2\mathbf{\Omega} \times \bar{\mathbf{u}} = -\frac{1}{\rho}\nabla \bar{p} + \mathbf{g} + (\text{friction})$$

and

$$\nabla \cdot \bar{\mathbf{u}} = 0 = \frac{\partial \bar{u}}{\partial x} + \frac{\partial \bar{v}}{\partial y} + \frac{\partial \bar{w}}{\partial z} = 0.$$

This small random deviation is of course unknown, but fortunately the nonlinear non-zero term is the only place in the equations where it occurs. It is called the Reynolds stress term and although it may look harmless, it actually contains nine terms consisting of all pairs of combinations of the components of $\mathbf{u}' = (u', v', w')$ with itself. As we have seen, one of the simplest assumptions to make is to follow Newton and to relate appropriate parts of this Reynolds stress to gradients in the mean flow. The Reynolds stress is in fact a tensor (strictly a second-order tensor) which is made up from the nine components and can be represented by a 3×3 matrix. It can be succinctly written τ_{ij} where the suffices i and j independently run from 1 to 3. In detail

$$
\begin{pmatrix} \tau_{11} & \tau_{12} & \tau_{13} \\ \tau_{21} & \tau_{22} & \tau_{23} \\ \tau_{31} & \tau_{32} & \tau_{33} \end{pmatrix} = \begin{pmatrix} -\rho \overline{u'^2} & -\rho \overline{u'v'} & -\rho \overline{u'w'} \\ -\rho \overline{v'u'} & -\rho \overline{v'^2} & -\rho \overline{v'w'} \\ -\rho \overline{w'u'} & -\rho \overline{w'v'} & -\rho \overline{w'^2} \end{pmatrix},
$$

where each τ_{ij} in fact measures the covariance (statistical measure of agreement, see section 1.4) between two of the fluctuating components $\mathbf{u}' = (u', v', w')$. If $i = j$ it is the autocovariance of u', v' or w' that is being measured. Both matrices are of course symmetric ($\tau_{ij} = \tau_{ji}$). This follows straight away from the right-hand matrix. In fact, if $i = j$ then the stress is called a normal stress. Since normal stresses act in a similar fashion to pressure, they can be safely overlooked (or more accurately absorbed by the pressure). When $i \neq j$ the stress is a shear stress and it is these that need to be modelled. It is of course the *gradients* of the stresses that contribute to the conservation of momentum in much the same way as does the gradient of pressure and not the pressure itself. Turbulence and friction have in common the transfer of momentum in a direction at right angles to the flow. The magnitude of the rate at which this transfer takes place in a turbulent ocean current is what is attempted to be modelled by eddy viscosity. The big assumptions are one that such a relationship is reasonable and two that the relationship is linear. Neither of these can be justified, except perhaps *a posteriori*. To home in on a particular component, let us choose τ_{13}, which of course equals τ_{31}. In this case, the eddy viscosity assumption, a turbulence equivalent to assuming a Newtonian viscous fluid, leads to

$$
\tau_{31} = -\rho \overline{w'u'} = -\rho \nu_v \frac{\partial \bar{u}}{\partial z},
$$

where ν_v is the eddy viscosity (sometimes in older texts called the Austauch coefficient). It is the turbulent equivalent to kinematic (not dynamic) viscosity. Not only is it much bigger than normal viscosity, usually by a factor of 10^6, but it is of course not a fixed property of the fluid. The gradient of the stress, and not the stress itself, gives the net force. No apology is necessary for reiterating that this is the same as gradients of pressure, not pressure itself, causing net force because it is so often misunderstood. Hence, if the overall force balance is to include shear stress effects then terms such as

$$\frac{\partial}{\partial z}\tau_{31} = \frac{\partial}{\partial z}\left(\nu_v \frac{\partial u}{\partial z}\right)$$

need to be included. If the eddy viscosity, ν_v, is a constant, then this can be written in terms of a second derivative

$$\nu_v \frac{\partial^2 u}{\partial z^2}.$$

The stress τ_{31} is a stress that represents the shear due to a current travelling in an easterly direction in the presence of either the sea bed or the sea surface. Other shear stresses, for example τ_{21}, would represent different current shears, in this case the shear due to an easterly current near a north south coast, and this can likewise be put in terms of gradients in mean flow as follows

$$\tau_{21} = -\rho\overline{v'u'} = -\rho\nu_H \frac{\partial \bar{u}}{\partial y},$$

but this time the eddy viscosity ν_H is representative of the *horizontal* transfer of momentum in the current. The horizontal eddies that effect this transfer are correspondingly larger than the vertical counterparts, hence $\nu_H \gg \nu_v$ (by a factor of about 10^4 in fact). From this point all reference to turbulent fluctuations denoted by primed quantities ceases. Let us examine the equation of motion (conservation of momentum) in the x direction. If we include advection, vertical and horizontal momentum transfer (eddy viscosity) then dimensional analysis can be performed on this equation. As all reference to primed (fluctuating) quantities has ceased, the overbar can be dropped and it is now assumed that the current is a (short) mean, enough to even out turbulent fluctuations but not enough to conceal swell waves or anything of longer period. The general x wise equation of motion then is rather daunting, until one remembers that is not actually going to be solved. Here it is:

$$\frac{\partial u}{\partial t} + u\frac{\partial u}{\partial x} + v\frac{\partial u}{\partial y} + w\frac{\partial u}{\partial z} - fv = -\frac{1}{\rho}\frac{\partial p}{\partial x} + \nu_v\frac{\partial^2 u}{\partial z^2} + \nu_H\left(\frac{\partial^2 u}{\partial x^2} + \frac{\partial^2 u}{\partial y^2}\right).$$

This equation is amenable to dimensional analysis of the sort mentioned in the last chapter. We can now do a detailed example.

The u, v and w above represent the flow but with turbulent fluctuations removed. We now non-dimensionalise as follows. Denote typical horizontal lengths by L, typical vertical lengths by D and typical horizontal currents by U. Denote non-dimensional variables by a prime. Again, a reminder that the old use of prime (to denote turbulent fluctuations) has long gone. Hence, we write:

$$x = Lx', \quad y = Ly', \quad z = Dz',$$
$$u = Uu', \quad v = Uv', \quad w = (UD/L)w'.$$

Also suppose that time is typically $t = t'/f_0$, where f_0 is a typical magnitude of the Coriolis parameter, so $f = f_0 f'$. What we have done is to segregate explicitly the magnitude of a quantity from its variation. The variation is between 0 and 1 (the primed quantities) whereas the dimensions (capital letters and f_0) are constant but may be any reasonable magnitude. Substituting for u, v, w, x, y and z into the Navier–Stokes equation gives

$$U f_0 \frac{\partial u'}{\partial t'} + \frac{U^2}{L} \left(u' \frac{\partial u'}{\partial x'} + v' \frac{\partial u'}{\partial y'} + w' \frac{\partial u'}{\partial z'} \right) - f_0 U (f'v')$$
$$= -\frac{1}{\rho} \frac{\partial p}{\partial x} + \frac{\nu_v U}{D^2} \left(\frac{\partial^2 u'}{\partial z'^2} \right) + \frac{\nu_H U}{L^2} \left(\frac{\partial^2 u'}{\partial x'^2} + \frac{\partial^2 u'}{\partial y'^2} \right).$$

We now drop the primes for convenience, and divide through by $U f_0$ to obtain:

$$\frac{\partial u}{\partial t} + \frac{U}{f_0 L} \left(u \frac{\partial u}{\partial x} + v \frac{\partial u}{\partial y} + w \frac{\partial u}{\partial z} \right) - fv$$
$$= -\frac{1}{\rho} \frac{\partial p}{\partial x} + \frac{\nu_v}{f_0 D^2} \frac{\partial^2 u}{\partial z^2} + \frac{\nu_H}{f_0 L^2} \left(\frac{\partial^2 u}{\partial x^2} + \frac{\partial^2 u}{\partial y^2} \right).$$

This is the second time we have dropped primes, and we are now left with a current (u, v, w) that has the turbulence averaged out and is non-dimensional, so has a magnitude of around unity. The coordinates (x, y, z) are also non-dimensional, also with magnitude of around unity. Notice that the pressure term has been left out of this process (it may be considered to be still dimensional if you like).

The usefulness of doing this lies in the three dimensionless groups that have emerged: they are

$$\frac{U}{f_0 L} = R_0 \text{ called the Rossby Number,}$$

$$\frac{\nu_v}{f_0 D^2} = E_V \text{ called the vertical Ekman number}$$

and

$$\frac{\nu_H}{f_0 L^2} = E_H \text{ called the horizontal Ekman number.}$$

If R_0 is small, the advection terms can be ignored; if E_V is small, the vertical friction term can be ignored, and if E_H is small, the horizontal friction terms can be ignored. Thus, this gives a way of assessing in a quantitative manner what terms to ignore and what terms to include in a dynamic model. If the Rossby number R_0 is around unity, then the nonlinear advection terms have to be included. This happens for models of streams such as the Gulf Stream crossing the Atlantic, or strong tidal streams around headlands where the radius of curvature of the headland is appropriate for L and again leads to a significant R_0. On the other hand, if E_V is around one then vertical gradients are large enough for D to be small, and this happens close to the sea bed or sea surface. Of course, a detailed model still has to be built, but this helps us eliminate unwanted terms before using methods to solve the dynamic balance. Nothing has been said here about density changes. There are two dimensionless numbers directly associated with changes in density that will be defined properly later in the book, these are the Richardson number — important when considering estuaries and other stratified flows — and the Burger number — also important when attempting to model stratified flow when Coriolis acceleration is not negligible. Both of these numbers will be met in Chapter 11.

The methods used to calculate currents and sea elevations are these days primarily numerical, and an introduction to these methods forms the subject of the next chapter.

The use of a constant given eddy viscosity for anything other than dimensional analysis still happens but is getting rarer. Once the choice has been made to use numerical methods for coastal or shallow sea modelling, then other options open up for the parameterisation of turbulence. It transpires that it is still convenient to keep eddy viscosity, but instead of it being a parameter to be prescribed, to use it as another variable. The introduction of another variable means that more equations are necessary, and these take the form of the conservation of turbulent kinetic energy at the sea bed together with an equation for the dissipation of the energy. These two equations constitute $k - \epsilon$ turbulence theory, which is briefly outlined at the end of Chapter 5, section 5.4.1. Before launching into numerical methods, let us derive a well known law that, despite modern complex parameterisations of sea bed friction, is still extremely useful.

3.8.1 *Parameterising bottom friction*

Before leaving dimensional analysis, let us return to the parameterisation of friction, which we called drag in Chapter 2, section 2.2. There we derived that drag should take the form

$$\rho L^2 U^2 \left(\frac{Lg}{U^2}\right)^e \left(\frac{\mu}{L\rho U}\right)^c,$$

which can be replaced by the more general

$$\rho L^2 U^2 f_1 \left(\frac{Lg}{U^2}\right) f_2 \left(\frac{\mu}{L\rho U}\right),$$

where f_1 and f_2 are general functions and not just powers. Let us look at this in the context of friction that is caused by the action of the sea bed on the water just above it. Now, this is only just beyond where we were in the last chapter, but when considering the motion of the sea over the sea bed it turns out that there are good physical reasons wrapped around pressure being force per unit area for letting $L^2 f_2(\mu/L\rho U)$ be a constant for a given geological regime. It has different values for sand, shingle, mud etc. but can be considered virtually constant over large areas. This constant is often called C_D, especially in coastal and hydraulic engineering. The other term, the function of Froude number $f_1(\frac{Lg}{U^2})$ is only important if surface gravity waves are present. In general, at the sea bed the effects of waves cannot be felt and this term is therefore taken to be unity. Hence, we arrive at the expression

$$F = \rho C_D U^2,$$

which is the quadratic friction law. This law has been widely used by civil and coastal engineers concerned with the flow of water over a river, estuary or coastal sea bed for over a century, and is still in use today. The D in C_D betrays its civil engineering origin in that it denotes the diameter of the pile about which drag is occurring. What we can deduce here is the restrictions for its validity. For very shallow water, surface waves may change this law, both because of Froude number effects and because, in addition, the changes in pressure will lead to a reappraisal of the other term $f_2(\frac{\mu}{L\rho U})$. Having said this, there is a severe reluctance in the community to relinquish the quadratic friction law, and in the presence of rippled beds the standard value of $C_D = 0.0025$ is simply replaced by a greater constant value $C_D = 0.0061$ in SI units (see Soulsby (1997) and Masselink, Hughes and Knight (2014)). Soulsby's book in particular goes into a lot more detail on this friction law, and we will meet some of these ideas after tackling finite difference modelling and the sea bed boundary condition in the next two chapters.

If the current is a vector, then in order to maintain that the drag and current must be in opposite directions, we write

$$\mathbf{F} = -\rho C_D \mathbf{u}|\mathbf{u}|.$$

This is the accepted form of the quadratic drag law.

3.9   Exercises

By the very nature of this chapter, the following exercises are quite mathematical in nature. They are, however, designed to help in the understanding of the material in the chapter so it is worth some effort to tackle these problems.

(1) If the current $\mathbf{u}$ is given by the expression $y\mathbf{i} - x\mathbf{j}$, calculate the two accelerations

$$\frac{\partial \mathbf{u}}{\partial t} \quad \text{and} \quad \frac{D\mathbf{u}}{Dt}.$$

Hence, calculate the pressure gradient in the absence of both Coriolis acceleration and friction, i.e. assume that

$$\frac{D\mathbf{u}}{Dt} = -\frac{1}{\rho}\nabla p.$$

By assuming hydrostatic pressure, determine the algebraic form of the sea surface.

(2) Confirm the final result of question 1 by using Bernoulli's equation.

(3) Show that the vorticity in a fluid has a value that is, in fact, twice the local angular momentum.

(4) Calculate the displacement x of an oscillating system in terms of frequency ω and time t if at $t = 0, x = 4$ and the speed is $1\,\mathrm{m\ s^{-1}}$.

(5) Starting with the geostrophic equation, which is the balancing of pressure gradients with Coriolisacceleration, in the form

$$2\mathbf{\Omega} \times \mathbf{u} = -\frac{1}{\rho}\nabla p,$$

take the curl to eliminate the right-hand side. By using the identity:

$$\nabla \times (\mathbf{A} \times \mathbf{B}) = (\mathbf{A} \cdot \nabla)\mathbf{B} - (\mathbf{B} \cdot \nabla)\mathbf{A} + \mathbf{B}(\nabla \cdot \mathbf{A}) - \mathbf{A}(\nabla \cdot \mathbf{B})$$

with $\mathbf{A} = \mathbf{\Omega}$ and $\mathbf{B} = \mathbf{u}$, show that only one of these terms is non-zero. Moreover, if only the z component of $\mathbf{\Omega}$ is non-zero show finally that the current must be independent of depth. This is called the Taylor–Proudman theorem; see section 10.4 for more on this.

(6) With

$$\phi = \frac{C}{k}\cosh k(z + h)\sin(kx - \omega t)$$

use equation (3.15) to calculate the rate of change of energy of progressive linear water waves. This result is used in section 8.2.

(7) Show that, by assuming hydrostatic balance along with linear wave theory, the speed of waves is $\sqrt{gh}$. Start with the

equations

$$\frac{\partial u}{\partial t} = -g\frac{\partial \eta}{\partial x},$$

$$h\frac{\partial u}{\partial x} = -\frac{\partial \eta}{\partial t}.$$

The first is the x wise momentum equation and the second the equation of continuity, both linearised and simplified ($h =$ constant) given that depth dependence is negligible.

By differentiating the first equation with respect to x and eliminating the current u via the second, derive a wave equation of the form

$$\frac{1}{c^2}\frac{\partial^2 \eta}{\partial t^2} = \frac{\partial^2 \eta}{\partial x^2}$$

and show that the celerity $c = \sqrt{gh}$.

(8) Forward referencing the next chapter, the Laplace tidal equations with constant depth take the form

$$\frac{\partial u}{\partial t} - fv = -g\frac{\partial \eta}{\partial x},$$

$$\frac{\partial v}{\partial t} + fu = -g\frac{\partial \eta}{\partial y},$$

$$\frac{\partial \eta}{\partial t} + h\frac{\partial u}{\partial x} + h\frac{\partial v}{\partial y} = 0.$$

Assume plane wave solutions of the form

$$\eta = Ae^{i(kx+ly-\omega t)}, \quad u = U_0 e^{i(kx+ly-\omega t)}, \quad v = V_0 e^{i(kx+ly-\omega t)},$$

substitute into the Laplace tidal equations to get three algebraic equations for A, U_0 and V_0. For consistency, the determinant of the co-efficients of these three unknown constant amplitudes has to be zero. Show that this leads to the dispersion relation

$$\lambda^2 = \frac{\omega^2 - f^2}{gh} = k^2 + l^2.$$

Deduce that plane waves cannot exist in an ocean of constant depth where $\omega = f$.

(9) In some large scale ocean flows f, the Coriolis parameter, can be assumed to vary according to the β plane approximation: $f = f_0 + \beta y$. Suggest an alternative dimensionless time, hence redefine both the Rossby and horizontal Ekman numbers, hence reappraise the importance of the nonlinear terms and (horizontal) turbulent dissipation.

(10) Determine the dimensions of the quantities c_V, κ and Q, as defined in section 3.6.

Chapter 4

Numerical Methods

4.1 Introduction

There is no doubt that these days it is not possible to do any serious detailed regional coastal sea modelling without the use of numerical methods. It is also the case, however, that much of the sophistication of the application of these methods lies hidden in proprietary software. It is only possible to understand this software if one understands the methods behind its operation. This in turn requires knowledge of the numerical methods themselves, and this is the main purpose of this chapter. Numerical analysis has been an established branch of mathematics for over 100 years, although it is only since the advent of the computer that it has taken on a central role in the application of mathematics to real problems. In ocean science and coastal engineering the numerical revolution began in the mid-1960s but was then only open to those with both the mathematical knowhow and the ability to program. In the late-1970s and throughout the 1980s, the development of the first mini-computers such as the VAX, and UNIX, based machines and then PCs meant that software became commercially available and user-friendly enough to be used by those not mathematically trained. Here, at the beginning of the 21st century, there are a generation of ocean scientists who are quite happy to run sophisticated software without knowing how it operates. Even coastal and civil engineers need not know the details these days. However, it can still be dangerous to do this when trying to model coastal and offshore seas for we still cannot model this environment with certainty. Given recent mathematical knowledge about chaotic systems, we may *never* be able to model this environment with certainty. The older generation of modellers still tend to be mathematically trained and able to program (usually in FORTRAN) and they are in a better position to understand how software works, in particular to realise its shortcomings.

This chapter will not, at a stroke, turn the mathematical novice into someone who can write and run complex computer programs. The whole point of modern commercially available software is that this is no longer a requirement. Rather, the emphasis here is to enable everyone to understand more about what goes on inside software; where the difficulties and inaccuracies are and how to use the software properly.

As we have seen in previous chapters, the equations that describe the behaviour of the sea are differential equations; that is they involve rates of change of quantities such as velocity, pressure, temperature, etc. These equations cannot be solved exactly, except in some very idealised cases. On the other hand, all modern computers are digital. That means that they can only deal with discrete bits of information. Everyone knows these days that computers work using instructions based on binary code (strings of ones and zeros corresponding to current/no-current passing along extremely small channels, now all inside silicon chips of greater and greater complexity). The continuous model described by the differential equations thus needs to be put into a form amenable to the digital computer, and this is done by adopting numerical methods. First of all, the governing differential equations are converted into a (large) number of algebraic equations. These can be millions of equations in millions of unknowns, but they can be solved using methods similar to the two equations in two unknowns (simultaneous equations) one meets at school. Examining methods of solution, their accuracy and quantifying errors are the province of numerical analysis. The most popular method to render our differential equations amenable to the computer is to use finite differences, and describing these forms the bulk of this chapter and the next. The other main method is to use finite elements, and a brief description of these is found here too.

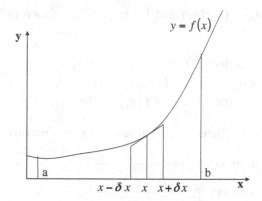

Fig. 4.1   The graph of the function $f(x)$ and the three finite differences.

For those who have met numerical methods before, the first part of this chapter will be familiar. It is there for those with a less sophisticated mathematical background.

Here is the basic idea behind finite differences. One wishes to replace a function $f(x)$ which is defined over a range, say $a \le x \le b$, by a finite sequence of values. The notation δx is introduced to indicate a very small (*infinitesimal*) length. This is not δ times x, δx is a single, very small quantity. If derivatives (i.e. rates of change) of $f(x)$ are required, then one of the three differences defined below can be used:

(1) forward difference: $\Delta f = f(x + \delta x) - f(x)$,
(2) centred difference: $\delta f = f(x + \delta x) - f(x - \delta x)$,
(3) backward difference: $\nabla f = f(x) - f(x - \delta x)$.

These are shown graphically in Figure 4.1.

Let us suppose that δx is the gap between each point on the x axis. This is called the step length and it is normal for this to remain the same throughout the calculation. This is because in adapting a quantity to be represented on a computer the step length is under our control and there is no reason why it should not be constant for our convenience. Let us home in on one value of x in the range $a \le x \le b$, say x_i; let the point immediately to the left of this be x_{i-1} and the point immediately to the right be x_{i+1}. The above definitions thus become:

(1) forward difference: $\Delta f = f(x_{i+1}) - f(x_i)$,
(2) centred difference: $\delta f = f(x_{i+1}) - f(x_{i-1})$,
(3) backward difference: $\nabla f = f(x_i) - f(x_{i-1})$,

so this leads to the three approximations to a derivative as follows

(1) forward difference: $\frac{df}{dx} \approx \frac{f(x_{i+1}) - f(x_i)}{\delta x}$,

(2) centred difference: $\frac{df}{dx} \approx \frac{f(x_{i+1}) - f(x_{i-1})}{2\delta x}$,

(3) backward difference: $\frac{df}{dx} \approx \frac{f(x_i) - f(x_{i-1})}{\delta x}$,

where we can write $\delta x = x_{i+1} - x_i$, the step length. It can be seen from Figure 4.1 that the three differences are approximations to the gradient of $f(x)$ at the point x_i. Closer examination reveals that the centred difference looks the best of the three approximations. However, this does depend on the behaviour of the function $f(x)$. In general, the centred difference is the most accurate; our definition masks this as it is based on a double step length (in order to avoid half steps). It is the symmetric difference using information from both sides of x_i in order to approximate the gradient and so ought to be the most accurate. The forward difference has a special place as it *predicts*. If x represents time, then the forward difference contains the value of f at the later time step, and an equation that contains a single time derivative may be able to be rearranged in order that this value is on the left whilst the right-hand side contains only values of quantities at the present time. As these are all known, we have a prediction for $f(x_{i+1})$. Perhaps not a very accurate prediction, but a prediction nevertheless. In the next Section we go into more detail about the methods used and how accurate they are.

4.2   Finite Differences

Fine mathematical detail on the kind of numerical methods we will use can be found in specialist texts (for example, Smith, 1965). This kind of specialist text will be concerned with the estimation of errors in a precise way, with convergence and with inconsistency. These subjects are at home in mathematics courses, but to cover these topics in any detail would be out of place here. Instead, various finite difference methods of the type used in marine modelling will be introduced and detailed discussion limited to assessing how each performs.

Partial differential equations that contain a single time derivative of the first order (e.g. $\frac{\partial u}{\partial t}$) and no other time derivatives are of the type called parabolic partial differential equations. This is the case no matter how many or what kind of derivatives with respect to x, y or z occur. One of the simplest parabolic differential equations, and the one all the books tend to use to illustrate the action of numerical methods on parabolic differential equations, is the diffusion equation. We shall meet this in Chapter 6 in its own right, but for now we state it as follows

$$\frac{\partial u}{\partial t} = \kappa \frac{\partial^2 u}{\partial x^2},$$

where κ is a constant called the diffusivity. If a forward difference in time but a centred difference in space is used, then at any particular location in the set of data points $x_i^s, i = 1, 2, \ldots, N; s = 1, 2 \ldots$, where N is a large integer, the discrete version of the diffusion equation is

$$\frac{u_i^{s+1} - u_i^s}{\Delta t} = \kappa \frac{u_{i+1}^s - 2u_i^s + u_{i-1}^s}{(\Delta x)^2}, \quad s = 0, 1, \ldots; \quad n = 1, 2, \ldots, N-1,$$

where we have written

$$u(i\Delta x, s\Delta t) = u_i^s,$$

Δx being the step length for space and Δt that for time. There are boundary conditions that will be considered later in the chapter and it is these that involve the end values x_1^s, x_N^s as well as the initial values x_n^0. The crucial point here is that it is possible for us to make u_i^{s+1} the subject of an explicit formula. In order to do this, it is necessary to calculate the finite difference form of the second derivative on the right-hand side. This is done by using the first-order centred difference, but in terms of derivatives as follows

$$\left[\frac{\partial^2 u}{\partial x^2}\right]_i \approx \frac{\left[\frac{\partial u}{\partial x}\right]_{i+1/2} - \left[\frac{\partial u}{\partial x}\right]_{i-1/2}}{\Delta x}.$$

Then the differences on the right are further expressed as centred differences in u

$$\left[\frac{\partial u}{\partial x}\right]_{i+1/2} \approx \frac{u_{i+1} - u_i}{\Delta x}$$

and

$$\left[\frac{\partial u}{\partial x}\right]_{i-1/2} \approx \frac{u_i - u_{i-1}}{\Delta x}.$$

Using these last three approximations gives the finite difference approximation

$$\left[\frac{\partial^2 u}{\partial x^2}\right]_i \approx \frac{u_{i+1} - 2u_i + u_{i-1}}{(\Delta x^2)},$$

which is the standard centred finite difference formula for the second space derivative. This leads to the following finite difference approximation for the diffusion equation

$$u_i^{s+1} \approx u_i^s + r(u_{i+1}^s - 2u_i^s + u_{i-1}^s), \quad s = 0, 1, \ldots; \quad n = 1, 2, \ldots, N-1,$$

where

$$r = \frac{\kappa \Delta t}{(\Delta x)^2}.$$

When it is possible to give an explicit expression for the value of the variable at the next time step, the method is, naturally enough, called an *explicit* scheme or *explicit* numerical method. The reason all practical numerical schemes are not explicit is that sometimes the explicit formula fails to converge. For example, in the formula just derived unless r is less than a specific value, this value $= 1/2$ for the diffusion equation, the scheme does not work. This is because it oscillates, often infinitely, instead of converging on a value close to that expected for $u(x,t)$. The name for this is *instability*, and there is no cure other than changing the finite difference scheme, either by making r smaller or by choosing a different scheme altogether. Making r smaller is often impractical as it would mean using such a small time step that forecasting would be too inefficient. Imagine having to use a 30 second time step for a one day forecast, this would take 172,800 steps! Not using an explicit method but instead using an alternative method that allows (say) a 7 minute time step would seem a better way forward. One example of this alternative method is that due to Crank and Nicolson (professor and research student, respectively, the Crank–Nicolson method). This discretises the diffusion equation as follows

$$\frac{u_i^{s+1} - u_i^s}{\Delta t} = \frac{\kappa}{2}\left[\frac{u_{i+1}^{s+1} - 2u_i^{s+1} + u_{i-1}^{s+1}}{(\Delta x)^2} + \frac{u_{i+1}^s - 2u_i^s + u_{i-1}^s}{(\Delta x)^2}\right].$$

The problem is now obvious. The predicted (*unknown*) values of u at the later time step occur on both sides of the equation and at three different points in space. There is therefore no explicit single equation for u_i^{s+1}. Instead, all the equations for all the discrete values of u have to be written down and a matrix equation set up and solved for all the values of u at the later time step. This is an example of an *implicit* scheme, but the inconvenience of having to invert matrices at each time step is outweighed by the method being unconditionally stable. Figure 4.2 gives a pictorial view of how the Crank–Nicolson method operates to solve the diffusion equation.

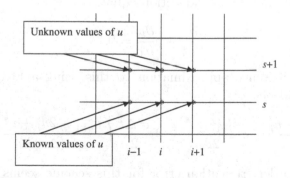

Fig. 4.2 The Crank–Nicolson method, shown schematically.

Of course, there are many different schemes possible just consider; the generalisation

$$\frac{u_i^{s+1} - u_i^s}{\Delta t}$$

$$= \kappa \left[\frac{\theta(u_{i+1}^{s+1} - 2u_i^{s+1} + u_{i-1}^{s+1})}{(\Delta x)^2} + \frac{(1-\theta)(u_{i+1}^s - 2u_i^s + u_{i-1}^s)}{(\Delta x)^2} \right],$$

where θ is a parameter that we, the user, are free to prescribe. The value $\theta = 0$ regains the explicit scheme, put $\theta = 1/2$ and we have the Crank–Nicolson scheme once more, but θ can be any value between 0 and 1. This generalisation is sometimes useful, but the Crank–Nicolson version is almost always best, certainly for the diffusion equation. If $\theta = 1$, then the scheme is called *fully implicit*. Fully implicit schemes have the merit of usually being the most stable. Unfortunately, they are prone to be inaccurate and exhibit large truncation error (see the next Section for an explanation of this terminology). There is a general class of numerical schemes, called Lax–Wendroff schemes, which are $O(\Delta t)^2$ in terms of truncation error, which is achieved through the cancellation of the $O(\Delta t)$ generated by the forward difference in time with the same term generated by the finite difference of the spatial derivative. These schemes are still explicit in time but incorporate centred space differences for accuracy. Lax–Wendroff schemes come into their own in problems that are two dimensional in space, but as a simple illustration consider the one dimensional advection equation

$$\frac{\partial u}{\partial t} + c\frac{\partial u}{\partial x} = 0.$$

The finite difference approximation to this using a Lax–Wendroff scheme would be

$$\frac{u_i^{s+1} - u_i^s}{\Delta t} + c\frac{u_{i+i}^s - u_{i-1}^s}{2\Delta x} = \frac{c^2 \Delta t}{2} \left(\frac{u_{i+1}^s - 2u_i^s + u_{i-1}^s}{(\Delta x)^2} \right).$$

The lowest order truncation error for this scheme seems to be due to the $O(\Delta t)$ error in the explicit finite time difference on the left.

However, this is exactly cancelled by the equivalent term on the right, since

$$\frac{\Delta t}{2}\frac{\partial^2 u}{\partial t^2} = \frac{\Delta t}{2}\frac{\partial}{\partial t}\left(-c\frac{\partial u}{\partial x}\right) = \frac{c^2\Delta t}{2}\frac{\partial^2 u}{\partial x^2}.$$

Precisely how these errors are calculated involves knowledge of Taylor's series and more is said about this a little later in this chapter. Many more elaborate schemes with useful properties have now been developed. For example, for problems that contain the nonlinear advective term, Lax–Wendroff, although accurate, can involve very lengthy computation. For this type of problem, one step algorithms give way to predictor–corrector methods whereby a rough first guess is used to recalculate the variables, the final answer being equal in accuracy to the Lax–Wendroff method but much simpler in terms of application. These predictor–corrector methods when applied to the equations of fluid mechanics, are sometimes called MacCormack's method or MacCormack's technique.

Another particularly useful formulation occurs when we utilise centered differences in time. These give rise to the same problems as using implicit schemes, but these difficulties are overcome by having more than one start time (or using a very accurate explicit scheme just once to generate a second start time given the first). Centred in time schemes are picturesquely called "leap-frog" schemes and the manner in which the computation proceeds does indeed resemble a line of leap-frogging children. The enhanced accuracy and stability of leap-frog schemes is well worth the extra trouble of their implementation, but actual details belong a little later in the chapter when we discuss the upstream (or upwind) schemes.

4.3   Errors and Instability

In this Section, the issues of error and instability will be addressed. The more technical aspects of stability have the next Section to themselves, so first let's examine error. The terms round-off error and truncation error define two very distinct types. The next paragraphs contains the definitions and examples that bring out the differences between them.

In fact, there are three types of error that occur in employing numerical methods. One of these is human error, which incorporates copying incorrectly, transposition of numbers, writing 774 instead of 744 etc. Apart from advice to use as many checks and balances as you can and to double check when possible, nothing further will be said about human error. The two errors we will consider are *round-off* error, and *truncation* error, and they are easy to distinguish in theory, but of course an error is an error, whatever the cause. Round-off error occurs because infinite decimals cannot be held in even the largest computer. The computer representation of the number $\pi = 3.14159265\ldots$ has to stop at some point, even if the decimal representation of the number itself does not. Of course, round-off error is not the problem it used to be with small capacity machines, but it cannot be ignored altogether. As an example, it is common in ocean science to be required to compute the value

$$\frac{1}{\omega^2 - f^2}.$$

Let us consider this expression with $\omega = 1.405 \times 10^{-4}$ and $f = 1.400 \times 10^{-4}$. The first is the frequency of the dominant (twice daily) lunar tide and the second a local value of the Coriolis parameter at latitude 74°N. The problem might therefore occur modelling the effects of tides at the top end of the Denmark Strait (between Greenland and Iceland), an area that is at the centre of global warming debates. Direct computation is of course possible, but undesirable as it involves the subtraction of two numbers that are very similar in magnitude. Using a calculator with seven significant figures, direct computation gives this quantity as infinity, as the difference between ω^2 and f^2 registers as zero. Far better to compute

$$\frac{1}{((\omega/f)^2 - 1)f^2},$$

which gives the answer 7.13×10^9 using the same number of figures on the same calculator. This answer is acceptable and is as accurate as the input data. It has to be said, however, round-off error is seldom a problem these days, even working on a calculator by hand, as long as computation is done sensibly. On a computer, if FORTRAN is the preferred high level language, then the use of Double Precision whereby twice the number of places of decimal, 16 instead of 8 is still recommended.

Truncation error comes about because a derivative has been approximated by a difference. It cannot be eliminated entirely but can be reduced by judicious choice of finite difference scheme. One crude way of reducing truncation error is to decrease the step length(s). This will virtually always give a more accurate answer, but at a price. The model will be slower to compute due to the increase in size of the number of unknowns. It will also take longer for any useful predictions to emerge as the time step will be shorter too. A more sophisticated way of reducing truncation error would be to change the scheme. As already mentioned, to those with a knowledge of Taylor's series, or Taylor's theorem as it is alternatively known, it is quite easy to estimate the order of the truncation error for a given scheme. Let us see how this can work. (If Taylor's series is a deep mystery, skip the next bit.) Taylor's series for a function of a single variable $f(x)$ about the point $x = a$ takes the form

$$f(a + h) = f(a) + hf'(a) + \frac{h^2}{2!}f''(a) + \frac{h^3}{3!}f'''(a) + \cdots,$$

where a dash denotes a derivative. This means that if $f'(a)$ is made the subject of this formula we can write

$$f'(a) = \frac{f(a + h) - f(a)}{h} - \frac{h}{2}f''(a) - \frac{h^2}{6}f'''(a) + \cdots.$$

Now, the first term is simply the forward difference approximation for the first derivative of $f(x)$ at the point $x = a$. Therefore the term

$$\frac{h}{2}f''(a)$$

is the leading error term for this difference. In this way it is possible to estimate truncation error. The forward difference (and the backward difference) have truncation errors that are proportional to h. It is similarly shown (by subtraction of Taylor series) that the truncation error of a centred difference scheme is proportional to h^2, hence centred differences are in general more accurate. The accuracy is of course dependent upon the size of h, the step length, as well as the behaviour of the derivatives of f. We are thus led to the unsurprising conclusion that approximating the derivatives of functions that vary rapidly leads to greater error than approximating the derivatives of smooth functions. In a realistic finite difference scheme, it is possible to use Taylor series expansions to estimate the leading term in the general truncation error for the whole scheme. This calculates what is termed the *local* truncation error. The *global* truncation error is the sum of all these local errors and is of course larger. Decreasing the step length in a scheme that is first order (truncation error $\sim h$) accurate will not decrease the global error as the more accurate value at each step will be exactly cancelled by the increase in steps. For a scheme that is accurate to $\sim h^2$ at each step the global error will be $\sim h$, and in general a scheme that is accurate $\sim h^n$ at each step will have a global error $\sim h^{n-1}$. Let's see why this is true: at each step

$$u_{i+1}^s = u_i^s + A_i(h^n),$$

assuming that the calculation is only concerned with spatial step lengths h. Here, A_i is a constant depending on the behaviour of u but not on h. The global calculation will be of the form

$$\text{Global Error} \sim \sum_i A_i(h^n).$$

The right-hand side will be equal to

$$C \sum_i h^n = CNh^n,$$

where we have written C for the average value of all the A_i's and N for the number of points. Now C does not depend on h, however $N = hL$, where L is a length, ostensibly the length of the one dimensional domain. So $N = L/h$ and we have

$$\text{Global Error} \sim CLh^{n-1},$$

as was stated earlier. So, it is confirmed that if a rather rough-and-ready finite difference scheme with local truncation error of order h is selected and found wanting, decreasing the step length will have no effect whatsoever on the global truncation error, as this will be of order h^0 (or 1) and independent of step length. This can be hard to detect in off-the-shelf software, where such details can be obscure and not emphasised to the purchaser.

The following may come as quite a surprise, but it is possible to select a finite difference approximation to a partial differential equation in all innocence, but for this approximation not to home in on the original equation once the step lengths have become infinitesimally small. The reason this can happen is that amongst the error terms, which are expressed as derivatives via Taylor's series (see above), there could be one or more that remains finite in the limiting process. There are many ways of letting the step lengths tend to zero, so you could be unlucky and choose a wrong one. When this happens, the scheme is called *inconsistent*. The only sure way of testing for consistency is to perform the Taylor's series expansions together with the limiting processes manually, as it were. The schemes outlined here will all be consistent.

As scientists and modellers became more used to using numerical methods, so many began to experiment with variations. One popular idea is to vary the details of the scheme depending on local conditions. A popular phrase encountered in the literature on finite difference techniques for geophysical problems is *upwind differencing* (a phrase arising from meteorology. *Upstream differencing* is the more standard terminology.) This is one scheme that is dependent on local conditions and is common enough to be worth taking a little space here to explain its meaning. Consider the advective equation met earlier

$$\frac{\partial u}{\partial t} + c\frac{\partial u}{\partial x} = 0.$$

The discretisation of this could be

$$\frac{u_j^{s+1} - u_j^s}{\Delta t} + c\frac{u_j^s - u_{j-1}^s}{\Delta x} \approx 0$$

or perhaps

$$\frac{u_j^{s+1} - u_j^s}{\Delta t} + c\frac{u_{j+1}^s - u_j^s}{\Delta x} \approx 0.$$

They have the same truncation error. If $c > 0$ the first approximation using the backward space difference is preferred, and if $c < 0$ the second approximation using the forward space difference is preferred. This is because each choice promotes stability. The choice is called *upwind* or *upstream* differencing. In fact, in practical applications the truncation error will probably be reduced overall as the mixture of backward and forward space differences over a sea area average out to be closer to the more accurate centred difference. But that's not really the point, the local increases in truncation error are worth tolerating to ensure stability. The next section is devoted to analysing stability; just enough is done here to enable the Courant number to be introduced. Let us examine the stability of an upwind scheme by looking at this nonlinear advection equation with $c = u$, which is called the inviscid Burgers equation, after Jan Burgers (1895–1981), a Dutch physicist:

$$\frac{Du}{Dt} = \frac{\partial u}{\partial t} + u\frac{\partial u}{\partial x} = 0.$$

In terms of fluid physics this equation indicates that u, the speed of the fluid, remains unchanged relative to a co-ordinate system that moves with the fluid. The second term on the right is nonlinear. The linear form of this is given by replacing the u in the nonlinear term $u\partial u/\partial x$ by c. So the equation

$$\frac{\partial u}{\partial t} + c\frac{\partial u}{\partial x} = 0, \qquad c > 0$$

is regained. The precise solution to this equation is any function of $(x - ct)$. It is definitely not the case that the nonlinear version is any function of $(x - ut)$. To solve

$$\frac{\partial u}{\partial t} + c\frac{\partial u}{\partial x} = 0, \qquad c > 0$$

one substitutes $\alpha = x - ct$ as a new variable and the equation transforms to

$$\frac{du}{d\alpha} = 0,$$

which integrates to

$$u = f(\alpha) = f(x - ct).$$

Replace c by u and this solution method fails as u is variable. The linear approximation is a one dimensional wave equation, c is the celerity or wave speed (a glance back at Chapter 3 gives you more about waves). Using the centred in space formulation, but keeping the time dependence continuous for now, yields the numerical scheme

$$\frac{\partial u_j}{\partial t} = -c\frac{u_{j+1} - u_{j-1}}{2\Delta x}$$

and for investigating numerical behaviour (anticipating the Fourier method of the next section) we assume

$$u_j = u(t)e^{ikj\Delta x},$$

where k is a constant. Here i reverts to meaning $\sqrt{-1}$, and j is an integer. Whence

$$\frac{\partial u}{\partial t} = -i\omega u,$$

where

$$\omega = \frac{c}{\Delta x}\sin(k\Delta x).$$

For the leap-frog scheme (centred in time) we approximate the time difference through

$$\frac{\partial u_j}{\partial t} \sim \frac{u_j^{s+1} - u_j^{s-1}}{2\Delta t},$$

whence we have

$$u_j^{s+1} \sim u_j^{s-1} - 2i\left[\left(\frac{c\Delta t}{\Delta x}\right)\sin(k\Delta x)\right]u_j^s$$

and stability requires that

$$\left|\left(\frac{c\Delta t}{\Delta x}\right)\sin(k\Delta x)\right| \le 1$$

or

$$\frac{c\Delta t}{\Delta x} \le 1.$$

The expression on the left is the (non-dimensional) Courant number, so this is the restriction that the Courant number must be less than one. For this equation, the forward in time, centred in space formulation is unconditionally unstable, and other schemes based on the leap-frog time difference are only conditionally stable. If we abandon the leap-frog scheme, we can still use the forward difference in time provided we use upstream or upwind differencing (in space) to restore conditional stability. Recall that upstream differences use a one sided difference in space. If we examine our advection equation using forward time differences and backward space differences, we get

$$\left(\frac{u_j^{s+1} - u_j^s}{\Delta t}\right) + c\left(\frac{u_j^s - u_{j-1}^s}{\Delta x}\right) \sim 0$$

from which, on rearranging, is

$$u_j^{s+1} = (1 - \mu)u_j^s + \mu u_{j-1}^s,$$

where

$$\mu = \frac{c\Delta t}{\Delta x}.$$

This next part looks rather like pure mathematics, but is very typical of the procedure that needs to be carried out in order to assess whether or not a given scheme is stable. We can take the modulus of both sides and deduce stability criteria directly here as follows. If $(1 - \mu) \geq 0$, then

$$|u_j^{s+1}| \leq (1 - \mu)|u_j^s| + \mu|u_{j-1}^s|,$$

so applying this at the point where $|u_j^{s+1}|$ is a maximum (over $j\Delta x$), call this value $|U_j^{s+1}|$, gives

$$|U_j^{s+1}| \leq (1 - \mu)|u_j^s| + \mu|u_{j-1}^s|$$
$$\leq |U_j^s|$$

and so the solution must be bounded, provided $\mu \leq 1$; this is the familiar condition that the Courant number does not exceed the local celerity. This expresses that the waves naturally occurring in the physics represented by the original equation must not travel faster than the "numerical wave speed". If this happens, the real waves cannot be captured by the numerical scheme and instability results. This inequality is called the Courant–Friedrichs–Lewy condition or CFL condition for short. It is a necessary condition for the stability of any given scheme. Sadly, it is by no means sufficient. The CFL condition is, of course, different for other schemes. A general interpretation of the CFL condition is that the numerical domain of dependence of a finite difference scheme must include the domain of dependence of the associated differential equation. This is the CFL criteria and more is said about these in the next chapter where there are practical examples.

Summarising, the upstream or upwind scheme can be explained as follows. Suppose a variable C is being modelled (it might be concentration, see Chapter 6) then the scheme whereby the speed $u < 0$

$$\frac{\partial C}{\partial x} = \frac{C_j^{s-1} - C_{j-1}^{s-1}}{\Delta x},$$

but if the speed $u > 0$ then

$$\frac{\partial C}{\partial x} = \frac{C_{j+1}^{s-1} - C_j^{s-1}}{\Delta x}$$

gives upstream or upwind differencing. The phrase stems from the fact that C can only be detected if one is downstream or downwind of it. The resulting scheme is stable, but the use of the one sided differences means that this could result in a large truncation error unless particular care is taken in the design of the overall scheme. Upwind schemes feature prominently in software packages of the type mentioned later in this chapter.

Using a Taylor series analysis the leading error term is

$$-\left(\frac{c^2 \Delta t}{2}\right) \frac{\partial^2 C}{\partial x^2},$$

which is only first-order in the time step. The use of upstream schemes for convenience that are only first order accurate has to be balanced against using schemes based on centred differences (for example, the leap-frog schemes and Lax–Wendroff or MacCormack schemes) that are more difficult to implement but more accurate.

Before moving to stability and then other numerical methods, the extension from one horizontal space dimension to two needs to be made. The notation used here in this introductory chapter is to put $(x, y) = (m\Delta x, n\Delta y)$; this is because at some stages complex numbers need to be used and $i = \sqrt{-1}$, which prevents us from using the more natural $(x, y) = (i\Delta x, j\Delta y)$, although this will be employed later in the book. In most books on finite difference methods, the first two-dimensional equation analysed is Laplace's equation

$$\nabla^2 F = \frac{\partial^2 F}{\partial x^2} + \frac{\partial^2 F}{\partial y^2} = 0.$$

Recall the centred difference formula for second derivative

$$\left[\frac{\partial^2 u}{\partial x^2}\right]_i \approx \frac{u_{i+1} - 2u_i + u_{i-1}}{(\Delta x^2)}.$$

With two horizontal dimensions and the change of notation to $F(x, y, t) = F(m\Delta x, n\Delta y, s\Delta t) = F^s_{m,n}$, we have

$$\left[\frac{\partial^2 F}{\partial x^2}\right]_{m,n} \approx \frac{F^s_{m+1,n} - 2F^s_{m,n} + F^s_{m-1,n}}{(\Delta x^2)}.$$

Similarly,

$$\left[\frac{\partial^2 F}{\partial y^2}\right]_{m,n} \approx \frac{F^s_{m,n+1} - 2F^s_{m,n} + F^s_{m,n-1}}{(\Delta y^2)}.$$

Adding these two and putting $\Delta x = \Delta y = \Delta$ for convenience means that Laplace's equation in finite difference form is

$$\frac{F^s_{m,n+1} - 2F^s_{m,n} + F^s_{m,n-1}}{(\Delta^2)} + \frac{F^s_{m+1,n} - 2F^s_{m,n} + F^s_{m-1,n}}{(\Delta^2)} \approx 0$$

or, on rearranging,

$$F^s_{m,n+1} + F^s_{m,n-1} + F^s_{m+1,n} + F^s_{m-1,n} \approx 4F^s_{m,n},$$

which is telling us that if Laplace's equation is valid at each internal point of a domain then the value of F there is approximately the average of the values F takes at the four surrounding points. The reputation of functions that obey Laplace's equation as smooth is thus well deserved; also it is theoretically possible to prove that if $\nabla^2 F = 0$ in a domain D then the maximum and minimum values of F must occur on the boundary of D. This might seem a strange result, but the numerical approximation just obtained immediately renders it obvious. The extension to two dimensions of upwind differencing is nothing special as upwind differencing is securely a one dimensional phenomenon, so we leave the theoretical treatment of these two dimensional differences here. They come alive in applications in the next chapter; for now we look at stability, then other finite difference schemes before moving on to different numerical methods.

4.4 Stability Analysis

It is worth being a little less *ad hoc* in the analysis of stability and giving two more general methods the first is called the Fourier method and generalises (slightly) that already performed in the last section when the Courant number was introduced. Most attribute its wartime development to that great 20th century mathematician John von Neumann (1903–1957) who was, amongst much else, one of the pioneers of computer science. Suppose we have a finite difference scheme derived using difference approximations from a partial differential equation. Denote by $u_{m,n}^s$ the approximate solution to this that has no round-off error. So the calculations are performed on a computer that holds all numbers exactly. This is impossible, but round-off error and truncation error have to be separated to establish this procedure. Denote by $N_{m,n}^s$ the same calculation but this time performed on a computer with finite places of decimals, so that there is round-off error. The difference $\epsilon_{m,n}^s = u_{m,n}^s - N_{m,n}^s$ is the round-off error. As both $u_{m,n}^s$ and $N_{m,n}^s$ will satisfy the finite difference form of the partial differential equation, so will $\epsilon_{m,n}^s$ provided the original equation is linear. All stability analysis breaks down for nonlinear equations. The guts of the method is to assume that this round-off error can be expressed as a Fourier series. In one space dimension this is

$$\epsilon_m^s = \sum_{p=0}^{P} A_p e^{\alpha s \Delta t} e^{i\pi m \Delta x p / L},$$

where $i = \sqrt{-1}, \alpha$ can be a complex number and the step length $\Delta x = L/P$. This series will grow if the real part of α is positive, decay if it is negative and oscillate if it is zero. So the technique is to substitute this into the finite difference form of the partial differential equation, and use algebra to find out which of these is true. This is simpler than it appears at first as α is not dependent on p, so the expression

$$e^{\alpha s \Delta t} e^{i\pi m \Delta x p / L}$$

can be inserted into the difference equation and algebra performed. The kind of algebra used is reasonably standard (separation of variables for example) but can get involved for elaborate schemes. Here is an example from history. Lewis Fry Richardson (1881–1953) was a pioneer of numerical weather forecasting, but here is a scheme he used on the one-dimensional diffusion equation that was unconditionally unstable. Let us prove this. Richardson used a leap-frog scheme in time, centred in space in order to reduce truncation error (this was 1910 when nothing was known about numerical instability). The difference equation is

$$\frac{u_m^{s+1} - u_m^{s-1}}{2\Delta t} \approx \kappa \frac{u_{m+1}^s - 2u_m^s + u_{m-1}^s}{(\Delta x)^2}$$

so the round-off error obeys the same equation

$$\frac{\epsilon_m^{s+1} - \epsilon_m^{s-1}}{2\Delta t} \approx \kappa \frac{\epsilon_{m+1}^s - 2\epsilon_m^s + \epsilon_{m-1}^s}{(\Delta x)^2}.$$

Write this as

$$\epsilon_m^{s+1} - \epsilon_m^{s-1} \approx 2K[\epsilon_{m+1}^s - 2\epsilon_m^s + \epsilon_{m-1}^s],$$

where $K = \kappa \Delta t / (\Delta x)^2$ and substitute the simplified Fourier expression

$$\epsilon = e^{\alpha s \Delta t} e^{i\pi m \Delta x},$$

which is acceptable as p/L is simply a scale factor that can be reintroduced when desired. It will not be desired in this section, but in specialist numerical texts it might be. Performing this substitution we get

$$e^{\alpha(s+1)\Delta t} e^{i\pi m \Delta x} - e^{\alpha(s-1)\Delta t} e^{i\pi m \Delta x}$$

$$\approx 2K[e^{i\pi(m+1)\Delta x} - 2e^{i\pi m \Delta x} + e^{i\pi(m-1)\Delta x}]e^{\alpha s \Delta t}.$$

Cancelling $e^{\alpha s \Delta t} e^{i \pi m \Delta x}$ from both sides gives

$$e^{\alpha \Delta t} - e^{-\alpha \Delta t} \approx 2K[e^{i \pi \Delta x} - 2 + e^{-i \pi \Delta x}].$$

Now, the right-hand side can be converted into trigonometry as equal to $-8K \sin^2[\frac{1}{2}\pi \Delta x]$, whence we have a quadratic equation

$$\{e^{\alpha \Delta t}\}^2 + 8K \sin^2 \left[\frac{1}{2}\pi \Delta x\right] e^{\alpha \Delta t} - 1 \approx 0.$$

Solving this using the usual formula for the solution of a quadratic yields

$$e^{\alpha \Delta t} \approx -4K \sin^2 \left[\frac{1}{2}\pi \Delta x\right] \pm \sqrt{16K^2 \sin^4 \left[\frac{1}{2}\pi \Delta x\right] + 1}.$$

Now, the interest is in whether the right-hand side grows or decays, so we take the absolute value of both sides but first expand the square root term, retaining only the largest term. In effect, writing $\sqrt{1 + X} \approx 1 + \frac{1}{2}X$ for small X to give

$$e^{\alpha \Delta t} \approx -4K \sin^2 \left[\frac{1}{2}\pi \Delta x\right] \pm \left\{8K^2 \sin^4 \cdot \left[\frac{1}{2}\pi \Delta x\right] + 1\right\}.$$

So, taking the minus option on the $\pm$ to maximise the right-hand side, we have

$$e^{\alpha \Delta t} \approx -4K \sin^2 \left[\frac{1}{2}\pi \Delta x\right] - \left\{8K^2 \sin^4 \left[\frac{1}{2}\pi \Delta x\right] + 1\right\}$$

or

$$e^{\alpha \Delta t} \approx -1 - 4K \sin^2 \left[\frac{1}{2}\pi \Delta x\right] \left\{1 + 2K \sin^2 \left[\frac{1}{2}\pi \Delta x\right]\right\},$$

which to the lowest order of approximation is

$$e^{\alpha \Delta t} \approx -1 - 4K \sin^2 \left[\frac{1}{2}\pi \Delta x\right]. \tag{4.1}$$

Taking the absolute value of this gives

$$|e^{\alpha \Delta t}| \approx \left|1 + 4K \sin^2 \left[\frac{1}{2}\pi \Delta x\right]\right|.$$

The right-hand side is always greater than 1, for any choice of K, so as the calculation proceeds the round-off error will increase until it gets unmanageably large. This is instability, and it has been shown that this particular scheme is unconditionally unstable so any results or predictions will be plain wrong. Notice that before the absolute value was taken, equation (4.1) has an exponential approximately equal to a quantity that is always negative. Some may be troubled by this. It indicates that $\alpha\Delta t$ could be written as $i\pi + \alpha_0\Delta t$, where α_0 is real and positive (since $e^{i\pi} = -1$). This confirms instability as time progresses in the calculation, in fact a classic increasingly oscillatory instability.

The second method for assessing stability uses matrices. All finite difference methods result in a number of algebraic equations for all the unknowns. For the matrix method, these are written down explicitly and manipulations are done on them. Although a mention was made of eigenvalues in Chapter 1 in relation to PCA, rather more knowledge of them would be required for an understanding of how to use matrices to assess the stability of a finite difference scheme. For that reason only, a brief outline is given here, with interested readers directed towards specialist texts on numerical analysis such as Smith (1965). Let us explain the procedure though the example of the Crank–Nicolson scheme for the diffusion equation. Here it is again

$$\frac{u_m^{s+1} - u_m^s}{\Delta t} = \frac{\kappa}{2}\left[\frac{u_{m+1}^{s+1} - 2u_m^{s+1} + u_{m-1}^{s+1}}{(\Delta x)^2} + \frac{u_{m+1}^s - 2u_m^s + u_{m-1}^s}{(\Delta x)^2}\right].$$

Rearranging it in terms of the three unknowns, this is

$$-Ku_{m+1}^{s+1} + 2(1+K)u_m^{s+1} - Ku_{m-1}^{s+1} = Ku_{m+1}^s + 2(1-K)u_m^s + Ku_{m-1}^s,$$

where K retains the previous value $\kappa\Delta t/(\Delta x)^2$. All the equations are now written down explicitly. Of course, there is now the problem of the end points, where there will be boundary conditions, and the value $s = 0$, the start condition. For stability, the start condition does not concern us and we assume that at the physical ends $u = 0$, hence all the equations written as a block will lead to the matrix equation

$$
\begin{pmatrix}
(2+2K) & -K & \cdots & 0 & 0 \\
-K & (2+2K) & \cdots & 0 & 0 \\
\vdots & \vdots & \ddots & \vdots & \vdots \\
0 & 0 & \cdots & (2+2K) & -K \\
0 & 0 & \cdots & -K & (2+2K)
\end{pmatrix}
\begin{pmatrix}
u_1^{s+1} \\
u_2^{s+1} \\
\vdots \\
u_m^{s+1} \\
u_{m+1}^{s+1}
\end{pmatrix}
$$

$$
=
\begin{pmatrix}
(2-2K) & K & \cdots & 0 & 0 \\
K & (2-2K) & \cdots & 0 & 0 \\
\vdots & \vdots & \ddots & \vdots & \vdots \\
0 & 0 & \cdots & (2-2K) & K \\
0 & 0 & \cdots & K & (2-2K)
\end{pmatrix}
\begin{pmatrix}
u_1^{s} \\
u_2^{s} \\
\vdots \\
u_m^{s} \\
u_{m+1}^{s}
\end{pmatrix}
$$

valid for $s = 0, 1, 2, \ldots$. The matrix on the left can be written as

$$
2
\begin{pmatrix}
1 & 0 & 0 & \cdots & 0 & 0 & 0 \\
0 & 1 & 0 & \cdots & 0 & 0 & 0 \\
0 & 0 & 1 & \cdots & 0 & 0 & 0 \\
\vdots & \vdots & \vdots & \ddots & \vdots & \vdots & \vdots \\
0 & 0 & 0 & \cdots & 1 & 0 & 0 \\
0 & 0 & 0 & \cdots & 0 & 1 & 0 \\
0 & 0 & 0 & \cdots & 0 & 0 & 1
\end{pmatrix}
+ K
\begin{pmatrix}
2 & -1 & 0 & \cdots & 0 & 0 & 0 \\
-1 & 2 & -1 & \cdots & 0 & 0 & 0 \\
0 & -1 & 2 & \cdots & 0 & 0 & 0 \\
\vdots & \vdots & \vdots & \ddots & \vdots & \vdots & \vdots \\
0 & 0 & 0 & \cdots & 2 & -1 & 0 \\
0 & 0 & 0 & \cdots & -1 & 2 & -1 \\
0 & 0 & 0 & \cdots & 0 & -1 & 2
\end{pmatrix}
$$

and those on the right as

$$
2
\begin{pmatrix}
1 & 0 & 0 & \cdots & 0 & 0 & 0 \\
0 & 1 & 0 & \cdots & 0 & 0 & 0 \\
0 & 0 & 1 & \cdots & 0 & 0 & 0 \\
\vdots & \vdots & \vdots & \ddots & \vdots & \vdots & \vdots \\
0 & 0 & 0 & \cdots & 1 & 0 & 0 \\
0 & 0 & 0 & \cdots & 0 & 1 & 0 \\
0 & 0 & 0 & \cdots & 0 & 0 & 1
\end{pmatrix}
- K
\begin{pmatrix}
2 & -1 & 0 & \cdots & 0 & 0 & 0 \\
-1 & 2 & -1 & \cdots & 0 & 0 & 0 \\
0 & -1 & 2 & \cdots & 0 & 0 & 0 \\
\vdots & \vdots & \vdots & \ddots & \vdots & \vdots & \vdots \\
0 & 0 & 0 & \cdots & 2 & -1 & 0 \\
0 & 0 & 0 & \cdots & -1 & 2 & -1 \\
0 & 0 & 0 & \cdots & 0 & -1 & 2
\end{pmatrix}
$$

or using matrix notation

$$
\{2\mathbf{I}_m + K\mathbf{M}_m\}\mathbf{u}^{s+1} = \{2\mathbf{I}_m - K\mathbf{M}_m\}\mathbf{u}^s.
$$

The matrix $\mathbf{M}_m$ is

$$\begin{pmatrix} 2 & -1 & 0 & \cdots & 0 & 0 & 0 \\ -1 & 2 & -1 & \cdots & 0 & 0 & 0 \\ 0 & -1 & 2 & \cdots & 0 & 0 & 0 \\ \vdots & \vdots & \vdots & \ddots & \vdots & \vdots & \vdots \\ 0 & 0 & 0 & \cdots & 2 & -1 & 0 \\ 0 & 0 & 0 & \cdots & -1 & 2 & -1 \\ 0 & 0 & 0 & \cdots & 0 & -1 & 2 \end{pmatrix}$$

Rewriting this as

$$\mathbf{u}^{s+1} = \mathbf{Q_m}\mathbf{u}^s, \tag{4.2}$$

where

$$\mathbf{Q}_m = \{2\mathbf{I}_m + K\mathbf{M}_m\}^{-1}\{2\mathbf{I}_m - K\mathbf{M}_m\},$$

enables the actual solution to be written by repeated application of equation (4.2) as

$$\mathbf{u}^{s+1} = [\mathbf{Q}_m]^s\mathbf{u}^0.$$

It is now reasonably clear that, in order to get a solution to this equation, the matrix $\mathbf{Q}_m$ needs to remain finite when raised to the power of s, the number of time steps. This could be a very large number, in the thousands perhaps. This is where things get technical, because most of you would not know what a matrix raised to a power is let alone be able to answer the question as to whether it remains finite, so we will not go much further. The eigenvalues of the matrix play a major part in answering this question. It is not too difficult to see that if the eigenvalues of $\mathbf{M}_m$ are λ_m then the eigenvalues of $\mathbf{Q}_m$ are given by

$$\mu_m = \frac{2 + K\lambda_m}{2 - K\lambda_m}.$$

What is more tricky to see is that $\lambda_m = -4\sin^2(m\pi\Delta x/2)$, where $m = 1, 2, \ldots, M - 1$. This leads to $|\mu_m| < 1$ and this is what is required for unconditional stability. The text by Smith (1965) has more detail, though not how to find the eigenvalues of an $m \times m$ matrix, for which books on linear algebra need to be consulted.

Although purists prefer the matrix method for assessing the stability of finite difference schemes, the Fourier series method is easier and, where it works, gives the same results.

4.5   Advanced Finite Difference Methods

Having just shown that the leap-frog method first tried all those years ago by Richardson cannot work, it's reasonable to try and find one that does. It turns out that it is best to make use of a predictor–corrector technique whereby a sensible guess at the next value is made, the *predictor*, and it is improved at the same time step, the *corrector*. Coming away from the diffusion equation, suppose an explicit scheme is given by the equation

$$u^{s+1} = F(u^s),$$

where the spatial dependence, usually denoted by the subscripts i, j (or m, n), is assumed for the moment, and the function F is all the advective and turbulence terms involving derivatives of u but now in finite difference form. In the fullness of time, there would be at least two equations here involving u and v but the principles are best done in one dimension. The Heun predictor–corrector scheme, for example, is given by the pair of equation

$$u_F^{s+1} = u^s + \Delta t F(u^s),$$

$$u^{s+1} = u^s + \frac{\Delta t}{2}[u^s + u_F^{s+1}].$$

The first is the Euler scheme, and the second a trapezoidal averaging, but it is not hard to see that there are plenty of possibilities; it is a question of doing some research and homing in on a good one. Leap-frog schemes are the order of the day in marine modelling, and so that is the direction in which we head. The specialist books by Kowalik and Murty (1993) and Durran (1999) give a lot more examples, but there's is scant time for that here, so let us cut to the chase and give a scheme due to Robert (1966) and Asselin (1972) which still has

currency (see the section on NEMO in Chapter 14). It is analysed in
the book by Durran (1999) and the basic equations are

$$u^{s+1} = \overline{u^{s-1}} + 2\Delta t F(u^s), \tag{4.3}$$

$$\overline{u^s} = u^s + \gamma(\overline{u^{s-1}} - 2u^s + u^{s+1}). \tag{4.4}$$

Here, is an explanation of the notation. This is a three level scheme
as it has times $(s-1)\Delta t, s\Delta t$ and $(s+1)\Delta t$. The quantity $\overline{u^s}$ is a
time filter that in recent years has been subject to some revision,
see Williams (2009). The value taken by the parameter γ varies,
dependent on the application. According to Durran (1999), a value of
0.25 completely eliminates oscillations of period $2\Delta t$ that are present,
however, values of 0.06 are used for atmospheric models and 0.2 for
cloud models. In NEMO (2012) a much smaller value of 10^{-3} is used,
see also Leclair and Madec (2009). The upshot is that this scheme
has an amplitude error of $O(\Delta t)^2$. Other alternatives are of course
possible; here is one that is very close to the Heun scheme mentioned
above

$$u^* = u^{s-1} + 2\Delta t F(u^s),$$

$$u^{s+1} = u^s + \frac{\Delta t}{2}[F(u^s) + F(u^*)],$$

which is a predictor–corrector scheme based on the first guess u^*, and
it is called the leap-frog trapezoidal scheme. This scheme is a good
one under most uses, but Magazenkov (1980) proposes combining this
with an Adams–Bashforth multi level predictor–corrector scheme.
If you have never heard of an Adams–Bashforth scheme (and why
should you?), consult a standard text like Ortega and Poole (1981).
According to Durran (1999), the errors are a lot less and there are
graphs to prove it.

4.6   Flux Corrected Transport

Finite difference numerical methods work best when the currents
being modelled are smooth. If there are fronts or other sudden
changes in current direction or magnitude then finite difference methods fail to perform well unless very carefully designed. Under these

circumstances, methods other than direct finite differences are worth considering. In 1973, Boris and Book published a method that, in addition, utilises bulk properties rather than just values at a point. In recent years this method, called *Flux Corrected Transport* (FCT), has become very popular and has led to commercial software. This method will now be outlined, and we shall follow the paper of Zalesak (1979). The easiest equation to use is the one dimensional advection equation that takes the form

$$u_t + p_x = 0,$$

where u and p may be any reasonable functions of t and x. Of course, this particular notation is consistent with our previous notation whereby u denotes the current in the x direction and p denotes pressure but it needn't, as we will here merely be illustrating the technique. Using a forward difference in time and centred in space gives

$$u_m^{s+1} = u_m^s + \Delta t \frac{P_{(m+1/2)} - P_{(m-1/2)}}{\Delta x_m},$$

where it will be noticed that the finite differences for p on the right-hand side do not have a time level specified. The letter m denotes the difference in the single space variable x (the letter i will be used for $\sqrt{-1}$, so is not available). On paper, this scheme looks workable if the right-hand side is evaluated at time step $s\Delta t$, however this is not the case. There are even more problems if the Coriolis terms are involved; problems of instability. The question is are there ways of avoiding this problem by judicious choice of time step(s) for the right-hand side? If one tries to be very clever and choose time steps that minimise the truncation errors, then it turns out that this causes "ripples", which can lead to instability. Being too conservative with the choice means a stable scheme but now, because of the one sided nature of the differences, the truncation errors becomes unmanageably large. The scheme introduced by Boris and Book (1973) and Zalesak (1979), an now called FCT, turns out to be an optimal choice for minimising both truncation error and instability. The FCT algorithm works by computing the flux of transport of p into and out of individual cells. This is done individually cell by cell and uses the quantity of p (its flux) that enters and leaves a particular cell. This flux depends upon the scheme adopted, and is changed by correcting the amount so that no p is created or lost overall. Hence the name "Flux Correction". Each FCT scheme is variable dependent to a certain extent, but in general terms the procedure is:

(1) Compute a set of fluxes $p^L_{(m+1/2)}$ which are low in order (first or second at most); this must be monotone, which means either all increasing or all decreasing with m.
(2) Compute a set of fluxes $p^H_{(m+1/2)}$ which are higher order, using a higher order scheme.
(3) Compute the so-called *anti-diffusive* fluxes

$$A_{(m+1/2)} = p^H_{(m+1/2)} - p^L_{(m+1/2)}.$$

(4) Compute a monotone estimate of the solution at $(s+1)\Delta t$, the desired next time step, also known in FCT as the transported or diffused solution through

$$u^{td}_m = u^s_m + \Delta t \frac{p^L_{(m+1/2)} - p^L_{(m-1/2)}}{\Delta x_m}.$$

(5) The $A_{(m+1/2)}$ is corrected so that the anti-diffusion does not generate any new maxima or minima. This can be expressed mathematically through defining

$$A^c_{(m+1/2)} = C_{(m+1/2)} A_{(m+1/2)}, \quad 0 \le C_{(m+1/2)} \le 1.$$

(6) Perform the anti-diffusion step

$$u^{s+1}_m = u^{td}_m - \frac{\Delta t}{\Delta x_m} (A^c_{(m+1/2)} - A^c_{(m-1/2)}).$$

There is a remaining technicality of how to compute $C_{(m+1/2)}$. If these were all zero, the solution would be the standard finite difference scheme; if all the Cs were one the scheme would be the chosen higher order scheme. For specific methods of finding $C_{(m+1/2)}$ see either the book by Durran (1999) or the paper by Zalesak (1979). One important factor that has to be kept in mind is that the scheme has to converge. In traditional finite difference schemes this can be assured reasonably easily, however when virtually each cell is treated differently things are not so simple. Using the above notation, we define the *total variation* in u as

$$\mathrm{TV}(u) = \sum_m \{u_{m+1} - u_m\},$$

although in oceanography or coastal engineering it will be a two or three-dimensional equivalent. It must be the case that the total variation diminishes as the FCT method progresses through time so we demand that at each time step $s\Delta t$

$$\mathrm{TV}(u^{s+1}) \leq \mathrm{TV}(u^s)$$

and this is referred to as *total variation diminishing* or TVD, although strictly it is total variation not increasing. Again there is more about this in specialist texts such as Durran (1999). Such details would be out of place here; we have the general idea of flux correction and TVD. The context — placing the method in the context of geophysical fluids — will have to wait until the next chapter when there will be examples.

4.7   Finite Volume Methods

To see flux in a different, physical context consider the x equation of motion in the form

$$\frac{\partial u}{\partial t} + (\mathbf{u} \cdot \boldsymbol{\nabla})u + (2\boldsymbol{\Omega} \times \mathbf{u})_x = -g\frac{\partial \eta}{\partial x},$$

where $(2\boldsymbol{\Omega} \times \mathbf{u})_x$ denotes the x component of the Coriolis acceleration; together with the continuity equation

$$\boldsymbol{\nabla} \cdot \mathbf{u} = 0.$$

So,

$$\boldsymbol{\nabla} \cdot (\mathbf{u}u) = (\mathbf{u} \cdot \boldsymbol{\nabla})u + (\boldsymbol{\nabla} \cdot \mathbf{u})u = (\mathbf{u} \cdot \boldsymbol{\nabla})u$$

because the continuity equation leads to the second term after the first equality being zero. As a consequence, the x equation of motion can be rewritten in what is called *flux* form

$$\frac{\partial u}{\partial t} + \boldsymbol{\nabla} \cdot (\mathbf{u}u) + (2\boldsymbol{\Omega} \times \mathbf{u})_x = -g\frac{\partial \eta}{\partial x}.$$

Similarly the y equation can be written in flux form as follows

$$\frac{\partial v}{\partial t} + \boldsymbol{\nabla} \cdot (\mathbf{u}v) + (2\boldsymbol{\Omega} \times \mathbf{u})_y = -g\frac{\partial \eta}{\partial y}.$$

Written out in full vector notation, the flux form of the equation of motion is

$$\frac{\partial \mathbf{u}}{\partial t} + \boldsymbol{\nabla} \cdot (\mathbf{uu}) + 2\boldsymbol{\Omega} \times \mathbf{u} = -g\boldsymbol{\nabla}\eta - g\mathbf{k},$$

where the strange looking term $\mathbf{uu}$ is an array of the nine possible combinations of the three components of the vector $\mathbf{u} = (u, v, w)$ (a second order tensor or, in old books, a *dyadic*) defined by the 3×3 matrix

$$\begin{pmatrix} u^2 & uv & uw \\ vu & v^2 & vw \\ wu & wv & w^2 \end{pmatrix}.$$

It is this version of the equation of motion that serves as the basis for a lot of coastal engineering software such as MIKE21. In this chapter, the concentration is on the introduction to the numerical methods; the applications are left for the next chapter. The principle of the method is to go back to physics. Consider the equation

$$\frac{\partial F}{\partial t} + (\mathbf{u} \cdot \boldsymbol{\nabla})F = \kappa \nabla^2 F,$$

where after a glance at Section 3.6 we see that F could certainly be either temperature T or salinity S. It could even be $\mathbf{u}$ as the inclusion of friction (see Section 3.7) but neglecting rotation in the form of the Coriolis acceleration gives rise to a diffusion of momentum term, and the balance takes the same form

$$\frac{\partial \mathbf{u}}{\partial t} + (\mathbf{u} \cdot \boldsymbol{\nabla})\mathbf{u} = \kappa \nabla^2 \mathbf{u}.$$

Using the continuity equation in incompressible form performed earlier, this is written in flux form

$$\frac{\partial \mathbf{u}}{\partial t} + \boldsymbol{\nabla} \cdot (\mathbf{uu}) = \kappa \nabla^2 \mathbf{u}.$$

This expression is used to show the physics behind the finite volume method. The last equation is written

$$\frac{\partial \mathbf{u}}{\partial t} = \kappa \nabla^2 \mathbf{u} - \boldsymbol{\nabla} \cdot (\mathbf{uu}) = \boldsymbol{\nabla} \cdot (\kappa \boldsymbol{\nabla} \mathbf{u} - \mathbf{uu}).$$

This is valid throughout the domain, which we will eventually assume to be two dimensional. We define what is termed a *control volume*, usually a rectangle but it may be a triangle or other closed area. Call the entire domain V and it is useful to think of it as three dimensional but vertically homogeneous and taking the form of a cylinder with axis vertical (in the z direction). Integrate the last equation over this volume to obtain

$$\int_V \frac{\partial \mathbf{u}}{\partial t} dV = \int_V \boldsymbol{\nabla} \cdot (\kappa \boldsymbol{\nabla} \mathbf{u} - \mathbf{uu}) dV = \int_S \{\kappa \boldsymbol{\nabla} \mathbf{u} - \mathbf{uu}\} \cdot d\mathbf{S}, \quad (4.5)$$

where S is the bounding surface. This last equality is Gauss' flux theorem, which states that the volume integral of a divergence must be equal to the flux of material integrated (summed) over the bounding surface. Put simply; stuff that's created in a closed volume must be the same as the stuff that flows out through the bounding surface of that volume. Much as we are used to the conservation of mass, whereby any mass created inside a closed volume must equal the flux of matter through the surface, this generalises to the creation of anything inside a closed volume being equal to the flux of that same quantity through the bounding surface of the volume. It is the conservation of 'stuff' and 'stuff' here equals $\kappa \boldsymbol{\nabla} \mathbf{u} - \mathbf{uu}$. Those familiar with vector calculus can prove the general result easily using integration. Physically, it is easier to see for T and easiest for salinity, S, where it expresses the conservation of salt.

We now turn to numerical evaluation and how this flux form facilitates this. The first point to note is that the flux form as represented in the integrated version equation (4.5) lends itself to a different kind of numerical evaluation. As the sum over a domain has taken place, we can evaluate the integrals numerically. This evaluation can lead to the *finite volume method* that will be described now. It also leads to the *finite element method* (FEM), a discussion of which is postponed for now but has its own section later. The right-hand side of equation (4.5) is an integral over the surface of the control volume, so to evaluate this the surface is divided up into small flat faces; call them ΔS_k, and when summed over integer k this generates the whole of the surface S. So we have

$$\int_S \{\kappa\nabla\mathbf{u} - \mathbf{uu}\} \cdot d\mathbf{S} \approx \sum_k \{\kappa\nabla\mathbf{u} - \mathbf{uu}\}_k \cdot \mathbf{n}\Delta S_k. \qquad (4.6)$$

The right-hand side of this equation is a sum over small areas, so to a good approximation, the integrand $\{\kappa\nabla\mathbf{u} - \mathbf{uu}\}$ takes the value at the centre of the small area so the use of Taylor's series, where the expansion is based around the centre, is certainly good way to improve the approximation. So, expand $\mathbf{u}$ as follows

$$\mathbf{u}_k(\mathbf{r}) = \mathbf{u}_k(\mathbf{r}_k) + (\mathbf{r} - \mathbf{r}_k) \cdot \nabla\mathbf{u}_k(\mathbf{r})_{\mathbf{r}=\mathbf{r}_k} + \cdots$$

in each cell S_k, retaining only a few terms each time. The summation on the right of equation (4.6) can then be evaluated to a good approximation. For many the vector notation makes this extremely obscure, so here is a one-dimensional version. The surface integral written in a single parameter α looks like

$$\int F(\alpha)d\alpha \approx \sum_k F_k \Delta_k$$

and Taylor's series about a value in the middle of the cell (α_k) is

$$F(\alpha) = F(\alpha_k) + (\alpha - \alpha_k)F'(\alpha_k) + \frac{1}{2!}(\alpha - \alpha_k)^2 F''(\alpha_k) + \cdots,$$

where this time the second-order term has been given (in the vector form it is, of course, more unwieldy). The expansion here is in terms of powers of $(\alpha - \alpha_k)$ and the value α_k will depend on k, just as the value $(\mathbf{r} - \mathbf{r}_k)$ will do too. The summation approximation to the integral is thus

$$\int F(\alpha)d\alpha \approx F(\alpha_k) \int d\alpha + F'(\alpha_k) \int (\alpha - \alpha_k)d\alpha + \ldots$$

$$= F(\alpha_k)\Delta\alpha + O(\Delta\alpha)^2.$$

It is now time to try and interpret this in terms of the original (vector) formulation. Consider the last integral which in terms of approximating equation (4.5) is

$$\int_S \{\kappa\nabla\mathbf{u} - \mathbf{uu}\} \cdot d\mathbf{S} \approx \sum_k [\{\kappa\nabla\mathbf{u} - \mathbf{uu}\}_k \Delta S_k] + \text{higher order terms}.$$

The first term $\kappa\nabla\mathbf{u} \cdot \Delta S_k$ is the diffusive flux across the surface S_k approximated using a value of $\mathbf{r}$ inside S_k, most likely the centre of mass of the small area, which is probably a square but could be another shape for adaptive grids or irregular grids (see later). The second term $\mathbf{u}\mathbf{u}_k \cdot \Delta S_k$, on the other hand, is the convective or advective flux approximated in the same way. The original surface integral to the right of equation (4.5) thus indicates the total difference between the diffusive and advective flux into the domain, and the numerical approximation of this is achieved by adding these up from a knowledge of each small surface, its centre of mass and its fluxes. This is the central theme of the finite volume method. The left-hand side of equation (4.5) is the time rate of change of the volume integral and is often either zero or can be easily computed from knowledge of flow into and out of the computational domain. It is also possible to evaluate the left-hand volume integral approximately through

$$\int_V \frac{\partial \mathbf{u}}{\partial t} dV \approx V_m \frac{\partial \hat{\mathbf{u}}}{\partial t},$$

where V_m is the mean value of the control volume and $\hat{\mathbf{u}}$ is the average value of the current taken spatially through the volume. This approximation only works well if the current is largely unidirectional. A practical example of using the finite volume method will be given in the next chapter.

4.8 Finite Element Methods

Finally, let us introduce FEMs. Perhaps the first thing to say is that there are several different types of FEM. If the problem is one that does not depend on time (rare these days) then one might try a finite element technique that is rooted in the calculus of variations. Here, the domain of the problem is divided into conveniently shaped small polygons, usually but not exclusively, triangles. The variables inside those triangles are assumed to be a simple form, say linear functions of x and y $(ax+by+c)$ where a, b and c are different for each triangle. To get mathematical for a bit, it is important that the choice of these functions is correct. That is, we need to be sure that our variables can be expressed in terms of them, and we can be certain of this if we choose what is called a *basis*. This is the name given to a correct set of functions, but a basis has two very important properties. First of all, any function can be expressed in a linear combination of them, and further, once this combination has been found it is unique. It is also possible for the functions that comprise the basis to be *orthogonal*; here this means the integral of a product of different members of the basis is zero. For those familiar with Fourier series this is not new, for others it might seem a tall order. Perhaps an analogy helps here. Colour TV works well and we all watch breathtakingly realistic scenes on our wide screen TVs these days. All the colour stems from combining just three colours by varying the intensity of each of the three basic components that are red, green and blue. These three components are the basis for all the amazing, contrasting varieties of colour on our TVs, so the creation of all colours from just three basic colours works. A basis for functions works similarly. A specific example is given in the next chapter. In the field of finite elements, the expression *shape function* is often preferred to the mathematical term basis function. Once the form of the functions has been decided, the continuity of the variables needs to be assured. This is done by matching the values at certain points (usually the vertices of the polygons) and the physical laws of conservation are applied by minimising what is called a functional by mathematicians, which might be energy or some other quantity. It is this minimisation principle that harks back to the calculus of variations which dates from the last years of the 17th century and was an early application of the calculus, at that time newly invented (or discovered, dependent on your philosophy). This version of the FEM is the Raleigh–Ritz technique, which can only really be applied to modelling tides of a single frequency in coastal engineering and coastal marine science. It gets an airing in the next chapter. Most of the interesting problems are time dependent and are either evolutionary, like the diffusion problems, or wave like. Numerical treatment of wave problems has not got a mention yet, but in principal they differ little from evolutionary problems when a particular sea is modelled. The theory of solving wave problems can be very different; housed in something called the method of characteristics, but let us not travel along that particular path as it is only useful under special circumstances. These are acoustics, which are outside our remit, and wave ray theory, which is only touched on briefly in Chapter 7 and then not modelled numerically.

A far more useful finite element technique is the Galerkin weighted residual method, although only a brief outline is given here. This is because the theory in all its detail would take too much space, and it is better to postpone such details until the next chapter, when an example is done. Let us start as in the finite volume technique and use as the starting point the underlying balance in the form of the partial differential equation. So suppose, as an example, the horizontal equation of motion written as

$$\frac{\partial \mathbf{u}}{\partial t} + \boldsymbol{\nabla} \cdot (\mathbf{uu}) + 2\boldsymbol{\Omega} \times \mathbf{u} = -g\boldsymbol{\nabla}\eta + \nu_H \nabla_H^2 \mathbf{u}$$

is considered. For those averse to vectors, this splits into the two components

$$\frac{\partial u}{\partial t} + \frac{\partial}{\partial x}(u^2) + \frac{\partial}{\partial y}(uv) - fv = -g\frac{\partial \eta}{\partial x} + \nu_H \left(\frac{\partial^2 u}{\partial x^2} + \frac{\partial^2 u}{\partial y^2} \right)$$

and

$$\frac{\partial v}{\partial t} + \frac{\partial}{\partial x}(uv) + \frac{\partial}{\partial y}(v^2) + fu = -g\frac{\partial \eta}{\partial y} + \nu_H \left(\frac{\partial^2 v}{\partial x^2} + \frac{\partial^2 v}{\partial y^2} \right).$$

Either the vector version or the two component equations are now manipulated. Consider the vector form first, and multiply this by a function $w(\mathbf{r})$, called a *weight function*, before integrating over the domain considered as a volume. Gauss' theorem is utilised when possible to turn volume integrals into surface integrals, then the boundary conditions (no flow through coasts) are used and the resulting volume integral remains.

If using the scalar version is preferred, take the first component and multiply this by a weight function $w_1(x, y)$ that is zero on the boundary of the domain of the problem. The integration under this regime is a multiple integral without vectors, and the form of the integrals is amenable to the technique of integration by parts. Whether the vector or scalar equation is used, it is important in the Galerkin technique that the boundary conditions are used through either use of Gauss' theorem or integration by parts. The choice of weight function is arbitrary, but it helps if this choice is made judiciously. Most applications designate the weight function to be a basis function (shape function) for the problem. This is because doing so greatly simplifies calculation. The reason is the orthogonality of the basis functions. If a basis function is multiplied by another different basis function, and the product is integrated over the domain, then this integral is zero. This is the definition of orthogonality and will be familiar to those who know about Fourier series. Fourier series are series of sine and cosine functions and the sine $(\sin(nx))$ and cosine $(\cos(nx))$ functions are orthogonal over either $(0, 2\pi)$ or $(-\pi, \pi)$.

If the original equation over the domain is directly solved by division of the domain into small areas then this is called a *strong* form of solution. Multiplying by a weight function and integrating over a domain then dividing into small areas before solving is called a *weak* form of solution. So this gives the basic idea behind the finite element technique. The application of finite elements to practical problems is delayed until the next chapter.

4.9   Adaptive Grids

In simple, straightforward terms, the use of adaptive grids is the ability of the grid points to move location as the computation proceeds. The term can also be applied to FVM and FEM where it is sometimes called moving finite volumes or elements. It dates back to 1981 and papers by Miller (1981) and colleagues (see the book by Baines, 1994). Originally, the nodes in a finite element mesh are allowed to move, giving an extra degree of freedom and allowing the calculation to become more accurate. However, later versions take into account the local speed of the flow and allow the nodes to concentrate along the paths of maximum shear, that is rate of change of flow with distance measured across the flow. Figure 4.3 shows how flow can distort the grid as it develops. Here, one can see the bunching of the grid points along the flow path that has a criss-cross pattern.

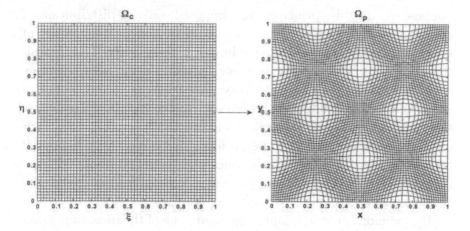

Fig. 4.3   An example of how an even grid can distort due to flow.

Writing in general terms, there are three types of adaptive method using finite elements, namely the *h-method* (mesh refinement), the *p-method* (order enrichment) and the *r-method* (mesh motion). The first two do not involve actually changing the nodes of the grid. The *h-method* is to create more nodes within the existing (fixed) grid so that better accuracy ensues. It is the finite element equivalent to using nested grids in finite difference methods. The *p-method* is to change the shape function, for example from linear to cubic or higher degree polynomial locally to improve accuracy, also locally. Again no actual movement of nodes takes place, and in this case no extra nodes either. In practice, the *h* and *p* methods combine to become the *hp-method*. It is fair to say now that it is the *r-method* that is proving central in providing the most innovative new adaptive grid finite element models. However, one cannot go moving nodes all over the place without some restraints. One of the most important is to be able to put some kind of error limits in place. This is done either by defining some limits on the velocity of the points themselves, or by establishing some restrictions on where the nodes can move to. Without getting into too many technicalities, the usual way to proceed is to define a principle that governs the distance between nodes. Maybe this has to be constrained to be within certain limits or the mean over subsections of the domain has to be within certain bounds. To do this, principles are imposed, on the distribution of nodes and monitor functions are defined that monitor the density of nodes, preventing either overcrowding or over sparsity. The imposition of such principles stops the algorithm for moving nodes running away with itself; it would be so easy for *r-methods*, in particular, to get into a feedback situation where ever faster speeds of flow cause finer meshes, leading to faster flow and eventually singularities. Tangling of the mesh also has to be avoided; this could easily happen if flow is fast but convoluted. Techniques such as the parabolic Monge–Ampère (PMA) technique can prevent this tangling (see Budd and Williams, 2009).

By common consent, adaptive grids are one of the most important advances in the numerical solutions of physical problems that has happened in the past ten or so years. Here, a few definitions and methods have been given, but this is an area of current research and so new improved techniques can be expected in the years after this book is published. At the moment, all basic papers are very mathematical as well as being very recent. Some examples of the power of these grids can be seen in section 10.4, later in the book.

4.10   Boundary Conditions

So far in this chapter very little has been said about the conditions imposed at the boundary of a domain of a model. Traditionally, conditions are imposed on the variable itself, on the rate of change of the variable or a combination of both. When a partial differential equation is solved in a closed domain, the domain D has a boundary labelled ∂D. If the differential equation is satisfied by a variable ϕ, and ϕ itself is specified on ∂D, it is called a Dirichelet boundary condition. If the derivative normal to the boundary is specified, then we have Neumann boundary conditions. If there is a mixed condition, such as the linear combination $\alpha\phi + \beta(\nabla\phi \cdot \mathbf{n})$, specified on the boundary, this is called a Cauchy (or Robin) boundary condition. For a linear second order differential equation, any of the three boundary conditions are enough to ensure a unique solution for ϕ. A mixture will also suffice, though in that case there is very little chance for an analytical solution. Here, there is no concern about that, but there is about uniqueness. Uniqueness is important as otherwise we would be unsure which of many possible solutions our numerical simulation was close to. The uniqueness can be assured for linear problems, it cannot for nonlinear problems. Most of the numerical solutions for nonlinear problems are, however, close enough to their linear approximations for uniqueness not to be an issue. It does remain an issue for strongly nonlinear problems. For practical problems in coastal and marine science, some — not to say all — of the boundary conditions are more tricky than these theoretical possibilities. At the sea bed, the physics of modelling friction is complex, and at the sea surface the boundary itself is moving due to waves. At the coast, there are beaches and sea walls, there are rivers flowing into the sea, estuaries or deltas. Then there are what are called open boundary conditions, where the part of sea we are interested in modelling stops but there is no physical boundary. This happens when modelling embayments or partially enclosed seas like the Irish Sea or North Sea around the United Kingdom. The theory indicates that either the variable or its rate of change (Dirichelet or Neumann condition) has to be imposed, so in practical cases these have to either be measured or estimated by other means. Having a numerical model, only the values at a finite number of points, rather than a continuously varying function, need to be specified. The inclusion of boundary conditions thus belong securely in practical applications; theoretical Dirichelet, Neumann or Cauchy conditions get no more attention.

4.11 Exercises

Due to the theoretical nature of this chapter, some of the exercises that follow are a bit technical. Try them, and if they stump you take a look at the solutions at the end of the book. This is not an examination, so it is not cheating. They will, it is hoped, help you to understand the concepts that have been introduced in this chapter.

(1) Consider the diffusion equation in two dimensions

$$\frac{\partial u}{\partial t} = \kappa \left(\frac{\partial^2 u}{\partial x^2} + \frac{\partial^2 u}{\partial y^2} \right).$$

Using forward differences in time and centred differences in space
with $\Delta x = \Delta y$ examine stability by letting

$$u_{m,n}^s = \lambda^s \exp{(ikx + ily)},$$

where $x = m\Delta x$ and $y = n\Delta y$, by finding an upper bound for
the parameter

$$\mu = \frac{\kappa \Delta x}{(\Delta t)^2}.$$

(2) By following the analysis of section 4.4 show that the one dimensional diffusion equation with the addition of a linearised advective term

$$\frac{\partial u}{\partial t} + U\frac{\partial u}{\partial x} = \kappa\frac{\partial^2 u}{\partial x^2}$$

is also unconditionally unstable with a leap-frog scheme. Suggest
how this can be overcome (without undertaking any elaborate
proofs or analysis).

(3) Using forward differences in time and centred differences in space
discretise the following partial differential equation, called the
telegraph equation or the equation of telegraphy

$$\frac{\partial^2 \phi}{\partial x^2} = a\frac{\partial^2 \phi}{\partial t^2} + b\frac{\partial \phi}{\partial t} + c\phi.$$

Without going into mathematical detail, suggest two stability
parameters. State the problem associated with starting, and suggest how it can be solved.

(4) Outline how the Crank–Nicolson semi-implicit scheme could be
used to solve the equation of telegraphy stated in the last question. By assuming $\phi = \exp{(ikx + \alpha t)}$, indicate how to calculate
stability criteria for this scheme.

(5) The scheme called the leap-frog trapezoidal scheme given in section 4.5 takes the form

$$u^* = u^{s-1} + 2\Delta t F(u^s),$$

$$u^{s+1} = u^s + \frac{\Delta t}{2}[F(u^s) + F(u^*)].$$

Outline a method that can be used to test this scheme for stability?

(6) Outline any difficulties there may be adapting Zalesak's flux corrected transport algorithm in a regime that contains a front.

(7) Explain the main distinctions between finite difference schemes and finite volume schemes based on rectangles. What are the advantages and disadvantages of each.

(8) Distinguish between the weak and strong form of the solution in the finite volume or FEM, and explain their function in maintaining conservation laws.

(9) Explain the difference between adaptive and nested grids. Explain when one is more desirable to use than the other, or whether adaptive grids have superseded nested grids.

Chapter 5

Applied Numerical Methods

5.1 Introduction

In the last chapter a number of different numerical methods were given. In the last few years the application of numerical methods to the solution of coastal sea and marine models increased and the kind of numerical techniques applied have considerably diversified. It used to be the case that only relatively straightforward finite differences were employed but now there are a wide variety of quite sophisticated methods of solution. One consequence of this is an increase in the technical requirements of the reader if they want to understand the subtle details of the methods and be able to distinguish between them in terms of estimating errors and comparing performance. In the paragraphs that follow, some of the methods outlined in the last chapter will be applied to real problems. These applications will be taken either from published papers or from commercially available software. Although there will be parts that are technically demanding, there will be enough explanation for everyone to be able to get information from these case studies. First of all, as the examples are from modelling coastal seas, there are general assumptions that apply. These are quantified first.

5.2 Defining the Environment

Here the modelling process will be followed, so some general assumptions will be made from knowledge of the environment. On continental shelves, tides are an important phenomenon; the length of a typical wave due to tidal forces is 1000 km and its amplitude is typically less than a few metres. The wave slope is thus around 10^{-5}. This means that tides are extremely long waves when considered as water waves. It is therefore acceptable to assume that the surface is sinusoidal in shape and to apply shallow water wave theory. Also, as the depth (around 200 m on the continental shelf) is very much less than the wavelength, the tidal currents at the surface will not have any room to decay before the sea bed is reached. This is because in linear water wave theory the currents decay with depth as $e^{-z/\text{wavelength}}$, where z is the distance beneath the sea surface (see Section 3.6.3). This expression is virtually e^{-0}, that is equal to one. To a first approximation therefore, the tide over a continental shelf can be modelled as a z independent sinusoidal oscillation. In Chapter 6 there will more to be said about tides and continental shelf sea models in general, but we need to state some of the basics now. As the sea is also approximately hydrostatic, (see section 3.3) the pressure $p(x, y, t)$ will obey the equation

$$\frac{\partial p}{\partial z} = -\rho g,$$

from which, upon integration with respect to z, gives

$$p = \rho g(\eta - z),$$

where η is the elevation of the sea surface above mean sea level, and zero pressure at the sea surface has been assumed. This can be replaced by the expression

$$p = p_a + \rho g(\eta - z),$$

where p_a is atmospheric pressure if desired, which can depend on x and y if this dependence is important in the model. It rarely is. If the hydrodynamic equations are considered, then the nonlinear terms are ignored, as are variations with respect to z and the vertical component of current. What results is a pair of simple partial differential equations in the *three* unknowns u, v the horizontal components of current, and η. The third equation that closes the system is given by the conservation of mass and is derived in full in Chapter 3, section 3.3. Including the Coriolis terms, the resulting set of three equations in three unknowns is

$$\frac{\partial u}{\partial t} - fv = -g\frac{\partial \eta}{\partial x},$$

$$\frac{\partial v}{\partial t} + fu = -g\frac{\partial \eta}{\partial y},$$

$$\frac{\partial \eta}{\partial t} + \frac{\partial}{\partial x}(hu) + \frac{\partial}{\partial y}(hv) = 0,$$

which are the Laplace tidal equations (sometimes abbreviated to LTE). Insight into the behaviour of the solutions to these equations in a tidal situation can be gained by assuming that the time dependence is sinusoidal, and although the principal aim of this chapter is finding numerical solutions, let us proceed in this way before using any numerical methods. Thus, it is assumed that

$$\eta(x,y,t) = A(x,y)e^{i\omega t}, \quad u(x,y,t) = U(x,y)e^{i\omega t},$$

$$v(x,y,t) = V(x,y)e^{i\omega t}.$$

Substituting these into the first two of the LTE gives explicit expressions for $U(x,y)$ and $V(x,y)$, namely

$$U(x,y) = \frac{1}{\omega^2 - f^2}\left[i\omega g\frac{\partial A}{\partial x} + fg\frac{\partial A}{\partial y}\right],$$

$$V(x,y) = \frac{1}{\omega^2 - f^2}\left[i\omega g\frac{\partial A}{\partial y} - fg\frac{\partial A}{\partial x}\right]$$

or multiplying by the exponential factor $e^{i\omega t}$ to regain the original variables,

$$u(x,y,t) = \frac{1}{\omega^2 - f^2}\left[i\omega g\frac{\partial \eta}{\partial x} + fg\frac{\partial \eta}{\partial y}\right],$$

$$v(x,y,t) = \frac{1}{\omega^2 - f^2}\left[i\omega g\frac{\partial \eta}{\partial y} - fg\frac{\partial \eta}{\partial x}\right].$$

The presence of the complex unit $i = \sqrt{-1}$ is confusing for some. It need not be. It represents that there are phase differences between the variables. In fact, substituting for u and v into the third of the

LTE gives a single equation for η (or equivalently its amplitude A). This equation is

$$\frac{(\omega^2 - f^2)}{g}\eta + \left[\frac{\partial}{\partial x}\left(h\frac{\partial\eta}{\partial x}\right)\right] + \left[\frac{\partial}{\partial y}\left(h\frac{\partial\eta}{\partial y}\right)\right] + \frac{if}{\omega}\left|\frac{\partial(h,\eta)}{\partial(x,y)}\right| = 0.$$

The last expression $\left|\frac{\partial(h,\eta)}{\partial(x,y)}\right|$ is the Jacobian of the parameter h and the variable η defined by the determinant

$$\left|\frac{\partial(h,\eta)}{\partial(x,y)}\right| = \begin{vmatrix} \dfrac{\partial\eta}{\partial x} & \dfrac{\partial\eta}{\partial y} \\[2ex] \dfrac{\partial h}{\partial x} & \dfrac{\partial h}{\partial y} \end{vmatrix}.$$

This expression is zero if the depth is constant. For the mathematically minded, we note that the complete story is as follows: provided neither h nor η are identically zero, the Jacobian term is only zero if there is a functional relationship between the parameter h and the variable η, which of course can never be the case. In the idealised case $h = $ constant, the equation obeyed by η is the Helmholz wave equation

$$\nabla^2\eta + \lambda^2\eta = 0,$$

where we have used the standard notation

$$\nabla^2 \equiv \frac{\partial^2}{\partial x^2} + \frac{\partial^2}{\partial y^2} \quad \text{and} \quad \lambda^2 = \frac{\omega^2 - f^2}{gh}.$$

Either of these equations are solved together with boundary conditions which are that there is no flow across coasts and that the value of η is prescribed at open boundaries.

In reality, the modelling of tides has to include the effects of the sea bed in terms of frictional effects. These produce a z dependence and complicate the equations. The details of this will have to wait, instead here the focus will be on the kind of numerical schemes that might be used in practical tidal models of particular seas. In the last chapter, we introduced some numerical methods that are useful for the modelling of coastal seas. Explicit finite difference methods are seldom used nowadays, but basic ideas and concepts are conveniently introduced through them. Their principal drawback is the time step has to be inordinately small to preserve stability. This has already been said in the last chapter. For example, a time explicit scheme based on centred space differences for the LTE introduced above would be

$$u_{m,n}^{s+1} = u_{m,n}^{s} + f\Delta t v_{m,n}^{s} - \frac{g\Delta t}{2\Delta x}(\eta_{m+1,n}^{s} - \eta_{m-1,n}^{s}),$$

$$v_{m,n}^{s+1} = v_{m,n}^{s} - f\Delta t u_{m,n}^{s} - \frac{g\Delta t}{2\Delta y}(\eta_{m,n+1}^{s} - \eta_{m,n-1}^{s}),$$

$$\eta_{m,n}^{s+1} = \eta_{m,n}^{s} - \frac{\Delta t}{2\Delta x}(u_{m+1,n}^{s}h_{m+1,n} - u_{m-1,n}^{s}h_{m-1,n})$$

$$- \frac{\Delta t}{2\Delta y}(v_{m,n+1}^{s}h_{m,n+1} - v_{m,n-1}^{s}h_{m,n-1}),$$

where m, n and s represent integers. However, if you try to solve these on a computer they will be unstable, no matter how small the time step. We will return to this particular set of finite difference equations after making some more general statements.

It can be proved that the error due to the round-off error and truncation error separately satisfy the chosen finite difference representation of the partial differential equations. In order to ascertain whether or not the scheme is stable, a common procedure is to use the method based on Fourier analysis (see section 4.4). This means assuming that the error $\epsilon(x, y, t)$ behaves like

$$\epsilon(x, y, t) = e^{at}e^{i(kx+ly)},$$

where we have assumed a two dimensional (x, y, t) equation for simplicity and k, l are integers. Substituting into the finite difference representation results in an expression for e^{at} in terms of $k\Delta x$, $l\Delta y$ and Δt, where Δx, Δy and Δt are, of course, the two space step lengths and time steps, respectively. Now, if the original partial differential equation contains solutions that are wave like with wave celerity or wave speed c, as seen earlier, then it is possible to define a non dimensional number, a two dimensional version of the Courant number C (see the end of section 4.3) where

$$C = c\frac{\Delta t}{\sqrt{(\Delta x)^2 + (\Delta y)^2}}$$

and, for stability, it is essential for this scheme that $C \leq 1$. This expresses that the waves naturally occurring in the physics represented by the original equation must not travel faster than the "numerical wave speed". We have already met this in one dimension, and its extension to two dimensions is straightforward. A general interpretation of the CFL condition is that the numerical domain of dependence of a finite difference scheme must include the domain of dependence of the associated partial differential equation; we have already stated the one-dimensional equivalent of this. For schemes that involve a large number of points the CFL condition can be hard to derive. More advanced books on numerical methods, e.g. Durran (1999) should be consulted for more information.

For the explicit scheme discretising the LTE, substituting the Fourier expressions

$$\eta^s_{m,n} = ae^{\alpha t}e^{i(kx+ly)},$$
$$u^s_{p,q} = be^{\alpha t}e^{i(k_m x+l_m y)},$$
$$v^s_{m,n} = ce^{\alpha t}e^{i(kx+ly)}$$

into the finite difference equations leads to an imaginary value for α. In these equations, it must be remembered that m and n are the integers elsewhere denoted by i and j but changed because of possible confusion with the i $(= \sqrt{-1})$ on the other side of the equation. The algebra is messy but not technically difficult, and we eventually deduce that the solution does not settle down. The inclusion of friction in the model helps it to be stable, but the truncation error and the friction term can be of the same order and, more importantly, indistinguishable. This is because the next term in Taylor's series that indicates the truncation error leading term involves the second derivative, precisely the form that the friction takes if the eddy viscosity formulation is used — see section 3.8.

5.3   Finite Differences; the Arakawa Grids

The history of the numerical modelling of ocean currents is not very long — it dates from the mid-1960s — but several types of finite

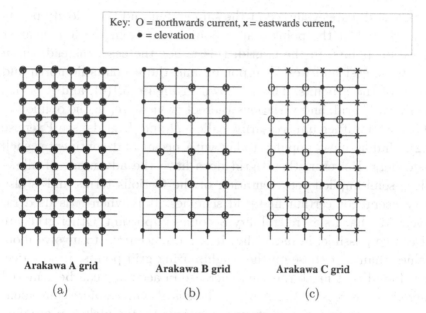

Key: O = northwards current, x = eastwards current,
• = elevation

Arakawa A grid

(a)

Arakawa B grid

(b)

Arakawa C grid

(c)

Fig. 5.1   The three Arakawa grids. • elevation (η), O northwards current (v), x eastwards current (u).

difference grid are still in use and we take this opportunity to introduce the main varieties. We do this even though new methods such as finite volume and finite elements do not require them. For the moment, we will only consider the horizontal discretisation of the variables u, v and η, which are the eastward current, the northward current and the surface elevation, respectively. There are in fact three grids (that is, ways of discretising these variables in the horizontal) still in use, and these are shown diagrammatically in Figure 5.1; (a) is the Arakawa A grid, (b) is the Arakawa B grid and (c) is the Arakawa C grid.

In the A grid, all variables are evaluated at the same location. At first sight this may seem logical, but bearing in mind that u and v are related to gradients of η, this turns out not to be very convenient. The B and C grids were developed so that points where the elevation was evaluated were always *between* points where the current was evaluated. This was first done by a meteorological modeller Akio Arakawa back in the 1960s. In the B grid, shown in Figure 5.1(b), both u and v are evaluated at the same point and the velocity points are situated at the point that is equidistant from the four nearest elevation points. In the C grid, this is not the case. Instead, the u points lie east and west of η points, and the v points lie north and south of the η points. This is shown in Figure 5.1(c), and it is this Arakawa C grid, and variations thereof (which are, in the opinion of this writer rather unnecessarily, called D and E grids, see Exercise 5.2), that is most popular today with ocean and continental shelf modellers. However, the B grid does have the advantage of allowing a semi-implicit representation of the Coriolis terms and is also very useful for certain fine grid schemes. It is when it is used for fine grids that its extra stability becomes important. Other grids are of course possible; in particular, it is more accurate to involve more values than just those on the neighbouring grid points. In practice, it is found that much the same increase in accuracy can be achieved merely by decreasing the step size. This is also much more convenient than complicating the difference equations by the inclusion of many more terms. So far, time has not been mentioned in the context of two dimensional schemes. Leap-frog schemes in time are popular, see section 4.5 in the last chapter. However, if they are used naïvely then there can be stability problems. In particular, the typical staggered grid involves three levels in time, $s-1$, s and $s+1$. If the same calculations in space are carried out at each value of s in a simplistic way then instability waves can propagate in much the same way as waves can travel on the C grid. These can be suppressed by changing the form of the space discretisation intelligently. The one proposed by Mesinger and Arakawa (1976) is shown in Figure 5.2 and is a combination of two B grids (also called the Richardson lattices) termed the Eliason lattice, after Arnt Eliason (1915–2000), the Norwegian scientist who pioneered the use of numerical analysis in the study of atmospheric physics. However, there are many alternatives. An interesting and more complex variation was used by Flather and Heaps (1975). They used mixed grids for the computations of storm surges; the details are in Kowalik and Mutry (1993). A more popular method these days is to use Alternating Direction Implicit (ADI) techniques. In this technique, the idea is to split the computation into two distinct

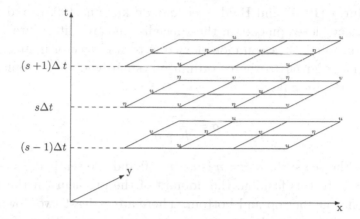

Fig. 5.2   A staggered grid in space that is also staggered in time using a leap-frog scheme. The notation is standard; (u, v) is the horizontal current and η the surface elevation. The vertical axis is time t.

parts. First the x direction is computed usually using semi-implicit methods, then the y direction is similarly calculated. Each direction is taken over a half time step, so that mathematically the ADI process is

$$\frac{u_{i,j}^{s+\frac{1}{2}} - u_{i,j}^{s}}{\frac{1}{2}\Delta t} = \text{RHS } (x \text{ deriv. at } s + 1/2) + \text{RHS } (y \text{ deriv. at } s),$$

$$\frac{u_{i,j}^{s+1} - u_{i,j}^{s+\frac{1}{2}}}{\frac{1}{2}\Delta t} = \text{RHS } (x \text{ deriv. at } s + 1/2) + \text{RHS } (y \text{ deriv. at } s + 1),$$

where RHS denotes the right-hand side of the equation. In applications to real coastal engineering or marine physics there will be equations involving the three variables u, v and η, but the ADI principle holds; the (six) right-hand sides will involve the computation of all three variables at the time step indicated. Analysing the stability and accuracy of such complex schemes is difficult and it is easier to conduct carefully designed numerical experiments. As a bonus, it is possible, in fact it emerges naturally that spatial boundary conditions are very easy to implement under ADI schemes. There are some more details in the specialist books, see for example Kowalik and Murty (1993) and Haidvogel and Beckmann (1999), and there will be new developments in the specialist research literature.

In the vertical, a common device is to use σ co-ordinates. This is the name given to the co-ordinate system whereby σ replaces z through the formula

$$\sigma = \frac{\eta - z}{h + \eta},$$

so that the sea surface $z = \eta$ is at $\sigma = 0$ and the sea bed $z = -h$ is at $\sigma = 1$. In this fashion, the domain of the problem (in the vertical) is flat at the top and bottom. There are at least two downsides to using the σ co-ordinate system. One is that derivatives of z have to be transformed into derivatives in terms of σ, which are more complicated. The second is more technical; if the slope is steep then there can be problems with the accurate representation of horizontal pressure gradients. The reason for this is wrapped up with the hydrostatic approximation. The pressure gradient term is split into a purely hydrodynamic part along the co-ordinate (x or y axis) and a correction due to hydrostatic balance; this last part being necessary to remove that bit of the pressure change due to hydrostatic variation, itself a consequence of the co-ordinate changing in height as it follows the undulations of the sea bed. It turns out that the truncation errors of the two terms are not the same and thus do not cancel, and a significant error can result. For fine resolution models, corrective numerical techniques can be applied (see the book by Haidvogel and Beckmann, 1999). There is no doubt that using these co-ordinates the sea bed and surface boundary conditions are greatly simplified, and modern software such as NEMO© uses σ co-ordinates extensively.

We now seem to be ready to implement finite difference schemes to marine and coastal problems, however before this can be done, there are some very important aspects that need investigation; they are the conditions that hold on the boundaries. The practical application of realistic boundary conditions is just a little more complicated than the idealised Dirichelet, Neumann and Cauchy conditions outlined in section 4.10.

5.4 Boundary Conditions in Practice

The laws of motion attributed to Sir Isaac Newton have for 300 years quantified the motion of physical objects. The sea is a fluid which is also governed by these laws of mechanics. The basic equations, outlined in Chapter 3, are these laws together with the conservation of mass. The mathematical theory that underlies the motion of a frictionless fluid is called potential theory. Although the equations that govern the movement of the sea contain nonlinear terms (namely $(\mathbf{u}\cdot\nabla)\mathbf{u}$), if these are small — and they usually are — then potential theory tells us that the motion is determined by what happens at the boundaries. The way numerical schemes work confirms this as was outlined in the last chapter; it is the start condition that is essential to kick start a numerical solution of a diffusion type equation, and the coastal boundary conditions are essential for any workable two dimensional finite difference scheme. Therefore, even with the nonlinear terms included, the linear theory of the last chapter still applies and the effects that are literally on the edge of models are in fact central drivers and play a major role in determining the resulting motion of the sea.

This chapter is concerned with numerical methods, but in order to formulate any numerical scheme, some space has to be devoted to considering how to model the processes that occur at the boundaries of the sea. First, let us distinguish between the different types of boundary. The sea bed and coastlines are of course solid to fluid whilst the sea surface is fluid to fluid. However, modelling the physics of sea bed as opposed to coastal boundaries is often very different. As for modelling what happens at the sea surface, so much has been done that this could be the subject of a separate book. It is one of the most complicated of boundaries and one is led to make drastic simplifications. However, these simplifications are more or less fully justified in terms of the specific phenomenon being examined. There are boundary conditions that have to be imposed because it is not possible for the domain to be the entire sea; these are called *open* boundary conditions. There is also a boundary condition in time, the start condition. All of these demand separate consideration so let us do this in the next few paragraphs.

So, although the layman may believe that what happens on the boundary appears not to be important, it actually drives the motion. It is all the more important, therefore, to understand precisely what is happening at the edges of regions. First, let us look at each boundary and distinguish a little more carefully between various types. Perhaps the most obvious boundary is the sea surface. It is also, as we have said, unfortunately one of the most complex. Another obvious boundary is the coast. The coast is a solid boundary, and the treatment of solid boundaries is certainly less controversial. Another solid boundary is the sea bed; but we need to be careful to know what we mean by "solid". Sea bed boundary conditions have occupied the attention of modellers for a very long time, and some quite sophisticated models are now in common use. The reason for this close attention is not hard to fathom; it lies in the role the sea bed plays in providing a sink for momentum. There is far more contact between the sea and the bed than the sea and its coastline, and a great deal of effort has been devoted to engineering problems associated with the sea bed, such as dredging, erosion and scour. Conditions imposed at the sea surface, the coast and the bed are called *closed* because there is a definite physical edge that dictates to some extent the type of equation valid at the boundary (relating the surface current to the wind perhaps, or imposing no flow through a coast). There are also open boundary conditions. These are not actual, physical boundary conditions, but arise because domains that are not closed lakes or the entire globe have to have edges that cross the open sea. Open boundary conditions are necessary to apply to the open edges of models of these regions.

Another very different boundary condition arises from having time as a variable. All models have to start, and the state of the variables at time zero, the initial condition of a model, is of course a boundary condition. In complex models that have significant nonlinear terms, the start conditions are not important because the sea soon 'forgets' how it started to move and becomes, so to speak, wrapped up in its own dynamics. Systems or models with short memories are the type that lead to chaotic behaviour which is deterministic (that is, it lacks a random or stochastic element), but exhibits behaviour on many very different length scales. This is a popular topic nowadays, but the modelling of strongly nonlinear systems is beyond the scope of this modest text. Having said this, in models that are not strongly non-linear, and that includes most of them, the start condition is very important. As examples, one can think of initial conditions that drive storm surge forecasts (see Chapter 6), diffusion models (see Chapter 7), and indeed weather forecasting itself, which would simply be a non-starter without accurate initial conditions. Using the language of systems (see section 2.5) the start condition may excite the natural frequencies but in general the long term response will mirror the characteristics of the input.

The most successful models begin by being simple and idealised so that the essential dynamics are present in glorious isolation, uncluttered by awkward boundaries or complex but unimportant effects. The rectangular ocean models of yesteryear provide particularly vivid examples of this. In very idealised models, the sides and bottom of the box-shaped ocean are flat, as is the surface. A rigid lid approximation is imposed, whereby the sea surface is assumed solid. The open boundaries are simplified to be lines of flow (streamlines), and time does not feature. In spite of these idealisations, important features pertinent to the understanding of the general circulation of the ocean can still be predicted (see Chapter 10). Moreover, because of the elementary nature of the model, the user can see precisely what causes a phenomenon such as the western intensification of ocean currents.

As models are made more realistic, boundary conditions need to be treated with greater care. In particular, the different treatments of horizontal and vertical boundaries which stem from their differing scales are very important to maintain in ocean models. In models of less horizontal extent, such as an estuarial model, the distinction between horizontal and vertical boundaries becomes less important. In an estuary, there are in fact only "water–air" and "water–solid" boundaries. Another recent development in modelling which fits naturally into discussion of boundary conditions is data assimilation. Data assimilation is the incorporation of data into a model *as it runs*. This was discussed under the umbrella of Kalman filers in Chapter 2, and we will discuss it again in the last section of this chapter. It finds a natural home in limited area ocean models where a neighbouring model, or perhaps a larger one, can input information into the given model as it is running by having both models running in parallel. A second possibility is to input observational data into a model as it is running. For example, if a limited area ocean model is running, then as it evolves it could be possible for an eddy to migrate into the model, even when the nature of the model makes the formation of such an eddy within the model impossible. This would be done by using as boundary conditions the velocity and elevation appropriate to an eddy along one of the model's edges.

Another obvious use of data assimilation is in storm surge modelling. In storm surge models, it is wise to update the weather input as the model is running in order for the enhanced elevation to be predicted with the greatest possible accuracy. The increase in availability of satellite data makes the improvement of model prediction by data assimilation a real possibility in areas where it has not been in use so far. However, for this text, data assimilation takes a relatively minor role as its main use is in obtaining answers that are a better fit to observations, and not answers that improve the understanding of the fundamental processes. Let us now return to the main subject of this particular section, namely the examination of the different types of boundary condition. First, let us look at the sea surface.

The surface of the sea can be a flat calm, it can be mountainous, or any condition between the two. In ocean models it is commonly taken as flat, because over reasonably short periods of time, say a few minutes, the up and down movement averages to zero. Since ocean models are concerned primarily with bulk movements through ocean currents, time steps longer that this are used, which implicitly implies averaging on the right time scale to eliminate the vertical movement of the sea surface. It is in Chapters 3 and 8 that simple models of the observed surface waves are introduced. These models are necessary for small scale motions such as wind rows (Langmuir circulation) and some estuarial dynamics (see Chapter 11). They must also feature in models that help the designer of offshore engineering structures; they are not relevant to oceanographic texts. There is, of course, an entirely different vertical displacement of sea surface due to astronomical forces called tides. These need special attention and are given it in Chapter 6.

The surface of the ocean can thus be assumed flat (apart from tides). It can still move, however; it moves horizontally like an airport travelator. To force the ocean surface to act as if a solid barrier were against it is nowadays almost always unacceptable. This was called the rigid lid approximation and flourished briefly in the 1950s and 1960s. It is much more usual to allow the wind to act on the sea and to move its surface tangentially through frictional forces. The momentum thus introduced into the surface is then transferred to the ocean underneath by the turbulence so created. A way to achieve this is to relate the velocity at the sea surface to the sea surface stress via a law which might be a stress rate of strain relationship (Newtonian eddy viscosity), or to use a more sophisticated law. The simple but successful Ekman model will be outlined in Chapter 6. The more complex laws at the sea bed are dealt with in the coming paragraphs, and overall models that deal with coastal sea problems that require sea surface boundary conditions form part of Chapter 8.

It is also possible to allow for other boundary influences through the sea surface. If the temperature variation of the ocean is being modelled, the surface can be allowed to heat or cool due to outside influences (night and day, or the different seasons) and this can be incorporated by imposing temperature or temperature gradient conditions at the sea surface itself. A model would incorporate this via a source (or sink) term in the equation governing the diffusion of temperature downward through the water column. Other sea surface effects, such as rainfall and evaporation, have not been the concern of the modeller and will not be considered here, however, they can be incorporated in principle using sea surface source or sink terms. In the newest ocean circulation models, the potential energy that arises from the rise and fall of the sea surface has an input into the salinity distribution. This is interesting in that models of the ocean circulation in the 1930s were entirely driven by the effects of precipitation, evaporation and freshwater inflow but these models were all but abandoned when the later wind-driven models seemed to be so much better at predicting the observed circulation. These latest models show us that we are wrong to reject completely such models. Instead, we must incorporate these effects alongside the more dominant wind driving in a more complete description of the surface (and coastal) boundary conditions. If it is necessary to incorporate the wind explicitly into a model, then it is common to use the following relationship between the wind speed at 10 m above the surface W_{10} and the stress due to the wind τ^w

$$\tau^w = \rho_a C_D |W_{10}| W_{10}.$$

Of course, both τ^w and W_{10} are vector quantities, but they have to be in the same direction so it is the above scalar relationship that is important. ρ_a is the density of air and C_D a drag coefficient obtained through experiment (0.003 is a typical oceanic value). The sea bed is the next boundary to consider; it is so important that it needs, and gets, its own special section.

5.4.1   *The sea bed*

The most obvious characteristic of the sea bed is that it is a solid barrier and that water must not be allowed to pass through it. This might seem obvious, but the smaller the scale of the model, the trickier such a criterion is to apply. A sea bed may first of all be steep, so that care needs to be taken that the boundary condition mentioned above involves the flow perpendicular to the sea bed, not to be confused with vertical flow. If the bed is flat, it may be sandy or muddy, in which case can it be assumed solid? Also, what about the cohesive properties (stickiness)? Mud is, in reality, a viscoelastic fluid with very complex properties that interact with the chemistry and biology of the sea but the elasticity and reactions will have to be ignored here. Viscosity is, by far, the simplest way that friction can be modelled. In a viscous fluid, all components of the velocity, including those parallel to the sea bed, must be zero at the bed itself. Although the ocean is not a viscous fluid in the accepted sense, its equations are similar enough for such a boundary condition to hold. Finally, if very detailed modelling is to be considered, then some techniques from mechanical and aeronautical engineering modelling can be used. In particular, the defining of a roughness length to represent the character of the bed (sand, gravel or rocks), a laminar sub-layer where the flow regime is viscous but not turbulent, then a transition to turbulence when one is clear of the bed.

In the 1980s, research workers developed a modelling approach to the sea bed boundary based on what has become known as $k - \varepsilon$ turbulence closure schemes (abbreviated to turbulent kinetic energy (TKE) schemes). The initial variation to the constant eddy viscosity model was to go back to the original work of Ludwig Prandtl (1875–1953), in particular his mixing length theory of the 1920s. Here, the eddy viscosity is deemed to be not constant but dependent upon the gradients in the mean flow. So, put $\nu = u_{\text{turb}}l$, where u_{turb} is some turbulent velocity scale and l a length scale. Further, Prandtl also assumed that

$$u_{\text{turb}} = l \left| \frac{\partial \bar{U}}{\partial y} \right|$$

so that

$$\nu = l^2 \left| \frac{\partial \bar{U}}{\partial y} \right|.$$

So, this is a model where l can be dependent on the nature of the flow as well as location. This is sometimes called a $k - l$ model. Many researchers used the $k - l$ model, but in general it lacked the flexibility desired by modellers, and full $k - \epsilon$ turbulence models gained favour. Although the models that utilise these schemes apply to the entire sea, it is the fact that this particularly accurate model is required to simulate turbulence in areas that are dominated by frictional effects which causes us to go into more detail. The TKE models themselves were first developed by modellers of aeronautical systems, where the ability to predict the detailed flow near critical parts of aeroplane wings is an essential requirement. In Chapter 3, we encountered the concept of eddy viscosity as a quantity that represented how stress is related to the rate of strain (or shear) of a turbulent flow. It is analogous to kinematic viscosity in laboratory viscous laminar flow, the principal differences being that turbulent viscosities are much larger than their laminar counterparts because turbulence is far more efficient at transferring momentum from one streamline to a parallel streamline. Also, a turbulent eddy viscosity is not a fixed property of a fluid. (In laminar flow, the viscosity is as fixed a property of the fluid as is, for example, density. In turbulent flow, eddy viscosity can vary from place to place and can change with time.) Eddy viscosities cannot, unfortunately, account for the observed behaviour of a flow adjacent to the sea bed. So eddy viscosity is hard to measure, varies with space and time and maybe changes with other variables too, and it does not model the current structure adjacent to a real sea bed; so it comes as no surprise that it is not a universally accepted quantity. Instead, modellers of recent times have used the above-mentioned, more sophisticated, turbulence closure models. What follows is a brief description of this more complex model of turbulence. By the very nature of these turbulence models, they involve complicated concepts. The mathematics is largely absent from the account here, although there is a little later. There is a fuller account in the original source papers: Blumberg and Mellor (1987) and Mellor and Yamada (1974).

The rationale behind most turbulence closure schemes is to start with the well known assumption of eddy viscosity but then to introduce other variables that can be related to length and velocity scales that represent scales of turbulence. To be specific, if K_q represents the eddy viscosity (or eddy diffusivity), then

$$K_q = lqS_a,$$

where l and q are appropriate length and velocity scales, respectively, and S_a is a factor that depends on the stability of the flow. This stability will, in turn, involve the density change with the vertical as well as the shear of the flow, $|\partial \mathbf{u}/\partial z|$. The velocity scale q and length l will themselves obey equations. For example, $\frac{1}{2}q^2$ represents the kinetic energy associated with the turbulence. In simpler models, S_a is simply a number and the complex structure of the turbulence model is carried in the two scales through the equations obeyed by them. It is quite usual these days for ostensibly simple two dimensional models to have quite complicated turbulence closure schemes attached to them which can mimic successfully the momentum transfer just above the sea bed. This kind of model is often called '$2\frac{1}{2}$ dimensional', and has the additional merit of being much cheaper to run (and in some cases more reliable) than more complex fully three dimensional models. One interesting feature of virtually all sea bed models that purport to be detailed models of dynamics is the presence of a layer adjacent to the bed where the velocity profile is logarithmic. That is, the speed U is related to the distance from the sea bed z by an expression such as

$$U = \left(\frac{\tau}{\kappa u_0}\right) \ln\left(\frac{z}{z_0}\right),$$

where τ is the sea bed stress, κ is a constant attributed to von Karman and usually given the value 0.4, and u_0 is a constant representative of the speed of the flow just above the roughness elements. These roughness elements are the sand, rock etc. that lie on the sea bed and interrupt the flow at the bed itself. The constant z_0 is a length that represents the average magnitude of the height of these roughness elements above the sea bed. At a height of z_0, the flow is no longer interrupted by sea bed debris. These ideas are displayed in Figure 5.3.

In addition to the correct modelling of the physics at the sea bed, at the bed itself some approximations are usually required because numerical methods which incorporate grid boxes are being employed. This means that one point will be above the sea bed, but the adjacent point must be in the bed itself.

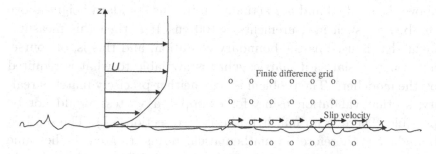

Fig. 5.3 The sea bed, showing roughness length (z_0), finite difference grid and slip velocity.

For this reason a *slip velocity*, whereby the velocity at the sea bed is not zero but simply some value, is a common proposal. The value chosen is consistent with an appropriate quadratic or linear friction law that relates frictional stress directly to velocity at the bed. The usual controversy over whether to use such a law or whether to use a no-slip law at the sea bed itself is precisely equivalent to the imposition of a linear friction law a little above the bed at the nearest grid point. Having said all this however, sophisticated turbulence closure schemes are now central to modelling the dissipation of momentum at the sea bed correctly, and numerical approximations of the sea bed boundary conditions merely alter the actual points at which these conditions are applied, not their application.

The specification of the velocity just above the sea bed, rather than actually on it, is tantamount to specifying a slip velocity at the bed itself or at the nearest grid point above it, simply because the resolution of the finite difference scheme does not allow them to be distinguished. A neat trick, already mentioned briefly, is to use something other than layers in the vertical or at least to use a continuous representation of velocity close to the bed itself. One alternative which has proved very popular is to use a series expansion of variables in terms of functions of the vertical co-ordinate (or the σ co-ordinate if this has been used).

If finite differences are used, then the sea bed boundary condition typically becomes the specification of stress at some point above the bed. Practically, however, this is not so much of a problem since our observational colleagues have to measure the stress at some point above the sea bed and not actually on it. The standard height above the bed for such measurements is 100 cm. It is then this measurement that is used as the boundary condition, and this is, of course, not controversial as it marries what is available to what is required by the modellers. The problem is that neither precisely matches reality, so that validation using, for example, a χ^2 test would not be possible. There is no null hypothesis (see section 5.7). The sea bed boundary is therefore, in mathematical terms, treated in the same way as any coastal boundary. The essential difference has already been alluded to; the sea bed is a crucial momentum sink and a correct interpretation of the physics there is extremely important to ensure a good model. "Good models" for the physics of the sea bed date from the 1960s when the aforementioned TKE models were first developed to describe what was observed in laboratory models of fluid flow. In the early 1970s, these were adopted for use first in the atmosphere and then the ocean. Researchers at the Princeton University's Geophysical Fluid Dynamics Laboratory in the USA then developed a hierarchy of models. Many similar hierarchies of models have now been developed elsewhere in the USA, Europe and the rest of the world. Indeed, a great deal of trouble has been taken in the last 15 to 20 years to develop physically accurate, but necessarily complex, TKE schemes. What follows are some details of the TKE scheme as proposed by Blumberg and Mellor (1987), which is now incorporated into proprietary software. The sea bed boundary conditions have been developed in the context of the General Ocean Turbulence Model, abbreviated GOTM. This is not a fixed model but a flexible generic code that can be modified. There is documentation at www.gotm.net, so only an outline is given here. GOTM is modular in construction, which takes advantage of the FORTRAN way of doing things. Inside the GOTM there is a water column model of some sophistication. In order to give a flavour of the complexity of the model, here is the version that this author has access to, though of course modifications will probably have occurred between the time that this is written and the time you read it. The Reynolds stress terms are related to gradients in the mean flow through eddy viscosities, but more generally than given in Chapter 3.

$$-\overline{u'w'} = K_{uu}\frac{\partial u}{\partial z} + K_{uv}\frac{\partial v}{\partial z} + K_{ub}\frac{k}{\epsilon}N^2,$$

$$-\overline{v'w'} = K_{vu}\frac{\partial u}{\partial z} + K_{vv}\frac{\partial v}{\partial z} + K_{vb}\frac{k}{\epsilon}N^2.$$

In these equations, K with various subscripts are the eddy viscosity coefficients, N is the buoyancy frequency, k is the turbulent kinetic energy and ϵ its dissipation rate. Note the inclusion of buoyancy effects here. These were ignored in Chapter 3. Here, we have that

$$N^2 = g \left(\beta_T \frac{\partial T}{\partial z} - \beta_S \frac{\partial S}{\partial z} \right),$$

where β_T and β_S are expansion coefficients. This formulation takes into account a linear relationship (equation of state) connecting temperature and salinity with density, otherwise it is the standard definition. Also within this parameterisation, we have that

$$\alpha_M = \frac{k^2}{\epsilon^2} \left[\left(\frac{\partial u}{\partial z} \right)^2 + \left(\frac{\partial v}{\partial z} \right)^2 \right],$$

$$\alpha_N = \frac{k^2}{\epsilon^2} N^2,$$

which together with

$$\frac{k}{\epsilon} \frac{\partial u}{\partial z}, \quad \frac{k}{\epsilon} \frac{\partial v}{\partial z} \quad \text{and} \quad \frac{k}{\epsilon} f,$$

turn out to be important stability parameters. In order to close the system of equations it is necessary to specify both k and ϵ. There are several ways that this can be done, but these days it is most usual to use what is called the two equation model

$$\frac{\partial k}{\partial t} = \frac{\partial}{\partial z} \left(K_k \frac{\partial k}{\partial z} \right) + P - K_h N^2 - \epsilon,$$

$$\frac{\partial \epsilon}{\partial t} = \frac{\partial}{\partial z} \left(\frac{K_k}{\sigma \epsilon} \frac{\partial \epsilon}{\partial z} \right) + c_{1\epsilon} \frac{\epsilon}{k} (P + c_{3\epsilon} G) - c_{2\epsilon},$$

where $K_h = S_k k^2 / \epsilon$ and

$$P = -\overline{u'w'} \frac{\partial u}{\partial z} - \overline{v'w'} \frac{\partial v}{\partial z} = K_m \left[\left(\frac{\partial u}{\partial z} \right)^2 + \left(\frac{\partial v}{\partial z} \right)^2 \right].$$

In the $k - \epsilon$ formulation, S_k is usually taken as

$$S_k = \frac{C_1 + C_2 \alpha_N + C_3 \alpha_M}{1 + C_4 \alpha_N + C_5 \alpha_M^2 + C_6 \alpha_M \alpha_N + C_7 \alpha_N^2},$$

where the constants $C_1 \ldots C_7$ are determined by fitting the model to higher order Reynolds averaged equations. This particular version of the $k - \epsilon$ turbulence closure scheme is by no means the only one, but its nature gives a flavour of the complexity of such models.

The numerical implementation of such complicated schemes as this has to be done very carefully. It would be nonsense to throw away all the sophistication by representing these complex schemes using ill-considered numerical approximations. As we have said, a straightforward difference scheme at the sea bed is precisely equivalent to imposing a slip velocity at the first wet grid point above the bed. Is there another way that this can be improved upon? One alternative which is simpler than refining the entire grid is to impose a quadratic friction law on the flow at this grid point and demand that the C_D (drag coefficient) be related to a logarithmic law in the boundary layer under the grid point. This kind of modelling was done by Blumberg and Mellor (1987). For those interested in some details, these follow but can be skipped if desired. At the lower boundary, the bottom stress (τ_{bx}, τ_{by}) is related to gradients in velocity through the standard law

$$\rho_0 \nu_v \left(\frac{\partial u}{\partial z}, \frac{\partial v}{\partial z} \right) = (\tau_{bx}, \tau_{by}),$$

where ρ_0 is the ambient density. In addition, the turbulent kinetic energy q^2 is related to the friction velocity $u_{\tau b}$ through the direct law

$$q^2 = B_1^{2/3} u_{\tau b}^2,$$

where $B_1 = 16.6$, a value derived empirically. There is the fact the $q^2 l = 0$, where l is a turbulence macroscale, and finally, the vertical velocity w_b must be given by

$$w_b = -u_b \frac{\partial h}{\partial x} - v_b \frac{\partial h}{\partial y},$$

where $z = -h(x, y)$ is the equation of the bottom topography. Assume a quadratic friction law at the bed of the form

$$\tau_b = \rho_0 C_D |\mathbf{u}_b| \mathbf{u}_b$$

with the drag coefficient C_D given by

$$C_D = \left[\frac{1}{\kappa} \ln \left(\frac{h + z_b}{z_0} \right) \right]^{-2}$$

with the suffix b denoting numerical evaluation at the grid point nearest the bottom and κ is von Karman's constant. Then it is possible to derive the standard logarithmic layer at the bed in the form

$$\mathbf{u}_b = \frac{\tau_b}{\kappa u_{\tau b}} \ln(z/z_0),$$

provided there is enough resolution. Of course, such elaboration is over the top if the sea bed boundaries are not well resolved. In this case, a numerical value of C_D, say 0.0025, will do.

5.4.2   *Coastlines*

In the ocean, the coast and the edge of the continental shelf can be taken as being virtually synonymous. This is because of the magnitude of the horizontal length scales. If a model is to cover the entire horizontal extent of an ocean, perhaps 5000 km, with 100 points, then each point has to be 50 km apart. This then means that a typical continental shelf is virtually lost between the coast and the first and second grid point. Under these circumstances, the coast itself can be assumed to be a shear vertical cliff. Exactly similar representations of the slip or no-slip boundary conditions at the sea bed are possible at the coast too when it is a vertical cliff. However, there is a subtlety that arises because in some models the Coriolis parameter varies with latitude which makes a no-slip condition dynamically distinct from a formulation which includes a slip velocity. However, in a text on coastal modelling this does not feature strongly. There is no need for the complicated turbulence closure schemes in these large-scale models, for they do not have the same energy extraction role.

For smaller-scale models, the continental shelf must be taken into account. Most models do not straddle the continental slope; they either lie entirely on it or it provides the location for boundary conditions of the deep sea models. However, now that finer resolution models are possible, it is expected that models that incorporate the shelf, the slope and the deep sea will become available. They are not here yet. Models of the continental shelf itself have open boundary conditions which need careful handling and deserve a section of their own.

At smaller scales still, say coastal inlets or estuaries, the coasts are treated in various ways. Often they are steep sided and a vertical face is allowable; on other occasions there may be mud flats or sandy beaches of minimal slope which demand that the depth of water itself is zero. There is a potentially serious problem if the sea dries inside a model region because, certainly in two-dimensional models, the depth occurs as a parameter in the denominator of various terms of the governing equations. One approach is to allow the sea to dry in cells in such a way as to compute the velocity as zero before (potentially) dividing by the zero depth. Research is still active, however, on the effects this has on the velocities elsewhere in the model domain. When there are drying regions present in a model, the horizontal domain of the problem changes with time (perhaps due to tides). Such a formulation is difficult to solve since the shape of the coast becomes an unknown of the problem. Dealing with moving coastlines can be technically difficult, but as a concept it (forgive the pun) holds water. At a coast, the normal component of the current must be zero. If the model includes a representation of horizontal diffusion of momentum, for example, eddy viscosity, then all components of the current must be zero at the coast. Let us now look at how a coastal boundary condition is dealt with in models that make extensive use of finite difference schemes. At a solid boundary, whether it be the sea bed or a coast in fact, the problem is essentially that a grid point that is in the sea and at which an equation is valid needs to be discretised in terms of differences, but these differences need to be calculated using values that are in dry land because that is where the adjacent point is. The point in question then has to be rejected insofar as being able to write down discretised versions of the governing equations is concerned. Instead, the boundary conditions need to be examined, and to be discretised appropriately. As an example, suppose that there is a straight line boundary as shown in Figure 5.4.

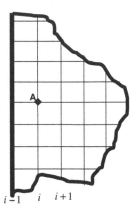

Fig. 5.4   The finite grid near a boundary.

At the point marked A, which is in the interior of the domain and therefore in the sea where various equations are indeed valid, the boundary condition that there is no flow through the boundary is discretised in the following way

$$\frac{\partial U}{\partial x} \approx \frac{U_{i+1} - U_{i-1}}{2\Delta x} = 0.$$

This replaces the full discretisation of the equations. Also, because U_{i-1} is due to be evaluated on dry land, it is taken as zero. Zero flow through the boundary therefore automatically leads to a zero value of U_{i+1} as well. This is obviously rather unrealistic, and various ways round this have been proposed, such as to stagger the grid such that velocity points are never actually on the coasts and to make a fictitious extension of the grid one point into the coast so that the domain is not artificially shrunk. The Arakawa C grid, outlined in section 5.3, is one modern finite difference grid that facilitates this. Backhaus (1982) in his semi-implicit scheme used the Arakawa C grid and at boundaries allowed for drying. The numerical scheme he employed was carefully designed such that as points "dried" no inconsistencies in terms of conserving transport of fluid occurred. Flather and Heaps (1975) give a good technical account of dealing with drying regions in a tidal model. In general, dealing with drying in a three-dimensional model is easier than dealing with drying in a two dimensional model, as in the latter it is a true singularity whereas in the former it is not.

In problems that are primarily concerned with the modelling of waves, there is another way of dealing with geometrically simple boundaries, which in the context of this text means straight boundaries. If perfect reflection of the waves occurs at coasts, then it is permissible to allow the domain to extend as if the coast was a mirror with an image domain on the other side. The method of images is very well known in various fields of applied mathematics and can be applied in oceanography, but only limited to cases of simple geometry. Because of this, it is not extensively used.

5.4.3 *Open boundaries*

In this section we shall look at the open boundary. Essentially, when an artificial boundary straddles the sea it is the job of the fictitious condition that needs to be applied to mimic the motion of the (non-existent) sea on the outside. In particular, the condition must allow waves to pass cleanly out of the region without reflection. It must also cater for any incoming information, typically an incoming tide. Nevertheless the term "radiation condition" is the usual phrase used for this open boundary condition. Unfortunately, in the situation where there are no incoming waves an ideal perfect radiation condition is only possible if the boundary is infinitely far from the source of any waves.

From a purely practical point of view, the two dimensional equation obeyed at the open boundary is usually a simple advection equation of the form

$$\frac{\partial u}{\partial t} + U\frac{\partial u}{\partial x} + V\frac{\partial u}{\partial y} = F,$$

where (U, V) is $(U_0 + c, V_0 + c)$ and (U_0, V_0) is a background flow, the local current near the boundary, and c is the appropriate wave speed. The right hand side F represents incoming waves, invariably tides in a continental shelf model, and is obtained through observations or via running another model. The celerity c is $\sqrt{gh}$ for tides and other long wavelength water waves (see Chapter 3). In finite difference form, this condition might be written as

$$U_{i,j}^{s+1} = U_{i,j}^s - \frac{U\Delta t}{2\Delta x}(U_{i+1,j}^s - U_{i-1,j}^s) - \frac{V\Delta t}{2\Delta y}(V_{i,j+1}^s - V_{i,j-1}^s) + F,$$

where we have used centred differences in space. Some kind of condition like this is essential to apply at open boundaries but perfection is not yet possible, as hinted at earlier (Durran, 1999).

As a specific example, let us look at the TKE model of Blumberg and Mellor (1987). In this paper, two types of open boundary condition are prescribed, the inflow and the outflow. Temperature (T) and salinity (S) obey the equation

$$\frac{\partial}{\partial t}(T, S) + u_n \frac{\partial}{\partial n}(T, S) = 0,$$

where n denotes the normal to the open boundary. Turbulence kinetic energy $(\frac{1}{2}q^2)$ and the macroscale turbulence quantity $(q^2 l)$ are calculated with sufficient accuracy at the open boundaries. It is sufficient, even considering the sophisticated nature of the rest of their model, to calculate these two quantities neglecting the nonlinear advective terms. Additionally, a radiation condition of the type mentioned above is also imposed, viz.

$$\frac{\partial \eta}{\partial t} + (gh)^{1/2} \frac{\partial \eta}{\partial n} = F(s, t).$$

Intrinsic co-ordinates (s, n) are used, where s is tangential and n is normal to the boundary. The forcing function $F(s, t)$ incorporates incoming tide, together with any other currents. Once again, in their model it is permissible to neglect the nonlinear advective terms in these open boundary conditions. Boundary conditions at the coast sea surface and sea bed can be considered *natural* in that they arise from a sea being surrounded by solid or air. It is the physics of the interface that by and large dictates the form that the boundary conditions take. Open boundary conditions occur simply because a model must end where there is no coast. Typical open boundaries are the edge of the continental shelf; Figure 5.5 shows a typical domain off northwest Europe.

In this example, there is an extensive open boundary which more or less marks the European continental shelf. Estuarial models and

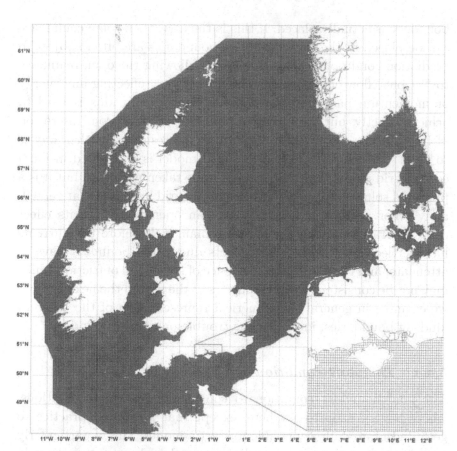

Fig. 5.5   The 1/60th of a degree by 1/40th degree resolution (approximately one nautical mile) grid CS20 from the National Oceanographic Centre website.

models of coastal inlets will have open boundaries with the sea. Even deep ocean models which span the Atlantic or Pacific Ocean from west to east often have equatorial and northern open boundaries. In this latter case, the open boundary is dealt with in the same way as a solid boundary in that it is assumed to be a streamline. (No flow passes through a streamline, although in fact all that is required is that there is no *net* flow; there can be some transfer of fluid, but it has to be the same in each direction.) For smaller models, however, the open boundary is treated differently. The essential feature a modeller wishes to retain is a lack of sensitivity as to where precisely the boundary is actually placed. If this is the case, then some confidence can be placed in the solution. Open boundaries are often subject to "radiation conditions". This is a way of allowing the open boundary to "let out" flow etc. from the domain without reflecting any energy or momentum back into the domain. Numerically, this is hard to achieve exactly, but the further a boundary is away from the region of interest, the better (more accurate) the results. Another way is to "nest" models such that a fine grid model is embedded inside a coarser grid model. Much care needs to be taken with this, but it is a preferred solution to simply extending the grid until it is "far enough away". For no matter how far away an open boundary, its effect can still travel through the entire domain well before the currents have settled down. In fact, the effects due to the boundary have to attenuate, and this implies the presence of some kind of friction which may not be consistent with the physics of the sea. In the next section, we examine in general the numerical representation of the different kinds of boundaries, including the troublesome open boundary.

5.4.4   *The start condition*

The final boundary condition to consider is the start conditions. There is very little to say here, except that it is good practice to get as accurate data as possible to start a model. This is particularly true for storm surge models and ocean models that are not strongly nonlinear. However, there is a class of models, (tidal models spring to mind) where the object of the exercise is to run the model in a predictive state to be as accurate as possible and *independent* of start conditions as soon after the start of the model as possible. For this kind of model, it is common to start the model by assuming a flat stationary sea and to allow the open boundary to input motion in order to *spin-up* the sea itself. This is called a *cold* start.

5.5   Finite Element Schemes

The basic ideas behind finite element schemes were outlined in the last chapter. Here, we go into a little more detail. To recap from the start, the domain of the problem is subdivided into small areas, usually triangles, but not exclusively so. These triangles are small enough for the values of the variables within each triangle to be approximated by a simple function, say a linear, quadratic or cubic polynomial function. At the internal boundaries between the triangles (these triangles are called *elements*, hence the name of the method) there has to be continuity of velocity, pressure etc. The principle behind the FEM is that, if the entire domain is considered — that is, the flow, temperature, etc. are each examined over the whole sea — then conditions can be imposed that correspond to basic physical laws such as the conservation of momentum and mass. This is done by forming a sum over all the elements and then applying each physical law. Boundary conditions, such as no flow through coasts or a specified tidal amplitude, have no place, directly, in such a scheme. In fact, what happens is this. The equations that arise from the imposition of these laws over all the elements lead to enough equations to be able to determine every simple function in each small area. In fact, it turns out that there are *too many* equations because the numbering of the locations (nodes) where the variables need to be determined does not distinguish between which is an internal mode and which is actually on the boundary. This problem is overcome by writing down *all* equations (even those that are false equations) and replacing these false ones by the appropriate boundary conditions, written in terms of the functions valid in these border elements.

Technically, the finite element representation of (say) tides is found by inverting a large matrix corresponding to solving a large number of simultaneous equations. Typically, there are about 1,000 equations with about the same number of unknowns. Most of these equations contain only four or so unknowns, corresponding physically to any given unknown being dependent only on those variables in its immediate vicinity. The matrix is, therefore, largely full of zeros with non-zero entries clustered around the main diagonal. Figure 5.6 is a cartoon of this matrix. The grey dots along the diagonal denote the non-zero entries, and there are other non-zero parts due to the implementation of the boundary conditions; however, the bulk of the

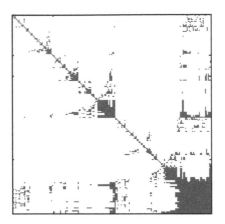

Fig. 5.6   A cartoon of a typical finite element matrix, size here is around 1000 × 1000. The grey dots are small squares and represent non-zero elements in the finite element matrix, a white dot represents zero.

square is white and therefore represents zero entries. Mathematicians call these banded matrices, and there are special methods for dealing with them efficiently. The introduction of boundary conditions leads to isolated off-diagonal entries which are potentially a nuisance, but not fatally so since judicious row operations usually restore the banded nature of the large matrix. In fact, it is usual only to store the non-zero elements of the matrix and their positions; every other position in the square is white and is thus assumed to be occupied by a zero. These days there are in-built routines that generate finite element meshes that have inside them ways of incorporating different types of boundary conditions. In order to get an idea of how finite element schemes work, in much the same way as the diffusion equation was used to introduce finite differences, a simple, rather idealised, equation is also used here. The LTE, in some circumstances, reduce to the Helmholtz wave equation

$$\nabla^2 \eta + \lambda^2 \eta = 0, \quad \text{where } \lambda^2 = \frac{\omega^2 - f^2}{gh}.$$

These circumstances are linearity, constant depth, a single frequency tide and no friction. There are many (often very thick) textbooks that explain the niceties of the FEM (see those by Zienkiewicz and Taylor, 2000a, 2000b, for example). It would be foolhardy to try and duplicate the entire contents of such books here; instead, a flavour will be given by running through the essentials. This should at least tell how the FEM works. As mentioned above, the domain of the problem is divided into small areas — let us assume, as is usually the case, that these are triangles. Consider a typical triangle. The variable, in our case the elevation η, has to approximated inside and on this triangle. First of all, suppose the vertices of this triangle are given by (x_1, y_1), (x_2, y_2) and (x_3, y_3). We then propose a function $N(x, y)$ that has the properties that it is zero at two of the vertices and one at the third. Suppose, for sake of argument, that $N = 0$ at (x_1, y_1) and (x_2, y_2) and $N = 1$ at (x_3, y_3). These conditions can be satisfied by assuming that

$$N = ax + by + c,$$

which is a general linear function, a, b and c constants to be determined. The precise form of $N(x, y)$ is actually

$$N(x, y) = \frac{(y_1 - y_2)(x - x_1) - (x_1 - x_2)(y - y_1)}{(y_1 - y_2)(x_3 - x_1) - (x_1 - x_2)(y_3 - y_1)},$$

but that is not particularly important just now. More important is that $N(x, y)$ is uniquely determined in this triangle. Note that in this case the numerator $= 0$ is the equation of the straight line connecting the points (x_1, y_1) and (x_2, y_2) so $N = 0$ anywhere on this line, not just at the vertices. Of course, the simple determination of N is due to its linearity; there are only three unknowns a, b and c to determine and prescribing $N(x, y) = ax + by + c$ at each of the three vertices does the job completely. The mathematical form that $N(x, y)$ takes is up to us, and we could make it a quadratic or a cubic or anything convenient. The more unknown parameters there are in $N(x, y)$, the more conditions required to fix its form in the triangle. For example, for a cubic ten conditions are required and these could be fixing its value at ten points in the triangle — say at the vertices, at points a third of the way along the sides and at the centroid. Or demanding a value plus continuity of derivatives with respect to x and y that's three conditions at each vertex, plus prescribing a value at the centre. This function $N(x, y)$ is called a *shape function* (mathematicians prefer the name *basis function*), and it is defined for each element in the finite element mesh. In setting up a FEM, one of the problems is bookkeeping, that is numbering the elements. Those brought up on differences are spoilt as the square or rectangular grid is its own Cartesian co-ordinate system. Faced with a mass of triangles, numbering is by no means obvious. Here, it is not appropriate to get involved with this problem (which has been elegantly solved via clever software). Instead, label the shape function associated with element i, N_i.

There are two principle techniques that are used under the broad heading of FEMs, and this same example can be used to illustrate both. With the single Helmholtz equation describing the dynamics the Rayleigh–Ritz method is applicable. This will be described first. When the original equation set is used without the assumption of all variables having a single (tidal) frequency, another more general method called the Galerkin weighted residual technique is used. Both Rayleigh–Ritz and Galerkin predate the FEM. The FEM originated in the 1930s and 1940s from structural engineering and really made a big impact when computing power became strong enough in the 1960s and 1970s. The use of old, hitherto theoretical, 19th century techniques Rayleigh–Ritz and Galerkin then became practical, and real complicated problems — not only in structural engineering but more generally (in, for example, fluid dynamics, heat transfer and electrodynamics) — became tractable. All the techniques are now labelled FEMs, perhaps a little unfairly.

5.5.1   *Rayleigh–Ritz method*

In order to explain the Rayleigh–Ritz method, let us start with the Helmholtz wave equation

$$\nabla^2 \eta + \lambda^2 \eta = 0, \quad \text{where } \lambda^2 = \frac{\omega^2 - f^2}{gh}.$$

One of the main attractions of this technique is that one is able to integrate once to remove second derivatives. Before getting to this, however, we need to guarantee that one has only a single solution and this useful fact arises from the Helmholtz equation, or more precisely the operator $\nabla^2 + \lambda^2$, being self-adjoint. This means that if it is assumed that η_1 and η_2 are two solutions then they must be the same. A self-adjoint operator means that

$$\int_D (\eta_1(\nabla^2\eta_2 + \lambda^2\eta_2) - \eta_2(\nabla^2\eta_1 + \lambda^2\eta_1))dS$$

$$= \int_{\partial D} \left(\eta_1\frac{\partial\eta_2}{\partial n} - \eta_2\frac{\partial\eta_1}{\partial n}\right)ds.$$

Here, dS is an infinitesimal area and ds an infinitesimal arc of its boundary. The result follows from Green's second theorem, but take this on trust if you have not met it before as to explain any more takes us into the realms of potential theory (see section 1.3). Since the left-hand side is zero throughout the domain D, the right-hand side is also zero and this ensures that $\eta_1 = \eta_2$ and the solution, once found, is the only one that there is. Uniqueness is a prominent feature of mathematical texts, and it is indeed important to know that even if you have only found an approximate solution, it is an approximation to the only solution that there is. Now comes the actual solution method. The Helmholtz wave equation is of course not exactly satisfied by the numerical approximation, in our case linear approximations over small triangular elements. However, what is certainly true is $\nabla^2\eta + \lambda^2\eta = R$, where η is now the numerical approximation rather than the actual elevation and R is a residual that we need to minimise for maximum accuracy. If this equation is multiplied by η and integrated over the domain D then

$$\int_D R\eta dS = \int_D \eta(\nabla^2\eta + \lambda^2\eta)dS$$

$$= \int_C \eta\frac{\partial\eta}{\partial n}ds - \int_S (|\nabla\eta|^2 - \lambda^2\eta^2)dS$$

$$= 0 + \int_S \lambda^2\left(\eta^2 + \frac{h}{g}\mathbf{u}^2\right)dS,$$

where the first term on the right vanishes due to boundary conditions (no flow through coasts); we have also used various theorems in potential theory plus that $\mathbf{u}^2 = u^2 + v^2 = -g^2|\nabla\eta|^2/(\omega^2 - f^2)$. The integral on the right-hand side is thus the potential plus kinetic energy (multiplied by a constant). Minimising the residual is thus seen to be the same as requiring this multiple of the energy to be minimised. For the interested reader this is connected to the Hamiltonian theory of dynamic systems outlined in papers by Ian Roulstone, e.g. Roulstone and Norbury (1994); there is a full list at his website. The next step is to minimise the difference between the real solution and the numerical solution, and this is done by expanding the variable in terms of the shape or basis functions:

$$\eta(x, y) = \sum_i \alpha_i N_i(x, y),$$

where the unknowns α_i are determined by minimising the integral

$$I(\alpha_1, \alpha_2, \ldots, \alpha_n)$$
$$= \int \int \left(- \left| \sum_i \alpha_i \nabla N_i(x, y) \right|^2 + \lambda^2 \left(\sum_i \alpha_i N_i(x, y) \right)^2 \right) dx dy.$$

This is done by insisting that

$$\frac{\partial I}{\partial \alpha_i} = 0, \quad i = 1, 2, \ldots, n,$$

which in fact gives n linear equations since I is quadratic in all the α_is. These equations do have a solution provided the equations behave themselves, and this can be shown to be the case here. Boundary conditions can be incorporated through deleting those equations involving the boundary nodes and replacing them with the correct conditions (no flow through a solid boundary which in terms of η is a mixed boundary condition, and prescribed elevation at open boundaries). The problem is then to solve a set of linear equations then to reconstruct $\eta(x, y)$. The Rayleigh–Ritz method will not be pursued further here as its applicability is limited.

5.5.2 *Galerkin weighted residual technique*

In general, equations satisfied by the sea elevation and associated currents are not self-adjoint, and the Rayleigh–Ritz variational method is not applicable. In these situations, another method, the Galerkin technique, comes to our rescue. As an example, let us use the following equation set, which are the LTE with linear friction.

$$\frac{\partial u}{\partial t} - fv + g\frac{\partial \eta}{\partial x} - \gamma u = 0,$$

$$\frac{\partial v}{\partial t} + fu + g\frac{\partial \eta}{\partial y} - \gamma v = 0,$$

$$\frac{\partial \eta}{\partial t} + h\frac{\partial u}{\partial x} + h\frac{\partial v}{\partial y} = 0,$$

where γ is a friction coefficient. These equations are formally replaced by a set of equations that are approximately true, but instead of the variables u, v and η we have the set $\tilde{u}, \tilde{v}$ and $\tilde{\eta}$ that are the same except they have been approximated by the shape functions (usually piecewise linear). The zero on the right of the three LTE with linear friction is replaced by a residual that is small but non-zero. So, we have

$$\frac{\partial \tilde{u}}{\partial t} - f\tilde{v} + g\frac{\partial \tilde{\eta}}{\partial x} - \gamma \tilde{u} = R_1,$$

$$\frac{\partial \tilde{v}}{\partial t} + f\tilde{u} + g\frac{\partial \tilde{\eta}}{\partial y} - \gamma \tilde{v} = R_2,$$

$$\frac{\partial \tilde{\eta}}{\partial t} + h\frac{\partial \tilde{u}}{\partial x} + h\frac{\partial \tilde{v}}{\partial y} = R_3.$$

The technique is to minimise the three residuals R_1, R_2 and R_3. The equations are written in matrix form as follows

$$\frac{\partial \tilde{\mathbf{u}}}{\partial t} + \mathbf{L}\tilde{\mathbf{u}} = \mathbf{R},$$

where

$$\tilde{\mathbf{u}} = \begin{pmatrix} \tilde{u} \\ \tilde{v} \\ \tilde{\eta} \end{pmatrix}, \tilde{\mathbf{R}} = \begin{pmatrix} \tilde{R}_1 \\ \tilde{R}_2 \\ \tilde{R}_3 \end{pmatrix}$$

and **L** is the operator

$$\mathbf{L} = \begin{pmatrix} \gamma & -f & g\dfrac{\partial}{\partial x} \\[2ex] f & \gamma & g\dfrac{\partial}{\partial y} \\[2ex] h\dfrac{\partial}{\partial x} & h\dfrac{\partial}{\partial y} & 0 \end{pmatrix}.$$

The procedure is now to write $\tilde{\mathbf{u}}$ as a weighted sum of the shape functions as follows

$$\tilde{\mathbf{u}} = \sum_{r=1}^{m} \mathbf{A}_r(t) N_r(x, y),$$

where

$$\mathbf{A}_r(t) = \begin{pmatrix} A_{1r}(t) \\ A_{2r}(t) \\ A_{3r}(t) \end{pmatrix},$$

the three components of $\mathbf{A}_r(t)$ corresponding to the coefficients of $\tilde{u}$, $\tilde{v}$ and $\tilde{\eta}$, respectively. The equation governing the coefficients $\mathbf{A}_r(t)$ can thus be written

$$\sum_{r=1}^{m} \frac{\partial \mathbf{A}_r}{\partial t} N_r + \sum_{r=1}^{m} \mathbf{A}_r \mathbf{L} N_r = \mathbf{R}$$

and this is now to be solved for the coefficients $\mathbf{A}_r(t)$. This is facilitated by multiplying this last equation through by an arbitrary shape function, say N_s, then integrating throughout the domain D

$$\sum_{r=1}^{m} \frac{\partial \mathbf{A}_r}{\partial t} \int_D N_s N_r dS + \sum_{r=1}^{m} \mathbf{A}_r \int_D N_s \mathbf{L} N_r dS = \int_D N_s \mathbf{R} dS$$

and the form of the shape functions is chosen so as to make the right-hand side as close to zero as possible. Although it might seem as if the integrals of products of shape functions are difficult to evaluate, this is not in fact the case. As seen earlier, typically

$$N_r(x, y) = \begin{cases} \dfrac{(y_{1r} - y_{2r})(x - x_{1r}) - (x_{1r} - x_{2r})(y - y_{1r})}{(y_{1r} - y_{2r})(x_{3r} - x_{1r}) - (x_{1r} - x_{2r})(y_{3r} - y_{1r})} & (x, y) \in \Delta_r, \\[3ex] 0, & (x, y) \text{ elsewhere,} \end{cases}$$

where Δ_r denotes the triangle with vertices (x_{1r}, y_{1r}), (x_{2r}, y_{2r}) and (x_{3r}, y_{3r}) and the suffix r runs through the integers $1, \ldots, m$ but in this illustrative example no attempt has been made to number all the nodes systematically. As was said, this is done via software these days. The value of $N_r(x, y)$ is thus zero, except over one triangle. This in turn means that integrals of products of the shape functions taken over the entire domain turn out to be the squares of expressions like the linear one for $N_r(x, y)$ (products with themselves) integrated over one triangle and these are straightforward to evaluate. Similarly, operating on a shape function with the differential operator **L** produces only constants, so these integrals are even easier to calculate.

The actual set of equations that determine the coefficients $\mathbf{A}_r(t)$ is solved using finite differences in time. A Crank–Nicolson formulation is often preferred to an explicit scheme, although because of the nature of the matrices, there are often stability problems, even with semi-implicit schemes, and these are still the subject of research.

As in the description of the Rayleigh–Ritz technique, the boundary conditions are incorporated by deleting the equations involving the boundary nodes and substituting the correct boundary conditions (no flow through a solid boundary; prescribed elevation at an open boundary).

5.5.3 *Finite element techniques in coastal sea modelling*

One of the prime reasons why finite elements were slow to be adopted by researchers in fluid flow in general and in ocean science in particular is that finite elements are not as natural a numerical method as finite differences. Structural mechanics is all about predicting stresses and strains in members and plates and this is the natural environment for finite elements — the physical elements, the ties, struts and plates could, and often do, coincide with the theoretical elements. In fluid flow this is not the case and most problems involve waves or evolution from an initial state. The complication of having time as a variable is a serious one, but is overcome by use of the weighted residual technique. In this section, the adoption of the FEM will be traced from its origins in the 1970s.

Early applications (mid-1970s) of the FEM to geophysical fluids contained unwanted oscillations in the solutions, numerical noise and instabilities. Research in the late-1970s and early-1980s showed why these instabilities occurred (Gray and Lynch, 1977; Lynch and Gray, 1979; Bateen and Han, 1981; Walters and Carey, 1983, 1984, and Walters, 1983 are good sources). The upshot of this was explicit substitution of modes proportional to $e^{i(\omega t + kx)}$ that took care of the instability by explicitly removing it from the numerical solution. Using such a combination that does not support spurious modes lies behind the successful finite element code QUODDY.

Note the removal of time by assuming that the response is a single (tidal) frequency. In FEM, the explicit presence of time can be a great influence on the solution method chosen; it is often a severe constraint due to the practical difficulties of dealing with time derivatives. Its absence is a real bonus and is the prime reason for the FEM being used extensively for tidal problems and not for other oceanographic problems such as the wind-driven circulation. Using Green's theorem can be a way to transform the problem into a form that can use the boundary element method (BEM), whereby the partial differential equations are transformed into integrals involving only values around the boundary together with terms that relate certain internal values to those on the boundary. BEMs now have a large literature of their own, see for example the thick tome in two volumes: Wrobel (2002) and Aliabadi (2002).

The finite element scheme used by Westerink *et al.* (1994) used highly unstructured graded grids based on a general wave continuity equation developed by Lynch and Gray (1979). The method is based on a clever combination of the basic equations which optimises the wave phase and amplitude characteristics. In general, finite element methods are still subject to instability, but the scheme used by Westerink *et al.* (1994) is stable. The authors quote a Fourier analysis in constant depth using linear interpolation which indicates that a tidal wave is resolved with 25 nodes per wavelength. It is still necessary for the Courant number, here defined as $\sqrt{gh}\Delta t/\Delta x$, to be less than or equal to one. The stability analysis is similar to that for finite differences on a B grid. In the intervening years a

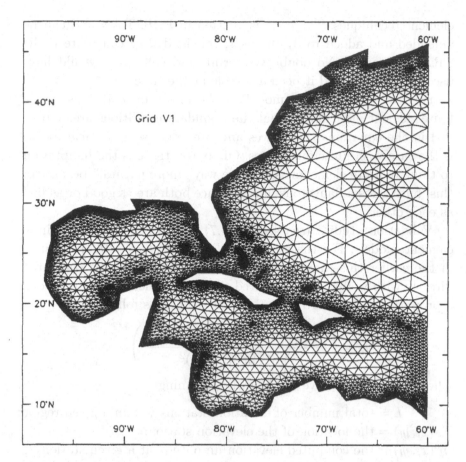

Fig. 5.7 An unstructured finite element grid. From Westerink *et al.* (1994) reproduced with permission.

system called TELEMAC has been developed by the French (originally by Électricité de France). It is a very powerful system based on a two-dimensional finite element unstructured mesh, but extended to three dimensions so each element is a triangular prism. It is nonlinear and can utilise a variety of models to simulate sea bed friction. The $k - \epsilon$ Mellor–Yamada scheme is available as well as simpler mixing length models based on Prandtl mixing length theory or even constant eddy–viscosity models. The latest manual on the web seems to be dated 2013 and references finite volume as well as finite element techniques. The code is written in FORTRAN and can be amended and added to by the user, provided they can write FORTRAN of course. No doubt Westerink and colleagues would have used TELEMAC had it been available at the time.

For the oceanographic modeller, the most convenient aspect of finite element modelling is that the boundary conditions are embedded in the equations themselves and are not a separate feature, as is the case with finite differences. Of course, there is the temptation to think of finite elements as in some way "more natural" because of this, but this is a dangerous illusion since both are as good or as bad as each other.

Validation studies (see section 5.7) have taken place for finite element models. For the study of Westerink *et al.* (1994) outlined in the earlier this chapter, the output of the model was compared directly with well known measured tidal constituents. In this model, a proportional standard deviation was defined as follows

$$E = \left(\frac{\sum_{l=1}^{L} (\eta^c(x_l, y_l) - \eta^m(x_l, y_l))^2}{\sum_{l=1}^{L} (\eta^m(x_l, y_l))^2} \right)^{1/2},$$

where the symbols have the following meanings:

L = total number of elevation stations within a given region;
(x_l, y_l) = the location of the elevation station;
$\eta^c(x_l, y_l)$ = the computed elevation amplitude at a given station;
$\eta^m(x_l, y_l)$ = the measured elevation amplitude of a given station.

In the model, a number of tidal constituents were simulated. In particular, the average error as measured by E was between 18.2% and 45.3%, where eight tidal constituents were considered and the entire domain was covered. Nothing statistical was attempted, but the regions of greatest error coincided with the amphidromic points (where the tide vanishes). Also, those stations with poor convergence properties (i.e. the model elevation only slowly converging to an answer) gave the poorest comparison with measurement.

For wind-driven flow, a statistical comparison between model output and observational data is hardly ever done since observations usually drive the model, rendering them dependent and rendering statistical approaches invalid. The future availability of satellite data could herald the onset of proper validation studies. For the model of Westerink *et al.* (1994), not only were validation tests in the form outlined above carried out, but sensitivity tests were also done. This took the form of splitting elements and examining the subsequent changes in both the amplitude and phase of the tide.

As both finite elements and finite differences can solve similar oceanographic problems, in the last 15 years or so there have been increasingly sophisticated inter-comparison studies. A good example is the paper by Jones and Davies (2005). There is no surprise that they conclude finite elements are very good for tidal computations, but the case is not done and dusted by any means.

5.6   Computational Fluid Dynamics

The use of computational methods for solving flow problems started in the 1960s and 1970s, but in the last ten years, the demands of a wide variety of industry and research for accurate modelling has increased markedly and has resulted in the birth of what is, in effect, a new science — the science of computational fluid dynamics (abbreviated to CFD). There are now textbooks and learned journals devoted to CFD, so only a summary will be given here. The numerical methods of this chapter form the basis of most CFD packages, but all the resources and demands placed in this area mean that it is fast moving and growing. At the moment, the website http://www.cfd-online.com is well worth a visit for the latest developments, but what are the recent technical advances? Large eddy simulation (LES) has been around as a specialist topic in numerical methods in fluids for over 30 years, but recently it has been taken up by those more concerned with practical applications. It is the modelling of actual eddies rather than their parameterisation. No matter how accurate the chosen parameterisation is, it is always bettered by actual modelling of the eddies. In LESs, models are getting close to including the actual Reynolds stress terms without parameterisation. Clever smoothing and interpolation routines minimise errors and suppress sub-grid scale unresolvable noise. This kind of modelling is now becoming incorporated into commercially available software such as the older FLUENT, FLOW-3D and PHEONICS, as well as larger codes in which these are embedded. Environmental finite element codes such as the already mentioned TELEMAC are built using a modular structure so that modules that are irrelevant to the user can be excluded whilst others can be included. A modern modular software system, NEMO, is explained in the last chapter due to its inclusivity. This kind of CFD software is being added to and improved all the time and no doubt future versions of most of them will include LESs too.

It remains the case, however, that as new developments are included and smaller and smaller scales are modelled, it is still essential to know the limitations of the model. It will be a long time before the ultimate CFD software exists that can solve all problems to any desired degree of accuracy.

Another development that has appeared in the last 15 years is adaptive grids. The essential feature of adaptive grids is the ability of the grid to change as the flow evolves. The grid could start off as uniform, but as the flow evolves, perhaps a jet appears. Adaptive grids then would concentrate the grid where there are changes in the flow, with passive areas having a sparser grid. The advantages of this are that computation time is saved by their concentration in areas where it matters at the expense of areas where it does not. The fine details of how this is done are outside the scope of this text, but the general idea is to weight the node of the finite difference or finite element scheme through tying its position to the magnitudes of the absolute values of the flow gradients, this is being done in such a way as to concentrate the grid where the changes are. There are technical difficulties here; done carelessly, grid lines could cross, producing new nodes and defying the basic laws of fluid mechanics such as mass or momentum balance. This is avoided by first by placing the repositioned nodes in a different space (called parameter space) then correcting as necessary. Adaptive grids are being used in engineering problems and are beginning to be used both in atmospheric and oceanographic modelling, but their use is in its infancy as yet. It is no doubt the case that to track truncation error and instability in an unstructured grid is more challenging than in traditional numerical modelling and that for problems with a natural time scale, there will be problems of interaction between these scales and any time scale associated with the adaption of the grid. Some new kind of Courant number will probably have to be defined. In the last few years, there have been a few advances. First of all, some promising commercial software have ceased due, perhaps, to the recession since 2008 but also to the growth of software such as NEMO, which is both comprehensive and highly advanced in terms of including the most up-to-date science but also by allowing the user inside the code to update and add (or subtract) features. This user-friendliness combined with great adaptability, tends to lead to its adoption ahead of its rivals, and there are many more rivals now. Let's leave this now and return to data assimilation.

5.7 Practical Data Assimilation

In Chapter 2, the idea of data assimilation was introduced as the introduction of data into a model as it runs. The Kalman filter is one technique that aids this process. For some models, particularly those that are simple and are built to demonstrate a straightforward principle, validation remains the right thing to do. The validation of a model is the comparison of model output with what can be termed current knowledge. This knowledge usually comes from observation (see Chapter 1, Section 1.2). It might seem as if nothing can be easier than to compare the output of a model with observations, which usually occur in the form of data. However, this is not the case, except perhaps for tides where there are very long and very accurate records (in most instances). If a model is run, and values of the variables are obtained, one is faced with the question of how accurate these results are. To take a specific example, a domain may be overlaid with a two dimensional grid (Figure 5.5 is typical) and values of surface elevation, eastward and northward velocity obtained as model output at all grid points. How good are the results? Perhaps an observational programme over the same area of a convenient cruise has occurred and instruments have produced some data. First, the data will not have been taken at precisely the same locations as the model grid points; second there will, in all likelihood, not be enough data; and third not all of the data will have been collected at the same time. All of these factors make comparison between data and model output difficult. In order for some values to be compared directly, interpolation has to take place. Interpolation is a numerical technique that enables values of a variable at some intermediate location to be computed from those that surround it. It is a generalisation of curve fitting and there is now quite an extensive library of software that can help with this process (terms associated with interpolation include splines and least squares, both of which are useful, and Lagrange interpolation, which should be avoided). This then deals with one problem, but if there are not enough data, or they were recorded at the wrong time, there is not much we can do. Satellite obtained data are proving to be very useful here and are beginning to provide a leap in the sophistication of model validation and data assimilation. Suppose that we have adequate data, what do we do about comparing these data with the output from the model? The easiest thing to do is to "eyeball" both and assess the significance of any differences. This is still often the only method of validation used in marine science. While not excusing this very unscientific methodology, it is perhaps understandable that oceanographers and ecosystem modellers should be unwilling to expend large amounts of time and energy on sophisticated statistical techniques when much of the data are so roughly hewn from the sea. Nevertheless, there are techniques that can be of some use. If a direct comparison between model output and observations is possible, then one can analyse the differences between them statistically. For example, using the statistics of sampling, it is possible to tell whether differences are significant by using the t-distribution and to place confidence limits on the significance of these differences. If, further, it is possible to postulate that, as a *null hypothesis*, there should be agreement between observations and model output, then one can define this measure of agreement via the χ^2 test. This is a statistic that is defined by the expression

$$\chi^2 = \sum \frac{(\text{Observed} - \text{Expected})^2}{\text{Expected}}.$$

The "observed" values are the model results, and the "expected" values are the corresponding observations. The $\sum$ sign denotes summation over all data points. Remember one technical point here: in order to use the χ^2 test the data have to be ranked, classified or otherwise rendered free of dimension. There are also questions of degrees of freedom. Consult Chapter 1 or statistics textbooks for further details. Of course, one could swap the role of model results and observations; this would in theory "test" the observations against the model, assuming the model to be correct. If both are reliable, then there is no problem. If neither is reliable, then we get the standard arguments between those who measure and do field-work, and those who model and do calculations. Once the χ^2 statistic has been calculated, it is simple procedure to look at a table (see for example, Murdoch and Barnes (1974) *Statistical Tables*) and ascertain whether or not departures are significant. Of course, it is tempting to regard this as a definitive argument for or against a particular model. In reality, standard statistical tables have built into them certain assumptions involving the normal distribution that particular observations or model output may disobey. As a general rule, it is most unwise to use statistical methods that are more complicated than is warranted by the veracity of the data. If in doubt, contact a professional statistician. Only a very brief introduction to some statistics was given in Chapter 1.

Let us now look at the combining of traditional validation with data assimilation. If it is assumed that neither the model nor the data are perfect, then perhaps the logical way forward is to combine the best features of both; but how can this be done? First of all, let us assume that there is a numerical model that predicts the state of the variables at the time $(s+1)\Delta t$ given the state at time $s\Delta t$. Without getting embroiled in notation, there will be many variables in a complex model, or the variables may be simply the tidal elevation. One fruitful thing to do, harking back to the last section of Chapter 1 and section 2.5 on Kalman filters is to pick out those variables (directions in model space to use the mathematical terminology) where there is most information. This is done through the use of PCA or the use of EOFs. Once the first few principal components are known, then it is possible to filter the data in such a way so as to be able to incorporate these new data into forecasts in an accurate and efficient manner. This has been done in several fields, most famously in geography; a branch called geostatistics which concerns assessing the Earth's resources and using satellite data to enhance model predictions. Here, the Kalman filter and associated interpolation and regression methods are called *kriging*, named after the South African Danie G. Krige who developed it for the gold mining industry. The problem with oceanography is the sheer richness of the data. For example, the ability of a model to simulate a real mesoscale eddy to enter and leave the model domain needs to be compared to the ability of the interpolation technique to maintain artificially generated features through the use of Kalman filters and their modifications. The time scales can go from under an hour to decades, and techniques developed for steady problems of geostatistics may be inappropriate for a highly time varying environment. One possible fruitful area of application might be ecosystems modelling where there are reasonably well-defined cycles and where particular data sets could be used for improving models simply because the parameters of such models are large in number. Enhanced Kalman filtering techniques can be used to filter out inappropriate data and reassign parameters to more reasonable values. The paper by Bertino, Evensen and Wackernagel (2003) is a good recent review; it is a little technical but the references are extensive.

Despite this enhanced sophistication, old arguments about the relative merits of model results as against those obtained through extensive observations persist. They are likely to persist for a long time.

5.8   Excrcises

(1) For the explicit scheme applied to the LTE

$$u_{m,n}^{s+1} = u_{m,n}^s + f\Delta t v_{m,n}^s - \frac{g\Delta t}{2\Delta x}(\eta_{m+1,n}^s - \eta_{m-1,n}^s),$$

$$v_{m,n}^{s+1} = v_{m,n}^s - f\Delta t u_{m,n}^s - \frac{g\Delta t}{2\Delta y}(\eta_{m,n+1}^s - \eta_{m,n-1}^s),$$

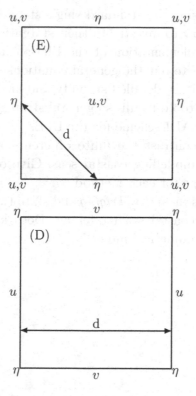

Fig. 5.8   The Arakawa D grid (bottom) and E grid (top), d is the grid spacing, η is the vertical displacement and (u, v) the horizontal current.

$$\eta_{m,n}^{s+1} = \eta_{m,n}^{s} - \frac{\Delta t}{2\Delta x}(u_{m+1,n}^{s}h_{m+1,n} - u_{m-1,n}^{s}h_{m-1,n})$$

$$- \frac{\Delta t}{2\Delta y}(v_{m,n+1}^{s}h_{m,n+1} - v_{m,n-1}^{s}h_{m,n-1}),$$

assume that the depth is constant, use the substitutions $\eta = A\exp{(ikx + ily + \alpha t)}$, $u = U_0\exp{(ikx + ily + \alpha t)}$ and $V = V_0\exp{(ikx + ily + \alpha t)}$ and take $\Delta x = \Delta y$ to show that this scheme is always unstable.

(2) Figure 5.8 shows the template for the Arakawa D and E grids. Write down the equation

$$\frac{\partial u}{\partial t} = \kappa\left(\frac{\partial^2 u}{\partial x^2} + \frac{\partial^2 u}{\partial y^2}\right)$$

for each grid using the standard single stage forward time step. Hence, show how to investigate their stability and accuracy.

(3) Discuss the implementation of the Eliason lattice (Figure 5.2) to the LTE. Write out the general equations of the scheme and discuss possibilities; detailed stability and analysis of truncation error would be too difficult, so general strategy only is required.

(4) Write down an ADI scheme for the LTE.

(5) Compare and contrast the finite difference and finite element techniques for modelling coastal seas. Give one advantage and one disadvantage for each method.

(6) Explain what is meant by *kriging* and state two problems applying kriging to coastal sea modelling. Does kriging help in the validation of a numerical model?

Chapter 6

Tides, Surges and Tsunamis

6.1 Introduction

In this chapter, we shall first discuss tides from a modelling viewpoint before going on to talk about wind-driven flow, tsunamis and finally seiches. By wind-driven flow we mean the currents due to the direct action of the wind on the water immediately underneath it, not the general circulation of the ocean (the Gulf Stream and the like). The goal will be to combine wind and tide to provide an explanation of practical storm surge modelling. One of the most distinctive features of continental shelf seas is the relative strength of the tidal currents compared with those that occur in the deep sea. From the point of view of fluid flow this feature is easy to explain. Just as flow through a large-diameter pipe accelerates to a faster value as the diameter decreases, so a slow tidal flow in the deep sea accelerates to fast tide over the shallow continental shelf; it is called the Venturi effect and is a consequence of applying the Bernoulli principle, which is the conservation of energy alongside the conservation of mass. The narrower the pipe, the faster the flow, or the shallower the water the faster the current. In shelf regions, tidal currents are usually ten times stronger than currents from other sources (wind or convection due to freshwater inputs from rivers). Let us begin with a look at continental shelf tides.

At the outset, let us declare that we shall be concerned with practical tidal dynamics and not with the kind of theoretical tidal modelling which was published 100 or more years ago, dating back to Laplace in 1776, involving periodic hydrodynamics on a rotating sphere.

Tides are due to the astronomical forces that arise from gravitational attraction between the Earth and the Moon and, to a lesser extent, the Earth and the Sun. These forces themselves act on the Earth's large bodies of water, the Pacific, Atlantic and Indian Oceans. Typically, the Moon pulls the surface of the Atlantic about half a metre from its mean level both when it is directly overhead and also when it is on the other side of the Earth. This can be thought of as arising because the same gravitational attraction of the Moon pulls the Earth itself closer to the Moon than the water which is furthest from the Moon thus resulting in a second high water on the opposite side of the Earth. These slight tides, and the equivalent depression when the Moon is on either horizon, causes a wave (two high waters a day as the Earth rotates) which propagates. This wave passes over the continental shelf and, once on the continental shelf itself, increases in amplitude because of the conservation of mass (the same reason as the enhancement of tidal velocities). Another feature of tides is their wavelength. This is the horizontal distance between two successive high waters at any fixed time. This distance is typically 1000 km (for a tide with two high waters a day in a continental shelf region). The amplitude of a tide is but a few metres or even less. The wave slope, a common dimensionless parameter which is often used as a characteristic measure, is the ratio of these two quantities and is thus only about 10^{-5}. This means that tides, considered as water waves, are very long waves indeed. Some elements of shallow water wave theory were given earlier in Chapter 3. Non-dimensional quantities such as the ratio of wave length to sea depth have already been defined in Chapters 2 and 3; these certainly can apply to tides. In order to see how shallow these water waves are, let us insert some typical values for the parameters. Since the depth of the continental shelf sea is only 200 m or so, and the ratio depth/wavelength is only 2×10^{-4}, we can consider the water as indeed very shallow. One consequence of this is the tidal currents should be virtually independent of the depth. This is certainly true if we are concerned with tidal height predictions.

However, tidal currents are influenced by bottom friction due to the roughness of the sea bed; this can be of crucial importance.

In recent years, tsunamis have hit the headlines due to the devastating 2004 Boxing Day tsunami and the equally disastrous 2011 Japanese tsunami. The modelling of tsunamis is interesting and important. Finally, there is a short section on the modelling of seiches, the oscillations in lakes and harbours, usually due to the passage of storms.

6.2   The Equilibrium Theory of Tides

As said above, the tides are of astronomical origin. As the Earth hurtles through space, the air and water that cling to its surface by gravitational force are themselves pushed and pulled by other heavenly bodies. It is the Moon that has the most influence, and it is the equilibrium theory of tides that calculates what this influence is. This theory assumes that the Earth is covered completely by water and that no other heavenly bodies (including the Sun) exist. So, accurate prediction of sea levels and high waters cannot be expected from the equilibrium theory, however it is necessary to calculate from this theory in order to find out the basic shape; it is from this basic shape when coasts and land masses are added that gets closer to how real tides behave. So good modelling practice is being followed; start simple and build on this to get an accurate model. Perhaps the best modern book on tides is Pugh and Woodworth (2014), which has a lot more detail than can be included here. Planetary bodies are subject to the inverse square law of attraction proposed by Newton in the 17th century. Two masses, m_1 and m_2, have a mutual attractive force

$$F = \frac{Gm_1m_2}{r^2},$$

where $G = 6.67 \times 10^{-11}$ Nm^{2}Kg^{-2} is the universal gravitational constant and r is the distance between the bodies. This is fine for particles where r is well-defined, but planets are not particles. If we want to calculate the pull of the Moon on the sea, then some extra trigonometry and a little potential theory is required. First of all, the gravitational force is really a vector that is conservative, that is it has what is called a potential. For the purist $\nabla \times \mathbf{F} = \mathbf{0}$ implies that $\mathbf{F} = -\frac{1}{m_1} \nabla \Omega_p$, where $\mathbf{F}$ is the conservative force and Ω_p is the potential referred to the mass m_1. In this application, however, such a vector treatment is unnecessary and the force and its potential are related through

$$F = -\frac{1}{m_1} \frac{d\Omega_p}{dr}$$

so, after integration we see that

$$\Omega_p = \frac{Gm_2}{r},$$

where by convention the constant of integration has been taken as zero. Constants when added to potentials do not change anything as they are always differentiated away. This is the definition of the gravitational potential Ω_p. Although it is true that the volumes of the Earth and Moon must be taken into account in order to calculate the shape the envelope takes under the gravitational pull of both, it is not as problematic as you might think as the forces of gravity stem from the centres of mass of the Earth and Moon. Thus we need to use simple geometry on the triangle connecting these centres of mass and a point on the surface of the water, Earth, Moon shows the set up. Figure 6.1 shows that r is in fact the distance between a point on the surface of the tidal envelope of the sea being pulled by the Moon and the centre of the Moon itself. The cosine rule can be used

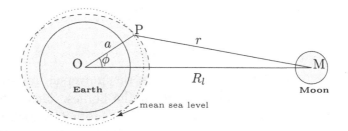

Fig. 6.1   The Earth–Moon system showing the water envelope.

to show that

$$r^2 = a^2 + R_l^2 - 2aR_l \cos \phi.$$

However, the distance between the Earth and the Moon (R_l) is much larger than the radius of the Earth (a). Inserting values, $a = 6.371 \times 10^6$ m and $R_l = 3.84 \times 10^8$ m giving the ratio a/R_l a value 0.017 to three decimal places. So, the expansion

$$\frac{1}{r} = \frac{1}{R_l} \left\{ 1 + \left(\frac{a}{R_l} \right)^2 - 2 \left(\frac{a}{R_l} \right) \cos \phi \right\}^{-\frac{1}{2}}$$

$$\approx \frac{1}{R_l} + \left(\frac{a}{R_l^2} \right) P_1(\cos \phi) + \left(\frac{a^2}{R_l^3} \right) P_2(\cos \phi)$$

$$+ \left(\frac{a^3}{R_l^4} \right) P_3(\cos \phi) + \cdots$$

needs only the first few terms for high accuracy. Although only the binomial expansion of a square root is demanded, it is a bit tricky. The final expansion is in fact written in terms of Legendre functions $P_n(\cos \phi)$ that have the values

$$P_1(\cos \phi) = \cos \phi,$$

$$P_2(\cos \phi) = \frac{1}{2}(3 \cos^2 \phi - 1),$$

$$P_3(\cos \phi) = \frac{1}{2}(5 \cos^3 \phi - 3 \cos \phi).$$

They are called spherical harmonics and the expression

$$\left\{ 1 + \left(\frac{a}{R_l} \right)^2 - 2 \left(\frac{a}{R_l} \right) \cos \phi \right\}^{-\frac{1}{2}} = \sum_{n=0}^{\infty} \left(\frac{a}{R_l} \right)^n P_n(\cos \phi)$$

can be used to generate the expansion to any order of accuracy. This expression, you will not be surprised to learn, is called the *generating function* for the Legendre polynomials $P_n(x)$ (writing x for $\cos \phi$; $P_0(x) = 1$ of course). Anyone who prefers differentiating could use the alternative formula

$$P_n(x) = \frac{1}{2^n n!} \frac{d^n}{dx^n} [(x^2 - 1)^n],$$

called the Rodrigues' formula, but this is usually more of an acquired taste. No matter, only the first four are ever required in the equilibrium theory of tides. It should also be mentioned that the tidal elevation is a few metres and negligibly small compared to the radius of the Earth so that the distance OP in Figure 6.1 is, to all intents and purposes, a. Thus, the potential Ω_P is given by

$$\Omega_P \approx Gm_2 \left[\frac{1}{R_l} + \left(\frac{a}{R_l^2} \right) P_1(\cos \phi) + \left(\frac{a^2}{R_l^3} \right) P_2(\cos \phi) \right.$$
$$\left. + \left(\frac{a^3}{R_l^4} \right) P_3(\cos \phi) + \cdots \right].$$

As Ω_P is a potential, the constant term Gm_2/R_l can be ignored as it gets differentiated out, however Ω_P is now a bit more complex than one dimensional. The potential at P can thus be written

$$\Omega_P \approx \left(\frac{Gm_2 a}{R_l^2} \right) \cos \phi + \left(\frac{Gm_2 a^2}{2R_l^3} \right) (3 \cos^2 \phi - 1)$$
$$+ \left(\frac{Gm_2 a^3}{2R_l^4} \right) (5 \cos^3 \phi - 3 \cos \phi).$$

This looks like a function only dependent on the angle ϕ, however this potential can vary with distance from the centre of the Earth as long as we stay within the approximations made. To be more precise, it is three variable dependent, just as the Earth itself is three dimensional, and we should be using spherical polar co-ordinates. To do this would be well over the top, and the letter a will continue to denote the distance out from the centre of the Earth and not just its radius as in this book not a lot more time will be spent on the equilibrium theory of tides. In general, the force due to this potential will be given by

$$\mathbf{F} = -m_1 \nabla \Omega_P = m_1 \left(-\frac{\partial \Omega_P}{\partial a}, -\frac{1}{a} \frac{\partial \Omega_P}{\partial \phi} \right).$$

To the lowest order, we get the result

$$\mathbf{F} = (-F \cos \phi, F \sin \phi),$$

so this term in the expansion, although dominant, is merely the force of attraction between Earth and Moon and does not contribute at all to the tide. Physically, this is not a surprise. Thus, it is concluded that the dominant tidal term is

$$\Omega_P = \left(\frac{Gm_2 a^2}{2R_l^3}\right)(3\cos^2\phi - 1).$$

This states that the equipotential surface Ω_P is given by the above expression to a good approximation. To see what kind of surface this is, let us use the following procedure. As Ω_P = constant on this surface, then its differential $d\Omega_P = 0$ on the surface. The easiest way to do the mathematics is to move into two-dimensional Cartesian co-ordinates and then realise that the three-dimensional picture is obtained by rotating the diagram Earth–Moon about the line OM. In x, y co-ordinates

$$a\cos\phi = x \quad \text{and} \quad a^2 = x^2 + y^2$$

so

$$\Omega_P = \left(\frac{Gm_2}{2R_l^3}\right)(2x^2 - y^2)$$

and

$$d\Omega_P = \frac{\partial \Omega_P}{\partial x}dx + \frac{\partial \Omega_P}{\partial y}dy = \left(\frac{Gm_2}{2R_l^3}\right)(4xdx - 2ydy) = 0$$

on the surface Ω_P = constant. The expression

$$\frac{dy}{dx} = \frac{\partial \Omega_P/\partial x}{\partial \Omega_P/\partial y}$$

is now used. This leads to the simple (but differential) equation

$$\frac{dy}{dx} = -\frac{2x}{y},$$

valid on the equipotential surface. Integrating this gives

$$x^2 + \frac{1}{2}y^2 = K,$$

where K is a constant of integration. This is an ellipse, with its major axis pointing towards the Moon. Rotating this ellipse about its major axis OM gives the prolate spheroid which is the shape of the equilibrium tide. For those familiar with such objects, this set of surfaces is orthogonal (at right angles) with the surfaces $\Omega_P =$ constant, as is always the case. It is the same procedure as outlined in section 3.6.3, but books on potential theory (e.g. Kellog (1929) is classic) give much more general detail.

6.3   Real Tides

Although the equilibrium theory outlined in the last section gives how the main tide arises, it is not good for describing the real tidal rise and fall of the sea. So, to obtain better predictions let us obey the modelling principles; first the tides as observed must be described. There are some reasonably general observations. There are usually two high waters each day, and this is consistent with them being due to the Earth's rotation under the Moon and the resultant equilibrium theory. There are also monthly tides with two high waters each month; when these coincide with the lunar high water they are called *spring tides* and there are two of these each month. They will happen when the Moon and Sun are aligned either as a new moon or as a full moon, but the timing is not exact. There is a delay between a new or full moon and high tide. When the Moon, Sun and Earth are positioned such that the Earth is at the right angle of the triangle formed, then the combined tides are at a minimum, apart from a similar lag. These minimum tides are called *neap tides* and there are also two of these each month. If this exactly described the tides then there would be little modelling to do, but this is not the case. At any particular coastal location the tides are usually more complicated and at a few places completely different. However, the most practical way of denoting the tides at any particular place is to let them consist of a sum of perfect oscillations. These oscillations are known as *tidal constituents* and Table 6.1 gives the 13 main ones.

In general, the semi-diurnal tides (that's two high waters each day) are due to the Earth revolving under the Moon or the Sun. The third tide of this period N_2, labelled the large lunar elliptic, is

Table 6.1   A table of principal tidal constituents.

Tidal component	Period (solar hours)	Description	Nature
M_2	12.42	Principal lunar	Semi-diurnal
S_2	12.00	Principal solar	Semi-diurnal
N_2	12.66	Large lunar elliptic	Semi-diurnal
K_2	11.97	Luni-solar	Diurnal
K_1	23.93	Luni-solar diurnal	Diurnal
O_1	25.82	Principal lunar diurnal	Diurnal
P_1	24.07	Principal solar diurnal	Diurnal
Q_1	26.87	Larger lunar elliptic	Diurnal
MF	327.90	Lunar fortnightly	Fortnightly
MM	661.30	Lunar monthly	Monthly
SSA	4383.00	Solar semi annual	Half-yearly
M_4	6.21		
MS_4	6.10		

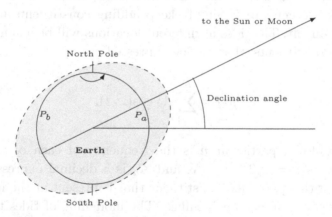

Fig. 6.2   Showing (at P_a and P_b) the uneven high waters as the Earth rotates.

entirely non-physical and is an attempt to model the ellipticity of the orbit of the Moon around the Earth. This is but one wobble; there are others. The most important of these, yet to be mentioned, are the changes in angle (or declination) of the Earth to the Moon and Sun. This declination changes and so would the position of the theoretical bulge of water due to the Moon's or Sun's gravitational pull (see Figure 6.2; the point P_a experiences high water, then 12 hours later the Earth has revolved this point to P_b where it experiences a lower high water due to the angle of declination). To simulate these, the tides K_2, K_1, O_1, P_1 and Q_1 are added to the physical tides; the details need not trouble us. Finally, there are longer period tides that compensate for other eccentricities such as the changes in declination not being precisely sinusoidal. M_4 is the first harmonic of the M_2 principal lunar tide and is due to nonlinear interaction. This gives a flavour of the empirical nature of this business. Table 6.1 lists 13 constituents, but the official government offices list well over 100 to capture every slight wobble. One is reminded of Ptolemy's model of the Solar System with its epicycloids; it worked very well for hundreds of years as epicycles can mimic the elliptical orbits very well provided the eccentricity of the ellipses is small. We now know the physics better of course, but we can still use epicycloids to predict eclipses without any astrophysical knowledge. In much the same way, we can simulate tides without any knowledge of the physics of how the tides are generated. Simply keep adding constituents until the observations fit. The tides at different locations will be a summation of these constituents, that can be expressed as

$$\eta(t) = \sum_{n=1}^{N} A_n \cos(\omega_n t),$$

where ω_n for a particular n is the frequency of each of the tidal constituents $n = 1, 2, 3, \ldots, N$, and A_n is a decimal expressing the amount of this particular constituent that is present at the location. In many cases, $N = 13$ will suffice. The modelling of tides this way is called the harmonic theory and, of course, the A_n are all location dependent as seen in Figure 6.2. It is the case that once the tide has been carefully measured over a year then it is predictable. Old Farmers Almanack (for example) has been publishing tide tables for major ports since the 18th century. So, as far as pure tides are concerned, they are predictable and there really is no need for sophisticated mathematical modelling. Why does harmonic theory work so well? It is because oscillations like the tide at a particular location are periodic, and periodic functions can be simulated by Fourier series.

A Fourier series is, however, infinite and is typically expressed as

$$\eta(t) = \frac{1}{2}a_0 + \sum_{n=1}^{\infty}[a_n \cos(n\pi t/L) + b_n \sin(n\pi t/L)],$$

where $2L$ is the period of $\eta(t)$ and all the a_n's and b_n's are constant. Although we shall not be using Fourier series much in this text, in the analysis of surface waves the notion of a spectrum is essential and spectra emerge from Fourier series as follows. If the period of $\eta(t)$ gets very large then the term $n\pi t/L$ is small unless n gets large too. As the sum is infinite, no matter how large L, the quantity n/L will eventually get finite and then large. It is just that the number of terms in the sum has to be correspondingly large. Eventually, such a sum becomes indistinguishable from an integral; and this is what leads to Fourier transforms. There is no mathematical rigour in these statements as they appear here but specialist books, for example Dyke (2014), supply this rigour. The Fourier transform has the definition

$$\eta(t) = \int_{-\infty}^{\infty} S(\omega)e^{-i\omega t}d\omega,$$

with

$$S(\omega) = \frac{1}{2\pi} \int_{-\infty}^{\infty} \eta(t)e^{i\omega t}dt,$$

where $S(\omega)$ is a function that provides information about the frequencies that are in the function $\eta(t)$. There are subtleties glossed over here, like the convergence of these improper integrals (those with ∞ of either sign in the limits) especially given the nature of the sea level $\eta(t)$ that never decays as $t \to \pm\infty$, but nor does $e^{i\omega t}$, which is $\cos(\omega t) + i\sin(\omega t)$ that simply oscillates. There is also the question of the factor 2π that can move within this definition of Fourier transform, but this is trivial if a little irritating. Taking the definition on trust, for tides there are only a finite number of frequencies (we called them constituents) so it is quite permissible to approximate the integral

$$\eta(t) = \int_{-\infty}^{\infty} S(\omega)e^{-i\omega t}d\omega$$

by a finite sum

$$\eta(t) = \sum_{n=1}^{N} A_n \cos(\omega_n t)$$

that regains the harmonic approximation we met earlier. In the trade, this harmonic series is called a Discrete Fourier Transform and is a well known device for analysing signals. So everything here is justified. Rigorous signal processing has been around for nearly 100 years now, tidal harmonics for considerably longer. It's good to see tidal harmonics justified in retrospect. Before actually modelling tides, it is necessary to have a way of displaying them. Currents are displayed by drawing streamlines, as pressure contours are streamlines for the wind in the atmosphere (well nearly). The up-and-down, to-and-fro motion of the tide has no obvious candidate to help display it, but to do so two sets of lines are defined: the *co-tidal line* and the *co-range line*. The co-tidal lines join points where high water occurs at the same time. It is analogous to a wave crest line in water waves, though the definition extends to joining any points where the tide has the same phase. Figure 6.3 shows a world map of the M_2 tide using co-tidal lines and it shows just how different it is from the idealised equilibrium picture. In this numerical model the co-tidal lines are shown in white, radiating from points where the tide is zero. These points where there is zero tide are called *amphidromic points*; they are equivalent to nodes in standard waves. The units for these co-tidal lines are, somewhat bizarrely, hours. As this is the M_2 tide, the period is 12 hours and, as there are six lines emanating from each amphidrome, they are two hourly intervals. The co-range lines join points that have the same tidal range. The range of the tide is the vertical distance between high water and low water. In Figure 6.3 the range is shown in colour with the code key given along the foot of the figure. One can see that the range of the tide is by far the largest at the coast, and the reason is the one outlined in the introduction — the Venturi effect. Figure 6.4 gives a good example of how tides are displayed using two sets of lines, full for co-tidal lines and dotted for co-range lines. Looking at this figure, one can see a wave

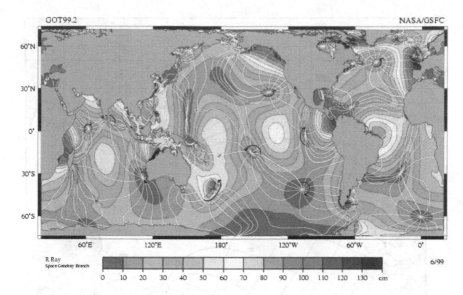

Fig. 6.3 The co-tidal lines for the world M_2 tide, from Accad and Pekeris (1978).

with crests virtually at right angles to the coast of the North East
UK propagating down the coast with land on the right as viewed
by someone riding on the wave pointing forwards. The largest tidal
range is at the coast and this range is virtually the same at similar
distances from the coast itself. This gives the square pattern with
the full co-tidal lines parallel to the coast and the dotted co-range
lines perpendicular to the coast. The co-tidal and co-range lines are
not always at right angles, but they are for these *Kelvin waves* (see
the next section). The kind of pattern exhibited in these co-tidal
and co-range diagrams has to be generated by a model; numerical
models will be looked at but first we develop more basic models that
help us to understand these patterns. Having a good working harmonic theory of real tides predicts future tides but does not help us
to understand them, in much the same way as Ptolemaic astronomical calculations would not lead us to Newtonian laws of gravity, let
alone Einstein's. For a better understanding, therefore, we return to
the equations of motion and some simple models that describe tidal
oscillations; this is the other use for modelling, to understand what is
going on.

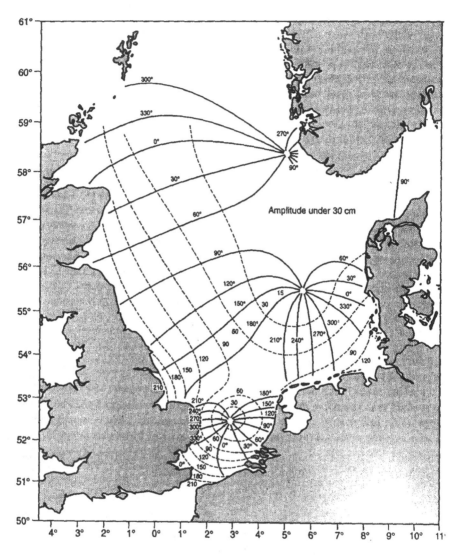

Fig. 6.4 Co-tidal and co-range lines, denoted by full and dotted lines, respectively. The associated numbers give the values of γ (in degrees) and of H (in cm).

6.4 Tidal Dynamics

The starting place for the understanding of the dynamics of tides must be the equations of motion. The simplest version of these equations has no friction and no nonlinear terms. Of course, we also assume hydrostatic balance. The results, when taken with the conservation of mass or continuity equation written in shallow water equation form, are the three equations (see section 3.3, repeated here for convenience)

$$\frac{\partial u}{\partial t} - fv = -g\frac{\partial \eta}{\partial x},$$

$$\frac{\partial v}{\partial t} + fu = -g\frac{\partial \eta}{\partial y},$$

$$\frac{\partial \eta}{\partial t} + h\left(\frac{\partial u}{\partial x} + \frac{\partial v}{\partial y}\right) = 0.$$

for the three unknowns u, v and η, where for simplicity at this stage, the depth h has also been assumed constant. The first two equations are geostrophic balance with the time dependent term included. These are sometimes called *quasi geostrophic balance*, but all three are called the LTE. In fact, they are due to Lord Kelvin (Sir William Thompson) (1824–1907); Laplace's 1776 version is in full spherical co-ordinates and is much less palatable. The first two equations are quasi geostrophic balance, where the horizontal pressure gradients have been put in terms of a sea surface slope via hydrostatic balance. The third equation was derived in Chapter 3 and is the conservation of mass in a sea of depth h and surface elevation $\eta(x, y, t)$, measured from mean sea level. The notation (u, v) for horizontal current, f for Coriolis parameter and (x, y) for Cartesian axes x-East and y-North has been retained. In tidal modelling it is reasonable to assume that all motion is at a single frequency. Although there is more than one frequency, the equations are linear, so working with one, calculating the outcome, then adding results from a similar calculation using a different frequency is valid. It is not easy to eliminate the variables u, v in favour of η unless you are mathematically minded, but if this is done using the first two equations, the third equation becomes

$$\nabla^2 \eta = \frac{1}{gh}\left[\frac{\partial^2 \eta}{\partial t^2} + f^2\eta\right], \qquad (6.1)$$

which is a wave equation with wave speed $c = \sqrt{gh}$. This equation also permits solutions of the form

$$\eta(x, y, t) = p_1(x, y) \sin(ft) + q_1(x, y) \cos(ft)$$

provided the amplitudes p_1 and q_1 satisfy $\nabla^2 p_1 = 0$ and $\nabla^2 q_1 = 0$. These are called *inertial waves* and only exist in an unbounded ocean where they are a special case of gyroscopic waves. They are of some theoretical interest but a deep discussion would be out of place here. Such waves are best discussed in the context of a stratified sea where there are density changes with depth; then these waves, along with internal waves, can be put in a better frame. Internal waves will be introduced later in Chapter 10, meanwhile those interested in a more formal treatment are directed to Chapter 4 of Greenspan's classic textbook (Greenspan, 1968) and the slightly more recent book by Leblond and Mysak (1978); be warned, however, that these texts are heavily vectorial. Typically of Coriolis dominated dynamics, a vector treatment is essential to capture the directions as nothing remains in a straight line. As far as we are concerned, they are definitely not tides, so we move on. For the modelling of tides we assume that the wave is at a single frequency ω, this time utilising the complex form $e^{i\omega t}$ rather than trigonometry. This is done not to be obtuse, but because not to do so more than doubles the quantity of algebra. Using the exponential form also means that any phase can be incorporated in a constant multiplier. This is not so if trigonometric functions and real quantities are used. With the exponential form, the time derivatives disappear as follows

$$\frac{\partial \eta}{\partial t} = i\omega\eta, \quad \frac{\partial u}{\partial t} = i\omega u \quad \text{and} \quad \frac{\partial v}{\partial t} = i\omega v.$$

Working with the first two of the original three LTE, this leads to the pair of equations

$$i\omega u - fv = -g\frac{\partial \eta}{\partial x},$$

$$i\omega v + fu = -g\frac{\partial \eta}{\partial y}$$

upon substitution. Whence, solving the algebraic simultaneous equations for u and v, we obtain

$$u = \frac{g}{\omega^2 - f^2} \left\{ i\omega \frac{\partial \eta}{\partial x} + f \frac{\partial \eta}{\partial y} \right\},$$

$$v = \frac{g}{\omega^2 - f^2} \left\{ -f \frac{\partial \eta}{\partial x} + i\omega \frac{\partial \eta}{\partial y} \right\}.$$

This actual calculation has already been performed in a different context in Chapter 5. Substituting into the third of the LTE gives a single equation for η, the Helmholtz wave equation

$$(\nabla^2 + \lambda^2)\eta = 0,$$

where $\lambda^2 = \frac{\omega^2 - f^2}{gh}$. The same partial differential equation is satisfied by u and v, and the same equation is obtained by inserting

$$\frac{\partial^2 \eta}{\partial t^2} = -\omega^2 \eta$$

in equation (6.1) which is gratifying and serves to check the mathematics. In some ways, the most obvious solutions to seek are the plane wave solutions

$$\eta(x, y, t) = A e^{i(kx + ly + \omega t)},$$

where k, l are wave numbers and thus real, but A the amplitude can be complex to accommodate phase. The velocity (u, v) will also be of this form and the Helmholtz equation will give

$$k^2 + l^2 = \lambda^2 = \frac{\omega^2 - f^2}{gh}. \tag{6.2}$$

These kind of waves are called Poincaré waves after the "monster" of mathematics, the Frenchman Henri Poincaré (1854–1912), said to be the last man to know the whole of mathematics. There is nothing special about these plane waves except perhaps their dispersion relation, which is worth some attention. Equation (6.2) which is the dispersion relation, gives the allowable wave numbers and thus wavelengths for a given frequency. In wave number space, this equation gives these curves as circles centred the origin of radius λ. Thus any point on this circle represents a possible Poincaré wave. When the

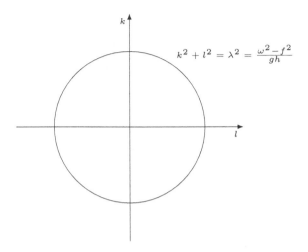

Fig. 6.5   The dispersion curve for Poincaré waves.

frequency of the tide ω is the same or very close to the value of the Coriolis parameter f, then $\lambda = 0$ and no waves are possible under this model. In reality, of course, the model breaks down and effects such as nonlinearity will become important.

6.4.1   *Kelvin waves*

A very special solution to the LTE are Kelvin waves. They are reasonably easy to derive: Kelvin waves are a gravity wave that is distorted by the Coriolis acceleration. It is an important model for coastal tides and, hence, certainly belongs in a text primarily written for offshore and coastal engineers. Kelvin waves are now derived in terms of a simple mathematical model. The basic equations remain as described above and symbolically are

$$\frac{\partial u}{\partial t} - fv = -g\frac{\partial \eta}{\partial x},$$

$$\frac{\partial v}{\partial t} + fu = -g\frac{\partial \eta}{\partial y},$$

$$\frac{\partial \eta}{\partial t} + h\left(\frac{\partial u}{\partial x} + \frac{\partial v}{\partial y}\right) = 0.$$

All nonlinear terms and friction terms remain ignored. The idealised perfect Kelvin wave has an infinitely long coast on its right as it propagates (left in the southern hemisphere). Thus, we first assume that the wave is at a single frequency ω, again utilising the complex form $e^{i\omega t}$ rather than trigonometry. Exactly as before — a single equation for η the Helmholtz wave equation

$$(\nabla^2 + \lambda^2)\eta = 0,$$

where $\lambda^2 = \frac{\omega^2 - f^2}{gh}$ — is obtained. The same partial differential equation is still satisfied by u and v.

Now we are ready to impose the fundamental Kelvin wave assumption. There is no flow perpendicular to the coast, which with our co-ordinate system means that $u = 0$. There are many ways to extract the mathematical form of the Kelvin wave; my particular favourite is to note that if $u = 0$, then

$$i\omega \frac{\partial \eta}{\partial x} = -f \frac{\partial \eta}{\partial y} \qquad (6.3)$$

or

$$-\omega^2 \frac{\partial^2 \eta}{\partial x^2} = f^2 \frac{\partial^2 \eta}{\partial y^2}$$

upon differentiation. Since $(\nabla^2 + \lambda^2)\eta = 0$, which written out in full is

$$\frac{\partial^2 \eta}{\partial x^2} + \frac{\partial^2 \eta}{\partial y^2} + \frac{\omega^2}{gh}\eta - \frac{f^2}{gh}\eta = 0,$$

the following two equations for η must hold

$$\frac{\partial^2 \eta}{\partial y^2} + \frac{\omega^2}{gh}\eta = 0,$$

$$\frac{\partial^2 \eta}{\partial x^2} - \frac{f^2}{gh}\eta = 0.$$

In effect, demanding that $u = 0$ forces us to look for solutions of $(\nabla^2 + \lambda^2)\eta = 0$ in the form of pairing off the four terms to give one equation in x (with exponential solution) and one equation in y (with sinusoidal solution). At this juncture, the calculations move from exponential with complex numbers to trigonometric and real exponential with real numbers. In subsequent calculation, therefore, only the real parts have significance, and we can thus write the Kelvin wave solution as

$$\eta_1 = C_1 e^{-fx/\sqrt{gh}} \cos\left(\frac{\omega y}{\sqrt{gh}} + \omega t + \phi_1\right),$$

where there is a coast at $x = 0$ (y axis), C_1 is a constant amplitude and ϕ_1 a constant phase. The Kelvin wave is shown in Figure 6.6.

If there is another coast at $x = b$ (for example, the width of the northern North Sea is about 450 km. So, in this case, $b = 450$ km.) then a Kelvin wave for that coast would be

$$\eta_2 = C_2 e^{f(x-b)/\sqrt{gh}} \cos\left(\omega t - \frac{\omega y}{\sqrt{gh}} + \phi_2\right),$$

where C_2 is another constant amplitude and ϕ_2 another constant phase. Writing $\phi_1 = \phi_2 = 0$ and $\eta_1 + \eta_2 = \eta$ gives a reasonable model for Kelvin waves in a canal. With $C_1 = C_2$ these Kelvin waves have the same amplitude and midway along the channel at $x = b/2$ there will be amphidromic points at

$$y = \frac{\pi}{2} \frac{\sqrt{gh}}{\omega} (2n + 1) \quad \text{for } n = 1, 2, \ldots,$$

since at these locations $\eta_1 + \eta_2 = 0$ for all t. The northern part of the North Sea and similar channels do exhibit this kind of tide, however the amplitudes of each Kelvin wave equivalent to C_1 and C_2 are seldom equal, so the amphidromes are typically far from symmetrically placed. They can, in fact, be absent altogether, especially if b is small; this is because for narrow gulfs the Coriolis effect can be neglected and the tide gets close to a standing wave of the form

$$C \cos\left(\frac{\omega y}{\sqrt{gh}}\right) \cos(\omega t).$$

Kelvin waves are a useful model for coastal tides, whereas Poincaré waves are not as they cannot easily accommodate coasts that demand zero flow perpendicular to a coast. Physically, it is the direction of the Earth's rotation that makes the sea pile to the right as the Kelvin wave travels. Mathematically, it is equation (6.3) from which it can be seen that if $f < 0$ (southern hemisphere) then

$$\eta = C_3 e^{fx/\sqrt{gh}} \cos\left(-\frac{\omega y}{\sqrt{gh}} + \omega t + \phi_3\right)$$

(C_3 a third amplitude constant and ϕ_3 a third phase constant) is the correct solution and the sea piles to the left as the wave travels.

6.5 Numerical Models of Tides

In this section, it is the numerical prediction of tidal elevation and tidal current over a sea that is being discussed, not the elevation of water level at a port that is successfully done through harmonic analysis. In order to predict tidal elevations and the currents that arise from tides, one builds a numerical model based on a grid designed to cover the desired basin, sea or coastal area and usually incorporating one of the Arakawa grids outlined in Chapter 5. Let us look at some of the processes it is essential to incorporate in a successful tidal model. Perhaps the best place to start is to see what can be ignored. First, it can be assumed that there is no weather. Wind and pressure effects that arise from the weather can always be added later, and it is much easier to validate a purely tidal model. Wind and pressure are not essential ingredients of a tidal model, although everyone is aware of their importance to understanding flooding during storm surges. The nature of a tidal model means that the effects of weather can be superposed later because the essential dynamics of tides are linear to the extent that there is very little dynamic interaction between wind and tide. Tides themselves, however, can exhibit highly nonlinear characteristics. One only needs to think of shallow water where drying can occur, or peninsulas where a tidal current can change sharply in direction. Both of these scenarios have length scales within them that render the advective term large enough to be important. When this term is important, the model is described as nonlinear. However, in the modelling of tides there is no non-linearity connected with wind or pressure. Second, unless internal tides are of concern, it is reasonably safe to ignore changes in density. In coastal seas, density changes are normally due to freshwater input via rivers rather than to temperature contrasts, although summer thermoclines do exist. Again, preliminary tidal modelling can be done accurately without involving these density fronts or pycnoclines. This is due principally to the difference in length scales associated with these density interfaces.

If the prime concern is modelling tidal elevation, certainly the most important factor for shipping, then the model must include the following factors: particle acceleration (because tide is a wave, albeit a very long one, and the waves have a to-and-fro motion which implies acceleration); Coriolis acceleration (because the length scale of the waves is, in general, large enough for the rotation of the Earth to be important); sea surface slope (this slope is alone responsible for the horizontal pressure gradients that act as a horizontal force); and friction. This last term needs some explanation. Astronomical forces generate the tides, but as far as the seas of the continental shelf are concerned, they are driven by the oscillations of a larger neighbouring body of water (the ocean). What one has, therefore, is the standard set of partial differential equations, the LTE perhaps modified by the inclusion of nonlinear terms and friction but discretised using finite differences. The friction is essential in a numerical model to dissipate the momentum at the bed, otherwise the solution to the equations will become too energetic — there is nowhere for it to go. At the open boundary, there is an incoming wave representing the oscillation of the neighbouring body of water. The actual scheme used is up to the user, but it is usually semi-implicit. The scheme used here dates from 1983 and is the classic numerical compromise between stability and truncation error. There are far more accurate numerical grids available — see Figure 5.5 in Chapter 5 — but this one is better for teaching purposes. Tidal models are, perhaps surprisingly, usually started from cold, that is a still region of sea is subjected to the open boundary oscillation and the incoming wave is allowed to propagate, refract and reflect and radiate out until such time as the model settles down to a "steady state". One can perhaps now appreciate more the importance of the radiation condition at the open boundary. Without one that works well, models like this will not operate. As mentioned, the only mechanism available to dissipate the energy that is being so liberally pumped in via the continental shelf is friction, and this frictional dissipation has to take place at the sea bed. Normally, a quadratic law relating the drag to the square of the speed provides an adequate model that can also be justified on dimensional grounds. If finite elements are chosen, then the procedure of integrating through each element and adding up contributions from all the elements means that the periodic forcing actually effectively acts as a body force. The act of integration has this effect; the limits of the integral are the boundary condition, and once the integration is performed, the variables evaluated at these limits are explicit in the equations.

Physically, the scenario is then rather like forced SHM, where a system is forced by an oscillation and the response exhibits the same frequency with the addition of harmonics. The harmonics will be due to non-linearities, either the advection terms from which M_4 tide stems from M_2 parent tide, or friction which can give rise to M_6. (M_2 is the semi-diurnal lunar tide.) In a purely linear model, no harmonics can arise and the model will give the M_2 generated currents and elevations throughout the modelled region and perhaps a natural frequency or two (see section 3.6). The natural frequencies of seas are too low to feature, however in a bay or inshore water they *are* important and are called seiches. Apart from very near the coast and where the sea is exceptionally shallow, linear tidal models with linear frictional dissipation at the sea bed are successful. Those that incorporate quadratic friction are even more so — so good in fact that they have not been surpassed by models with more elaborate sea bed friction. The model of the North Sea is a good example of real life numerical modelling as it combines tide with wind and pressure in a successful attempt to model actuality.

Figure 6.4 shows the accepted picture of co-tidal and co-range lines for the M_2 tide in the North Sea. Figure 6.7 shows the numerical simulation of this for the North Sea, and surrounding area. The net picture is, loosely, a wave entering the North Sea, keeping the coast on its right and exiting along the Norwegian coast. The wave is often very close to the theoretical Kelvin wave, especially along the Scottish

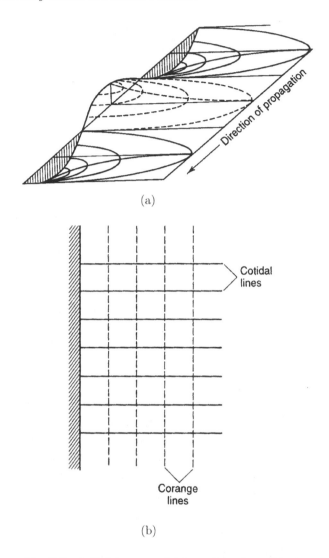

(a)

(b)

Fig. 6.6   A Kelvin wave: (a) isometric view; (b) plan view.

and Northumbrian coasts. The wave gets distorted, however, as it travels around the North Sea, both by irregularities in the sea bed topography which cause refraction and by irregularities in the coast which cause reflection, and of course by frictional effects. A purely tidal model does predict the observed tide in the North Sea, as can

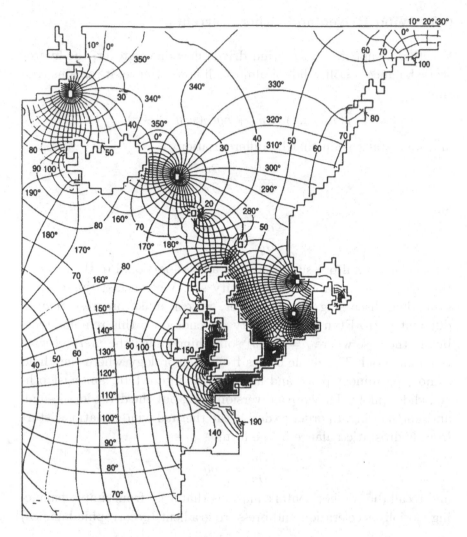

Fig. 6.7 The output from a numerical tidal model.

be seen by this figure. The numerical calculations were done on a grid of spacing $\frac{1}{3}^{\circ}$ latitude by $\frac{1}{2}^{\circ}$ longitude and the amphidromes are reproduced with reasonable accuracy. We return to this model of tides when considering storm surges later. Other non-tidal currents (mainly, but not entirely, due to wind) are discussed next.

6.6   Wind Driven and Other Currents

Whether or not it is tides, wind-driven flow or whatever, the hydro-static balance is still valid for almost all flows. Hence, it remains true that

$$p(x, y, z, t) = p_A + \rho g(\eta(x, y, t) - z),$$

and we restate again that by differentiation

$$\frac{\partial p}{\partial x} = \rho g \frac{\partial \eta}{\partial x},$$

and

$$\frac{\partial p}{\partial y} = \rho g \frac{\partial \eta}{\partial y}.$$

Hence, horizontal pressure gradients are proportional to the slope of the sea surface in a constant density sea. We reached this conclusion when discussing tides, however, we can make progress along a different path. Even if the density depends on z this remains true, but in the case where $\rho = \rho(x, y)$ something called the *thermal wind equations* hold. The name comes from meteorology where they hold a more prominent place and are of great interest to, amongst others, glider pilots. They are a diversion here, but the diversion is small and worth doing. In order to derive the thermal wind equations, start from hydrostatic balance in the form

$$\frac{\partial p}{\partial z} = -\rho g$$

and recall the two horizontal equations that arise from simply balancing Coriolis acceleration and pressure gradient (geostrophic balance)

$$-fv = -\frac{1}{\rho} \frac{\partial p}{\partial x},$$

$$fu = -\frac{1}{\rho} \frac{\partial p}{\partial y}.$$

Assuming that the density ρ depends on x and y, carefully differentiating each equation with respect to z gives

$$-f \frac{\partial v}{\partial z} = -\frac{\partial}{\partial z} \left(\frac{1}{\rho} \frac{\partial p}{\partial x} \right)$$

$$= -\frac{1}{\rho}\frac{\partial}{\partial x}\left(\frac{\partial p}{\partial z}\right) - \frac{\partial p}{\partial x}\cdot\frac{\partial}{\partial z}\left(\frac{1}{\rho}\right)$$

$$= -\frac{1}{\rho}\frac{\partial}{\partial x}(-\rho g) - \frac{\partial p}{\partial x}\left(-\frac{1}{\rho^2}\frac{\partial \rho}{\partial z}\right)$$

$$= \frac{g}{\rho}\frac{\partial \rho}{\partial x} + \frac{1}{\rho^2}\frac{\partial p}{\partial x}\frac{\partial \rho}{\partial z}$$

and substituting for $\partial p/\partial x$ gives

$$f\frac{\partial v}{\partial z} = -\frac{g}{\rho}\frac{\partial \rho}{\partial x} - \frac{fv}{\rho}\frac{\partial \rho}{\partial z} \approx -g\frac{\partial}{\partial x}(\ln \rho).$$

Similarly

$$f\frac{\partial u}{\partial z} = \frac{g}{\rho}\frac{\partial \rho}{\partial y} - \frac{fu}{\rho}\frac{\partial \rho}{\partial z} \approx g\frac{\partial}{\partial y}(\ln \rho).$$

The approximation can be avoided by differentiating on surfaces of constant pressure, but this is firmly meteorology and outside the realm of this text.

The most important implication of these equations is that a horizontal density gradient (salinity or temperature usually) induces a change in velocity with depth. Even more importantly perhaps, if both u and v are independent of depth, then there can be no horizontal density gradients. In the atmosphere, the presence of horizontal temperature gradients can produce updrafts which are particularly sought after by glider pilots. The thermal wind equations neatly express this. In the ocean, they can have local importance but it is the wind driven currents that are dominant.

So, let us assume a constant density but a driving wind that acts at the sea surface. The balance valid near the sea surface is one between the Coriolis acceleration and friction. This is called Ekman circulation; it is both useful and has historic importance and is derived below. The main influence of the wind obviously occurs close to the surface of the sea. However, the dynamics of the sea are complex and the fact that the sea is moving under the influence of the wind can, in turn, drive deeper flows. Thus, wind can cause upwelling which is of importance in some coastal regions. Coastlines are a barrier to currents and all motion is forced to flow parallel to them. Modelling currents along coasts tends to be specific; there is very little to be said of a general nature, except that again the overall dynamic balance remains between frictional forces and pressure gradient forces. When there are bends in the coastline on a scale of hundreds of kilometres, then it is usually prudent to consider, additionally, the effects of advective acceleration. An effective way of assessing whether or not various terms should or should not be included in a particular model of a specific sea is to utilise dimensional analysis, as outlined in Chapter 2. The first models of surface layer physics consisted of a simple balance between Coriolis acceleration and vertical transfer of momentum. From the dimensional considerations discussed at the end of section 3.8, this implies that the vertical Ekman number

$$E_v = \frac{\nu_v}{f D^2}$$

must be of order one. The values $f = 10^{-4}$ s^{-1}, $D = 10^3$ m and $\nu_v = 10^2$ m^2 s^{-1} are appropriate, which does indeed give $E_v = 1$. When actually solving equations, as distinct from using them for dimensional analysis, one has the choice of either working with dimensional or dimensionless variables. Both have their merits, but in the present context the use of dimensional variables is less confusing. The equations for steady Ekman flow, called the Ekman equations are

$$-fv = \nu_v \frac{\partial^2 u}{\partial z^2},$$

$$fu = \nu_v \frac{\partial^2 v}{\partial z^2}.$$

Before presenting the solution, one should ask what has happened to the pressure terms. If one writes the total current as $(u + u_g, v + v_g)$ then we have

$$-f(v + v_g) = -\frac{1}{\rho}\frac{\partial p}{\partial x} + \nu_v \frac{\partial^2}{\partial z^2}(u + u_g),$$

$$f(u + u_g) = -\frac{1}{\rho}\frac{\partial p}{\partial y} + \nu_v \frac{\partial^2}{\partial z^2}(v + v_g)$$

and we set (u_g, v_g) to balance the pressure gradient. Therefore, since (u_g, v_g) does not depend on z, (u, v) satisfy the Ekman equations as desired.

As the Ekman equations are differential equations, boundary conditions are required for a complete solution. Let us look at what these might be. Ekman flow is driven by the wind, therefore far beneath the sea surface (u, v) will tend to zero. Remember this is as z tends to $-\infty$ as the z-axis points upward. At the sea surface, one makes the assumption that the stress is the same as the wind stress. So, if the wind stress is given by the horizontal vector (τ^x, τ^y) then

$$\tau_{31} = \tau^x, \quad \tau_{32} = \tau^y.$$

Hence, at the sea surface

$$(\tau^x, \tau^y) = -\rho \nu_v \left(\frac{\partial u}{\partial z}, \frac{\partial v}{\partial z} \right).$$

With the other assumptions already made (neglecting the nonlinear advection terms arising from assuming small Rossby number), the sea surface can be taken as $z = 0$. This is sometimes called the "rigid lid" approximation and amounts to the suppression of surface waves. Any sea surface slope due to geostrophic effects is included in (u_g, v_g) and may be added later. The Ekman equations plus boundary conditions form a boundary value problem with a unique solution. In particular, a unique solution which can be written in closed analytical form as follows

$$(u, v) = V_0 e^{\pi z / D_0} \left(\pm \cos \left(\frac{\pi}{4} + \frac{\pi}{D_0} z \right), \ \sin \left(\frac{\pi}{4} + \frac{\pi}{D_0} z \right) \right),$$

where the constants V_0 and D_0 have been introduced for convenience and are defined as follows

$$V_0 = \frac{\pi}{\rho D_0 |f|} \sqrt{2(\tau^x)^2 + 2(\tau^y)^2},$$

$$D_0 = \pi \sqrt{2\nu_v / |f|}.$$

$|f|$ is the absolute value (magnitude) of f which accounts for whether we are north or south of the equator. The plus sign is taken north of the equator, the minus sign south of it. The constant V_0 is termed the "total Ekman surface current" and the constant D_0 is called the "depth of frictional influence". Both terms are reasonably self explanatory. Here, we have more or less followed Pond and Pickard

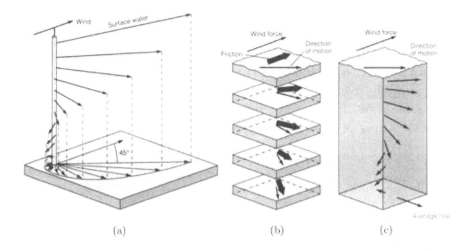

Fig. 6.8   The Ekman spiral.

(1983) and the Ekman spiral is shown in Figure 6.8. There are equally valid solutions that differ in detail (positions of $\sqrt{2}$ or π) but the essentials are the same. Now that we have a solution, we can validate it via various observations.

The classical reason for the development of the Ekman layer model was in response to observations made by the Norwegian biological oceanographer, Arctic explorer, statesman, humanitarian and now national hero, Fridtjof Nansen (1861–1930) whilst he was icebound during a three year (1893–1896) voyage. It was an heroic attempt to be the first to reach the North Pole. His ship (the Fram) was specifically designed to withstand the pressures of the ice as it froze; in fact the freezing water lifted the Fram so that the ship rode on its frozen surface. He got close but did not succeed in reaching the pole. The Fram is now preserved in a museum across the water from the city of Oslo with easy access by ferry and is well worth a visit. One observation that surprised Nansen was that the surface ice movement did not follow the driving wind but seemed to move to the right of it. Nansen suggested between 20° and 40°. The Ekman model gives 45°. It also gives a net flow 90° to the right (left) of the wind in the northern (southern) hemisphere. This feature can be used to explain why a flow along a coast can give rise to vertical currents at the coast (upwelling), which is important for biological productivity. For a west facing coast, a wind along the coast from the south in the southern hemisphere or from the north in the northern hemisphere will induce the surface flow to be offshore according to the Ekman model. In order to replace this surface water that moves out towards the sea, water which is usually colder has to rise from the depths. This is the upwelling and this cold water, being nutrient rich, provides food for local plankton which in turn feed the fish. This model of upwelling first saw the light of day in 1908 and, although there are considerably better, more complex models, the underlying reason behind upwelling in places like Nigeria and Peru is still very much the Ekman picture. The depth of frictional influence (D_0 above) is of the order of 10 or 20 meters in the mid latitudes, so Ekman flow (if it is a valid model) will describe flow very near and at the sea surface. Ekman spirals are seldom observed; in fact they were discounted as just an ideal picture, almost fictional, for years until the more sensitive non-invasive instruments (laser doppler anemometry) of the 1970s could pick them out. In general, the surface waves and unsteady nature of the wind will destroy Ekman layer structure. A three-dimensional diagram of the solution to the Ekman equations in the northern hemisphere will reveal the current to be a spiral, in a direction 45° to the right of the wind at the sea surface and spiralling away from the direction of the wind the deeper one goes. All the while, the amplitude of the current diminishing until, at a depth corresponding to D_0, the current has all but vanished. The solution is often called the Ekman spiral for this reason. South of the equator $f < 0$ and so the spiral starts off 45° to the *left* of the wind at the sea surface and spirals away. One simple alternative is to propose a depth dependent eddy viscosity. This could be justified as representing the change in turbulence characteristics the further one gets from the sea surface. One simple model using a quadratic dependence was used by Dyke (1977) and predicted a smaller angle of deviation for the surface stress (10°). The total depth integrated current is always 90° for eddy viscosity models, no matter how the eddy viscosity depends on depth; see the following subsection. One should ask about different models of wind-driven currents. There are many. Oceanographers recognise the existence of a well mixed surface layer virtually always present in cooler climes and virtually always due to the action of the wind on the sea surface, and it is tempting to equate this with the Ekman layer. However, to do so would invariably be wrong as the base of the mixed layer, called the thermocline (more correctly, the pycnocline), is very stratified. It is an interface between well mixed surface water, and less well mixed deeper water, and since Ekman layer physics ignores all density changes it could not possibly say anything about where this interface might be. Commonly, if the Ekman layer occupies the top 20 meters of the sea, then the well mixed layer extends to about 100 or 200 meters. Models of the well mixed layer will include unsteady effects (Ekman's original model included the time dependent terms), more sophisticated turbulence models as well as thermal forcing and possibly advective acceleration. First, here is a simple bulk model that deduces the bulk flow, is consistent with Ekman dynamics, but does not require eddy viscosity assumptions.

6.6.1   *Ekman bulk flow*

The stress on a small volume of fluid was considered in Chapter 3, and as for pressure, it is the *gradient* of shear stress that exerts a net force, not the stress itself; therefore, retaining Ekman's assumptions apart from eddy viscosity, the steady horizontal equations of motion are

$$-f(v+v_g) = -\frac{1}{\rho}\frac{\partial p}{\partial x} + \frac{\partial}{\partial z}\left(\frac{\tau_{zx}}{\rho}\right),$$

$$f(u+u_g) = -\frac{1}{\rho}\frac{\partial p}{\partial y} + \frac{\partial}{\partial z}\left(\frac{\tau_{zy}}{\rho}\right),$$

where (τ_{zx}, τ_{zy}) is the horizontal shear stress anywhere throughout the depth of the sea. Consider that geostrophic balance takes care of the (u_g, v_g) and pressure terms; this was done above. It leaves us with

$$-fv = \frac{\partial}{\partial z}\left(\frac{\tau_{zx}}{\rho}\right),$$

$$fu = \frac{\partial}{\partial z}\left(\frac{\tau_{zy}}{\rho}\right),$$

where as before it is understood that this is just the Ekman flow. Other texts might call this (u_E, v_E). Let us now assume that the sea is very deep, deep enough for the wind induced shear stress to vanish far below the sea surface. Define

$$(U, V) = \int_{-\infty}^{0} (u, v) dz,$$

where (U, V) is the bulk flow, sometime called the net transport. Integrating these new Ekman equations from deep sea $(z = -\infty)$ to sea surface $(z = 0)$ gives

$$-fV = \left[\frac{\tau_{zx}}{\rho} \right]_{-\infty}^{0},$$

$$fU = \left[\frac{\tau_{zy}}{\rho} \right]_{-\infty}^{0}.$$

In the deep sea, the stress is zero, and at the sea surface we can equate it to the wind stress; therefore, these equations give

$$(U, V) = \left(\frac{\tau^y}{\rho f}, -\frac{\tau^x}{\rho f} \right).$$

This result has been arrived at without the use of eddy viscosity, but agrees with the result obtained by integrating the Ekman equations. The bulk transport has magnitude $|\tau|/\rho f$ and is 90° to the right of the wind in the northern hemisphere (left of the wind in the southern hemisphere). $|\tau| = \sqrt{(\tau^x)^2 + (\tau^y)^2}$. Often the density ρ is included in the definition of (U, V), in which case

$$(U, V) = \int_{-\infty}^{0} (\rho u, \rho v) dz$$

and we get the slightly simpler

$$(U, V) = \left(\frac{\tau^y}{f}, -\frac{\tau^x}{f} \right).$$

This latter definition is perhaps better as the dimensions of (U, V) are now $ML^{-1}T^{-1}$ and not just $L^2 T^{-1}$. Most think the net transport represents the transport of matter and, therefore, the inclusion of mass in its dimensions is more appropriate. This is a simple conceptual model and helps us to understand overall movement due to wind stress. Let us now turn to modelling a real phenomenon — one that is responsible for flooding, loss of property and, sadly, loss of life.

6.7   Storm Surge Modelling

In this section, we shall see how to put modelling to the very practical use of forecasting storm surges. The tides of the North Sea are largely M_2, that is they are due to the Earth revolving under the Moon. There are two high waters a day. The accepted pattern of co-tidal and co-range lines is shown in Figure 6.4. This pattern was established through measurement in the 1920s. The three points where the tide vanishes are called *amphidromic points* or *amphidromes*. The tide along the Northumbrian coast is very close to a perfect Kelvin wave, but distortion is introduced through the variation in the depth and the deviation of the coastline from straight. Frictional effects are also important and these are also absent from the pure Kelvin wave of course. A purely tidal model of the North Sea tides, based on a uniform grid of spacing $\frac{1}{3}^{\circ}$ latitude and $\frac{1}{2}^{\circ}$ longitude, gives the co-tidal and co-range lines of Figure 6.7.

This model uses a semi-implicit finite difference scheme, whereby a mesh is overlaid on the North Sea and variables computed according to an Arakawa B grid. Later models used the C grid. The model is essentially two dimensional as the surface elevation is of prime concern and not how the currents vary with depth. The predicted pattern of co-tidal and co-range lines in the North Sea is certainly acceptably accurate. In these tidal models, there must be a friction term in order to extract the energy of the tidal forcing that is continually being fed in at the open boundary. This friction usually takes the form of a quadratic law which we have met as an example of dimensional analysis. Alternatively, for three dimensional models one could use eddy viscosity, but a vertical integration can still preserve depth dependent information by the careful use of transformations, as done by Norman Heaps in 1971. Even with these sophisticated additions, a purely tidal model is of limited practical use, although an accurate tidal model is certainly an essential starting point. The one used to produce this figure is accurate in both amplitude and phase to 10%. In order to include meteorological effects, stress terms are fed in at the sea surface. The surface elevation predicted by the output of such models is, therefore, due to both tide and meteorology (pressure and wind). In order for the artificial open boundary at the shelf edge to have minimal effect, a *radiation condition* is imposed that ideally allows waves to exit freely from the domain of the numerical model with zero reflection. Any reflection at the open boundary would eventually render the solution over the entire domain inaccurate. Fortunately, the friction in the system prevents this, but effects can be minimalised by careful application of an optimal radiation condition. Finding such an optimal condition is difficult and still taxing the brains of numerical analysts. One of the first successful three dimensional models of the North Sea was built by Norman Heaps in the mid-1960s for the specific purpose of predicting storm surges from the knowledge of current tide, wind and pressure. With later developments, this model of storm surges in the North Sea can be considered successful and is used operationally — for example, in decisions as whether or not to employ the Thames barrier. The practicalities of storm surge prediction are summarised in the flowchart of Figure 6.9.

This particular chart is due to Norman Heaps. In the last 30 years or so there have been modifications to the storm surge model, principally by Roger Flather and his colleagues at the Proudman Oceanographic Laboratory. Present day models include a representation of nonlinear processes. There was an attempt to use statistical hindcasting with a view to longer term prediction, but this proved unreliable and is no longer operative. When the meteorological situation and the state of the tide (spring tide) combine in such a way as to make a storm surge even a remote possibility, the storm surge warning system swings into action. The UK Meteorological Office computer produces weather forecasts routinely and these are 24 hour, 48 hour and 72 hour predictions of the synoptic weather pattern over the UK. If the weather pattern seems likely to develop favourably, i.e. a low pressure centre deepening and becoming slow moving with its centre somewhere between the Shetland Islands and Norway, then the storm surge model is run. The spring tide high water is modelled

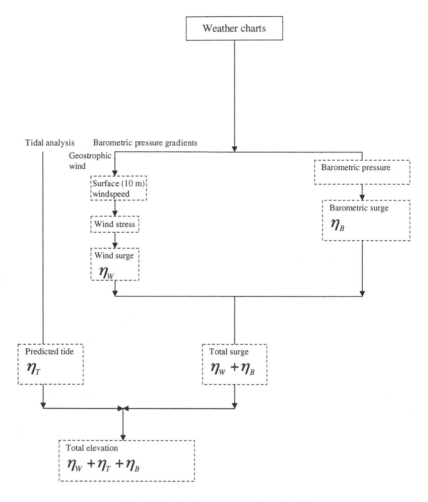

Fig. 6.9   A storm surge flowchart.

as it travels down the UK coast as a distorted Kelvin wave. It takes
about 12 hours to move from Orkney to the Thames so there is
reasonable notice for any action, which might include operating the
Thames barrier, distributing sandbags, issuing various warnings to
yachtsmen via coastguards etc. As the first model runs, later models
are implemented which have more up to date information on wind
and pressure and these have a smaller margin of error. Eventually, a
decision will have to be made as to whether the likely magnitude of

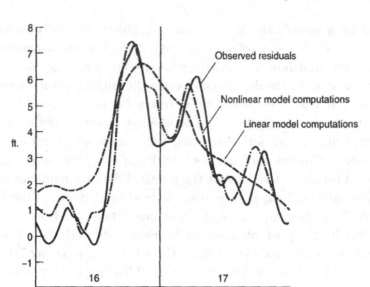

Fig. 6.10 Wind induced surge levels at Southend, 16–17 February 1962. From Banks (1974). Reproduced with permission.

the flooding is enough to warrant the expense and inconvenience of implementing the most severe flood prevention measures. It is always better to be safe than sorry. Figure 6.10 shows an example of how well such models can work. Storm surge modelling is a good, possibly unique, example of successful modelling.

6.8   Tsunami Modelling

Before the Boxing Day tsunami caused devastating damage in 2004, the word tsunami was restricted to specialist textbooks, but overnight suddenly everyone knew the word, if not the precise definition. Just as tsunamis were drifting out of the vocabulary, along came the equally devastating tsunami off the coast of Japan in 2011 and back the word came to swamp the news once more. The (Japanese) word "tsu-nami" actually means "harbour wave", implying a wave that was felt in the harbour but not further off-shore. This does indeed describe what a tsunami is, but we add that it is a wave caused by a seismic shock to the ocean. This is normally an earth-quake, but could be a volcano and might be, on rare occasions, an underwater landslide. Essentially what happens is this; the seismic disturbance under the deep ocean causes the entire several kilometres of ocean water to be displaced either upward or downward. This leads to a wave which when travelling to shallow water at the coast gets amplified and becomes the tsunami, which manifests itself as a wall of water. In Chapter 8 of the book by Pugh and Woodworth (2014) there is a lot more detail about the physical damage tsunamis cause. The phrase "tidal wave" was used instead of tsunami — probably due to the rather superficial resemblance a tsunami has to a tidal bore but it is a poor phrase that is inaccurate and should not be used. Figure 6.11 shows general relations between wave period, wave velocity and wavelength for ocean waves. This figure shows where the tsunami is on this spectrum, which will be seen again in Chapter 8. We can certainly use the linear wave model described in Chapter 3 to estimate the speed of travel of a tsunami when it travels over the deep sea. Of course, as it nears the coast and rears up to form a wall of water, linearity fails to describe it. Never mind — at least the time taken for a tsunami to get from the source to just offshore, to the edge of the continental shelf perhaps can be calculated using linear wave theory. Thus, we can use the formula $c^2 = gh$ to esti-mate the wave speed c where $g = 9.81$ m s^{-2} is acceleration due to gravity and h is an estimate of the local depth. Without being over precise, the depth of the ocean is usually between 3 km and 4 km so, using the formula, the speed c will be approximately 180 m s^{-1} or around 400 mph. This is very fast. If a seismic shock happens, say, 1000 km away then the wave travelling at this speed will only take 1.5 hours to reach you. Agreed, it will be slowed down by the decrease in depth at the shelf break, but it will still hit in under 2 hours. As earthquakes and the like are still unpredictable, around 2 hours is all the warning that can be expected. If you are on the beach, just get out of there and get to higher ground. You've got 2 hours. It is worth looking at the process that the tsunami goes through when it meets the continental shelf. This is best done in terms of energy as energy is not dissipated by such long waves; it is conserved. At

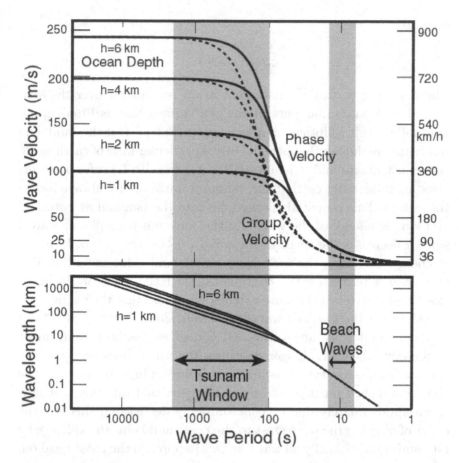

Fig. 6.11 *Top panel* Phase velocity $c(\omega)$ (solid lines) and group velocity $u(\omega)$ (dashed lines) of a tsunami wave on a flat earth covered by oceans of depths 1, 2, 4 and 6 km. *Bottom panel* Wavelength associated with each wave period, the tsunami window is marked. (Steven N Ward, University of Santa Cruz)

the site of the earthquake (say) the ocean surface is uplifted. For the Sumatran earthquake of 2004 that caused the devastating Boxing Day tsunami, the energy has been calculated to be 5×10^{15} J (J = Joules), see Pugh and Woodworth (2014). This is large enough but is only 0.5% of the available strain energy released by the quake itself. The tsunami then races away from the site at breakneck speed. The energy of the wave is initially entirely potential. It is given by the integral

$$\frac{1}{2}\rho g \int_S \eta^2 dS,$$

where η is the vertical displacement of the sea surface over the area S, and S denotes that part of the sea surface that is lifted up. S is usually a thin elliptical shape covering that part of the fault line where the earthquake happened. Only a particular kind of earthquake causes a tsunami and it is unusual to get one. To be safe, however, after an underwater earthquake, tsunami warnings are always issued these days. This potential energy goes into the tsunami as potential and kinetic energy of the wave. As the wave meets shallower water, since h decreases the speed of the wave $\sqrt{gh}$ decreases too. In terms of energy, therefore, the wave loses kinetic and gains potential. Put in layman's terms, on the continental shelf the front of the wave slows and the rest of the wave catches up, converting the energy and transforming the low, fast wave into a tall, slower wave — the "wall of water" that is so devastating. Of course, as has been said, linear water wave theory no longer describes the wave. However, consider the following scenario. A tsunami is approaching and has hit the continental shelf break at an angle. The part that hits first will slow down approximately due to $\sqrt{gh}$ suddenly decreasing; this has the effect of aligning the wave crest to the coast and helps to explain why tsunamis (and virtually all waves in fact) approach the coast head on. Wave ray theory can be used to be more precise. It is quite easy to use software to simulate the generation and propagation of tsunamis because the important aspects of them are dominantly linear. Precise modelling of tsunami run up, or getting the height correct to the nth degree is really fiddling while Rome burns. Figure 6.12 shows the output from a numerical model of the 2011 Japanese tsunami. The model is a simple two-dimensional shallow water equation solver which is quite satisfactory to simulate the propagation of the tsunami and its contact with the coast. It uses the equations in radial co-ordinates

$$\frac{\partial u_r}{\partial t} - fu\theta = -g\frac{\partial \eta}{\partial r},$$

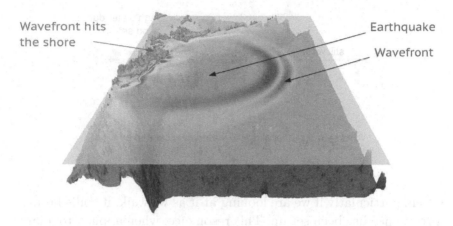

Fig. 6.12 The 2011 tsunami propagating from the earthquake, showing the impact at the Japanese coastline.

$$\frac{\partial u\theta}{\partial t} + f u_r = -\frac{g}{r}\frac{\partial \eta}{\partial \theta},$$

$$\frac{\partial \eta}{\partial t} + \frac{\partial}{\partial r}(hu_r) + \frac{1}{r}\frac{\partial}{\partial \theta}(hu\theta) = 0,$$

perhaps with $f = 0$, and analytically, the solutions are cylinder functions — the kind you get from throwing a stone into a still pond. Mathematicians express this solution is terms of Bessel functions, but no matter, solved numerically they look like Figure 6.12 and an analytical treatment with constant depth runs this solution close except at the coast itself. The analytical model would not precisely simulate the way the wave crashed against Japan and caused havoc; the numerical model does better, but is such precision necessary when the emphasis is on first getting clear then later cleaning up? Exercise 6.10.6 shows how to estimate how long a tsunami takes to arrive after the earthquake occurs, surely the priority for modellers.

6.9   Seiches

The strange word seiche is actually something quite familiar. When we carry a bucket of water or a cup of coffee — particularly if we do so up or down a flight of stairs — the liquid rocks back and forth.

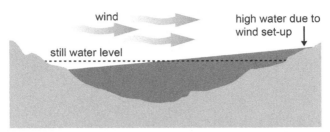

Wind set-up is a local rise in water level caused by wind.

Fig. 6.13   A seiche.

Often, particularly if we are looking at it as we walk, it spills because a resonance has been set up. This resonance, when applied to a large body of water such as a lake or a semi enclosed harbour, is called a seiche. How, one may ask, are seiches generated? The answer is by the wind. In particular, a low pressure centre moving across a lake can easily have a speed that exactly sets up a resonance. Figure 6.13 shows an example of this. Consider an idealised rectangular lake that has constant depth h and its sides are a and b with $a > b$. The smallness of the dimensions means that rotational effects (the Coriolis effect) can be ignored. Let us also assume the simplest friction free linear model. Define x and y as the usual axes through the middle of the lake that occupies $-b \leq x \leq b$, $-a \leq x \leq a$, and the following equations are the linear shallow water equations without rotation. η is the surface elevation and (u, v) the current:

$$\frac{\partial u}{\partial t} = -g\frac{\partial \eta}{\partial x},$$

$$\frac{\partial v}{\partial t} = -g\frac{\partial \eta}{\partial y},$$

$$\frac{\partial \eta}{\partial t} + h\left(\frac{\partial u}{\partial x} + \frac{\partial v}{\partial y}\right) = 0,$$

so straightforward elimination gives the two dimensional wave equation

$$\frac{\partial^2 \eta}{\partial t^2} = gh\nabla^2\eta = gh\left(\frac{\partial^2 \eta}{\partial x^2} + \frac{\partial^2 \eta}{\partial y^2}\right).$$

We now have the reasonably standard problem of solving the wave equation in two dimensions inside a rectangle. Although there are no

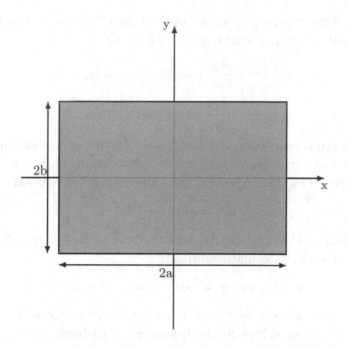

Fig. 6.14   The rectangular lake.

rectangular lakes, this is one of those excellent examples of modelling along the same lines as Stommel's 1948 model of ocean circulation. Even lakes that are only vaguely rectangular will exhibit resonances and lay themselves open to the possibility of seiches, so by considering the perfectly rectangular lake, see Figure 6.14, the mechanics of the seiche are isolated in the same way the mechanism for western intensification is isolated in Stommel (1948) with his idealised rectangular constant depth ocean. The solution to the problem is dictated by the domain and boundary conditions. Usually, for example when Kelvin waves were analysed earlier in this chapter, the exponential form of the solution was employed. This will not be done here as at the sides of the rectangular lake there has to be zero perpendicular flow. To solve this problem, therefore, choose instead the form

$$\eta = A \sin \omega t \cos kx \cos ly,$$

where ω is a frequency and k, l are wave numbers (see Chapter 3). This is ideal for standing waves and it is standing waves that produce seiches. This expression, when inserted into the top two equations above and integrated with respect to t, yields

$$u = -\frac{Agk}{\omega} \cos \omega t \sin kx \cos ly,$$

$$v = -\frac{Agk}{\omega} \cos \omega t \cos kx \sin ly,$$

where there is no constant of integration as this only contributes to the phase of the solution, which is not important when considering resonances. The wave equation gives the dispersion relation

$$\omega^2 = gh(k^2 + l^2).$$

At the sides of the rectangular lake there is no flow perpendicularly through the sides, so mathematically

$$u = 0 \quad \text{at} \quad x = \pm a; \quad v = 0 \quad \text{at} \ y = \pm b$$

so that $ka = n\pi$ and $lb = m\pi$, where m, n are integers. This gives a possible set of values to the frequency ω through the dispersion relation. Defining the period $T = 2\pi/\omega$, these values are

$$T = \frac{2}{\sqrt{gh}} \frac{1}{\sqrt{\frac{n^2}{a^2} + \frac{m^2}{b^2}}}.$$

Of course, the integers m and n can take on any integer value — well almost — they cannot both be zero else this violates continuity. If $m = 0$, then the solution becomes independent of y and one dimensional in x. The case $n = 1$ then gives the solution

$$\eta = A \sin \omega t \cos \frac{\pi x}{a}, \quad u = -\frac{Ag}{\omega} \cos \omega t \sin \frac{\pi x}{a}, \quad v = 0,$$

with the period

$$T = \frac{2a}{\sqrt{gh}}.$$

This is the fundamental period of this oscillation and is the seiche. The formula known as Merian's formula, after a paper by J. R. Merian in 1828, gives the frequency that can be estimated for any lake from its length ($2a$) and its depth (h). Of course, most lakes

Fig. 6.15   The shape of Loch Earn, Central Highlands of Scotland, UK.

are not such a convenient geometrical shape, but any long lake will have a resonant period dictated approximately by Merian's formula. The next question is how likely is it? For Loch Earn, see Figure 6.15, the rectangular approximation using Merian's formula gives seven minutes, whereas the resonant period from observation is 8.1. All the other oscillations are of course possible, and in a misshapen only vaguely rectangular lake there will be others, but it is the case that most of the energy will be in the fundamental mode as given by Merian's formula. The beauty of this kind of modelling is that it helps us understand why seiches exist. These fundamental oscillations are usually excited by a low pressure system passing over the lake (or loch) in an ideal way, so that the wind is along the long axis of the lake in one direction and reverses in direction just as the natural oscillation is also taking the oscillation in the same direction. It is the same phenomenon as pushing a child on a swing; if the push occurs at the same point of the child's oscillation then your push has the same frequency as the swing and resonance is enforced.

Away from lakes, seiches on the coast are also common. The harbour of Port Stanley in the Falkland Islands, South Atlantic, is virtually enclosed and subject to seiches that are meteorologically driven (see Pugh and Woodworth, 2014). Elsewhere, all around the Indian Ocean the inlets and harbours are subject to seiches, and they are not all due to meteorology. Some will be due to the local topography interacting with the tide and swell waves. Ocean swell is certainly the cause of short period oscillation in some Australian harbours that can harm the moored ships. Larger bodies of water are also not immune to seiche motion — the Adriatic has a seiche period of 21 hours and the Baltic a seiche period of 26 hours. Both are quite well duplicated by merian's formula; the discrepancy can be accounted for by the neglect of the Coriolis effect, certainly in the Baltic.

6.10　Exercises

(1) Starting with the equations

$$g\frac{\partial \eta}{\partial x} = -\frac{\partial \Omega}{\partial x} - \frac{1}{\rho}\frac{\partial p_a}{\partial x},$$

$$g\frac{\partial \eta}{\partial y} = -\frac{\partial \Omega}{\partial y} - \frac{1}{\rho}\frac{\partial p_a}{\partial y},$$

where η is the equilibrium surface, Ω is the tidal potential and p_a is atmospheric pressure. Find equations valid for (a) purely tidal effects and (b) purely atmospheric effects over the entire Earth.

(2) From the equation that describes a Kelvin wave, deduce an off-shore decay scale. Research the appropriate values to calculate the magnitude of this scale for the North Sea.

(3) Numerical models of tides in a coastal sea usually start from "cold". Explain what this means, and why is this the case when tides are clearly periodic?

(4) If in an Ekman driven flow the wind is from the south west, in what direction is the surface current and in what direction is the net transport (in the northern hemisphere)? If the wind stress has magnitude 0.3 N m^{-2}, the Coriolis parameter is 1.12×10^{-4} s^{-1}, the depth of frictional influence is 40 m and the density of seawater is a constant 1027 kg m^{-3}, calculate the magnitudes of the surface current and the net flux.

(5) What aspects of storm surges are not well represented by linear models?

(6) A tsunami results from an earthquake that takes place 200 km from the Japanese coast. The depth of the sea is 3 km but there is a continental shelf 50 km wide with a depth of 200 m. Estimate the time before the tsunami hits the coast, and comment on any errors there might be in your calculation.

(7) Given that Lake Baikal in Siberia has a length 665 km. and an average depth of 680 m, calculate an approximate resonant period for the lake. It is measured to be 4.64 hours, what is the percentage error in your approximation?

(8) How would Merian's formula be changed for a harbour of approximate length $2a$, rectangular and closed on three sides but open on one short side?

Chapter 7

Modelling Diffusion

7.1 Introduction

One of the clearest indications that there is human habitation on this planet is the presence of artificially produced material in the world's oceans. For mankind, this is of course of central importance — also of more immediate concern is the presence of such material in coastal seas and estuaries. Sadly, much of this material is often poisonous to some degree, as has already been mentioned in the first chapter of this book. The name *pollution* has been coined to describe foreign, often toxic, material in the sea. Once pollution is present, it does not remain unchanged but is pulled and pushed around by the currents and waves of the sea, spreading and (usually) diluting. This spreading takes place even if the sea were to be quiescent. The name of this process is *diffusion*. Molecular diffusion can be observed if a grain of potassium permanganate (purple) is placed in still water. A purple patch gradually grows. Of course, this growth is enhanced if there are currents present. The school experiment that demonstrates convection by dropping a crystal of potassium permanganate ($KMnO_4$) in a beaker of water being heated from beneath by a bunsen burner clearly shows thermally driven circulation, as well as enhanced spreading. Similarly, in the environment pollution can be made to spread effectively by the action of strong and usually variable currents. A word must be said here about the use of the word *dispersion*. Quite rightly, dispersion is used as a synonym for diffusion since diffusion is nothing if not the dispersion of foreign matter by waves and currents. However, dispersion has come to have a special meaning for physicists and applied mathematicians; it means the change (increase) in the wavelength of a wave, or group of waves, as it propagates and we have met this in Chapter 3. This spreading of the waves is dictated by the dispersion relation that gives the relation between wavelength, frequency and other relevant physical parameters. In view of this and subsequent possible confusion, the word dispersion will not be used as meaning general diffusion.

7.2   The Process of Diffusion

Although diffusion can and does take place at the molecular level at sea in the same way as in the laboratory, far more important is the diffusion of pollution through turbulence. Turbulence, the random commotion of water, is ideal for diffusing pollution and it can do it reasonably successfully. It can act in a similar way to molecular diffusion but at far larger scales and the effect can be up to 1000 times greater. As you may suspect, however, the story is by no means a simple random spreading but a combination of all kinds of different and complex mechanisms. We shall be trying to model some of these mechanisms in this chapter. Other effects, not in themselves diffusive, can greatly enhance or, in some cases, inhibit diffusion. At a shear, which is often present near a boundary, the unidirectional current varies from zero on the boundary itself to a large value only a short distance from the boundary. A shear is particularly efficient at diffusing any pollution. It is far more efficient, for example, than turbulence, which is suppressed near boundaries (solid ones at least). On the other hand, there are convergence zones in some flows (for example, Langmuir circulation). These convergence zones can act anti-diffusively, especially for buoyant contaminants. Their ability to re-concentrate hitherto dilute toxic substances is one of the main reasons for their study.

As far as modelling diffusion is concerned, one rather different species of model is the *particle tracking* model. Particle tracking or Lagrangian models are well suited to modelling the diffusion of particles because their *modus operandi* is to follow individual particles. Since the pollution can be simulated as a collection of marked fluid particles, these can be tracked by the model and various parameters (size of patch, location of its centre, etc.) can form the output at appropriate stages.

To the fluid mechanist, diffusion is merely a consequence of applying the laws of fluid motion. Granted, they need to be applied with precision, so that effects mentioned above such as turbulence and current shear near boundaries are adequately represented, nevertheless diffusive effects should appear as consequences. To the practical modeller of real diffusion of real pollution this is not useful. In real life, too much is happening for everything to be included, even in the most advanced of models. For example, an oil slick is one liquid (oil) interacting with another (seawater) in a complicated way. Even the oil is probably composed of several varieties which over time separate into liquids which have a wide variety of properties; tar is very different from light crude oil. Then there is all the chemically induced flow caused by reactions, the heat generated by these reactions and various combinations of these two effects. Biology, especially the microbiology of bacteria and cells, provides yet another possible source of flow. Accurate hydrodynamic modelling of this is still very far from possible.

However, the fluid mechanist's view is useful in one respect. That is, diffusion is a turbulent process that happens once a parameterisation of turbulence is included in the dynamic balance. The simplest model of diffusion is one in which diffusive transfer of material (through action) is directly related to the gradient of the concentration. The constant of proportionality governs how quickly the diffusion occurs and is called the diffusion coefficient.

7.2.1 *Fickian diffusion*

This kind of simple model can be used to predict the diffusion of outfall material due to sewage or industrial waste spillage in a river, provided the spillage is large enough, homogeneous enough, and provided the river flow is reasonably uniform. This model is termed Fickian (after Adolf Fick (1829–1901), who developed the idea in the mid-19th century).

Let us derive from first principles the effects of turbulence on a contaminant (pollutant). A much simpler dimensionally based argument will be given a little later. Suppose a unidirectional flow can be split into a mean flow and a fluctuation due to turbulence. In terms of symbols, let

$$u = \bar{u} + u',$$

where $\bar{u}$ is a mean and u' is a fluctuation about the mean ($\overline{u'} = 0$, of course). The pollutant has concentration c, and because this foreign matter is caught up in the local motion of the fluid, it too will have a mean and fluctuating part, so

$$c = \bar{c} + c'.$$

Of course, nothing can be said about the relationship between c' and u', $\bar{c}$ and $\bar{u}$ without further assumptions. All we know is that $\overline{c'} = 0$ too by *a priori* assumption.

For simplicity let us assume that the flow is uniform across an area A. This is not essential, but by doing so we avoid ugly looking and potentially confusing multiple integrals. Thus, the flux of contaminant across this area, which is perpendicular to the direction of the current u, will be Auc. The flux per unit area averaged over a suitable time is thus $\overline{uc}$. Some simple algebra reveals that

$$\overline{uc} = \overline{(\bar{u} + u')(\bar{c} + c')}$$
$$= \bar{u}\bar{c} + \overline{u'c'}.$$

The term $\bar{u}\bar{c}$ is called the *advection* of contaminant and might be what the intelligent layman would expect the flux of pollutant to be. The term $\overline{u'c'}$ is the *diffusion* of the contaminant and is due to the interaction between the two turbulent fluctuations. It is not surprising that such diffusion takes place, and for those with some knowledge of statistics, or who have read and understood Section 1.4, a parallel with covariance of the two random variables can be drawn. Having introduced advection and diffusion through a simple one dimensional model, we now move into three dimensions. If a cube of sea (or estuary for that matter) is considered, then the total amount of pollutant entering it must be the same as the total amount of pollutant that

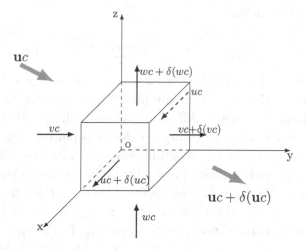

Fig. 7.1 The conservation of pollutant concentration c through a box of infinitesimal dimensions $\delta x, \delta y$ and δz.

leaves it. Calculus type arguments in a box of dimensions δx, δy, δz (see Figure 7.1 and Proudman (1953), Chapter 6) give, on taking appropriate limits

$$\left(\frac{\partial}{\partial t} + u\frac{\partial}{\partial x} + v\frac{\partial}{\partial y} + w\frac{\partial}{\partial z}\right)c = 0.$$

Now, writing $u = \bar{u} + u'$, $v = \bar{v} + v'$, $w = \bar{w} + w'$ together with $c = \bar{c} + c'$ and also noting that

$$\frac{\partial u}{\partial x} + \frac{\partial v}{\partial y} + \frac{\partial w}{\partial z} = 0$$

shows that in order that actual flows into and out of the cuboid are the same we must have

$$\left(\frac{\partial}{\partial t} + \bar{u}\frac{\partial}{\partial x} + \bar{v}\frac{\partial}{\partial y} + \bar{w}\frac{\partial}{\partial z}\right)\bar{c} + \frac{\partial}{\partial x}(\overline{u'c'}) + \frac{\partial}{\partial y}(\overline{v'c'}) + \frac{\partial}{\partial z}(\overline{w'c'}) = 0.$$

We now write

$$\frac{D}{Dt} = \frac{\partial}{\partial t} + \bar{u}\frac{\partial}{\partial x} + \bar{v}\frac{\partial}{\partial y} + \bar{w}\frac{\partial}{\partial z},$$

which is the usual notation for the total derivative (also called differentiation following the motion or Lagrangian derivative — outlined .

before). We also define

$$\overline{u'c'} = -\kappa_x \frac{\partial \bar{c}}{\partial x}; \quad \overline{v'c'} = -\kappa_y \frac{\partial \bar{c}}{\partial y}; \quad \overline{w'c'} = -\kappa_z \frac{\partial \bar{c}}{\partial z},$$

where κ_x, κ_y and κ_z are the diffusion coefficients. These three relationships are of the type originally proposed by Fick in 1855, but he was considering molecular diffusion. When dealing with turbulence induced diffusion it is necessary to define separate quantities for, as we have already seen in defining eddy viscosity (see Chapter 3), turbulence induced diffusion is usually by no means independent of time or space. We have defined three of them in the three different directions. The vertical diffusion coefficient κ_z will always be distinct from the two horizontal coefficients κ_x and κ_y. In the open sea, there may be arguments to support putting $\kappa_x = \kappa_y$, but in an estuary, for example, the diffusion along the line of the estuary will be much larger than the diffusion across the estuary.

With these Fickian assumptions, the equation for $\bar{c}$ can be derived as

$$\frac{D\bar{c}}{Dt} = \frac{\partial}{\partial x}\left(\kappa_x \frac{\partial \bar{c}}{\partial x}\right) + \frac{\partial}{\partial y}\left(\kappa_y \frac{\partial \bar{c}}{\partial y}\right) + \frac{\partial}{\partial z}\left(\kappa_z \frac{\partial \bar{c}}{\partial z}\right).$$

For some more detailed sensible discussion, see the textbook by Lewis (1997). With κ_x, κ_y and κ_z (unjustifiably) equal and constant, this reduces to what is known as the "advection–diffusion equation"

$$\frac{D\bar{c}}{Dt} = \kappa \nabla^2 \bar{c}.$$

Finally, if the advection terms are ignored, which may be justified for very slow flow, we obtain the standard diffusion equation

$$\frac{\partial \bar{c}}{\partial t} = \kappa \nabla^2 \bar{c}.$$

We shall return to look at the solution of the advection–diffusion equations a little later. However, let us first pause for some much simpler dimensional analysis.

If discussion is centred on the diffusion of momentum, the diffusion coefficient is our old friend eddy viscosity. The dimensions of the quantity are $L^2 T^{-1}$ (or perhaps density times this, $ML^{-1}T^{-1}$, which is the dynamic version). If, on the other hand, discussion focuses on a passive quantity, say temperature, salt or in fact any contaminant, then the amount of this passive quantity in the volume considered in Figure 7.1 must be conserved in its simplest which is considered free of sources and sinks, must be conserved. In its simplest form, this means that the time rate of change of the concentration of this passive quantity at any particular point must balance the spatial gradient of, not the concentration itself, but the agent that causes the diffusion of the concentration. In an open ocean or sea, this agent is usually the turbulence induced eddies. This agent is usually parameterised as being proportional to a constant times the concentration gradient (the Fickian assumption), as explained above. The balance is thus

time rate of change of (concentration of passive quantity)

= gradient of [agent that causes the diffusion of the

passive quantity].

In turn, the square bracket is a constant times the concentration gradient. If the passive quantity is labelled C, then in purely dimensional terms we have the dimensional version of the diffusion equation

$$\frac{1}{T}C = \frac{1}{L}k\frac{C}{L},$$

where k is the constant diffusion coefficient arising out of the Fickian assumption. Once again, therefore, it is seen that the dimensions of k are L^2T^{-1}. This is now seen to be independent of the identity of the diffused quantity.

We have seen the simplifications that have to be made in order to derive the diffusion equation itself. The standard textbook for the solution of the diffusion equation in all its attendant forms and under all kinds of boundary conditions is Crank (1975). The often quoted solution to the one dimensional diffusion equation in the form

$$\frac{\partial c}{\partial t} = \kappa\frac{\partial^2 c}{\partial x^2},$$

(where the overbar has been dropped for convenience), is

$$c = \frac{A}{\sqrt{t}}e^{-x^2/4\kappa t},$$

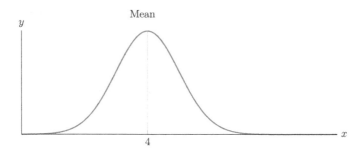

Fig. 7.2   Gaussian profile with mean at $x = 4$.

where A is an arbitrary constant. This solution describes the diffusion away from the source, which is concentrated at $x = 0$. The diffusion is one dimensional, hence is either a streak along the centre line of an estuary, or more realistically, a cross-stream average diffusion. The solution is not valid at time $t = 0$, except that it describes a ficti-tious infinitely strong source of infinitesimally small extent. (Those familiar with Dirac-δ functions may recognise an impulse when they see one.) At subsequent times, however, the (mean) concentration c has a Gaussian profile, see Figure 7.2.

Of course, this is exceptionally idealised. Even if the diffusion equation is valid, it is additionally necessary for idealised boundary conditions to hold. Nevertheless, there are certain useful deductions one can make. Certain advanced mathematical methods can be used to deduce that, no matter what initial distribution of concentrate, the spread is still at the same rate. The Gaussian model is valid in one dimension, in cylindrical geometry in two dimensions and in spherical geometry in three dimensions. The analytical solution in two dimensions is complicated, but the three dimensional solution is quite straightforward. The equation is

$$\frac{\partial c}{\partial t} = \kappa \left(\frac{\partial^2 c}{\partial r^2} + \frac{2}{r} \frac{\partial c}{\partial r} \right),$$

assuming spherical symmetry, with solution

$$c = \frac{A}{r\sqrt{t}} e^{-r^2/4\kappa t}.$$

In other words, there is simply an extra r in the denominator. Here, we have written $r^2 = x^2 + y^2 + z^2$; r is the distance away from a point source at the origin. There is little direct marine application of such point sources in three dimensions. It is background material. Let us get back to the general advection–diffusion equation

$$\frac{Dc}{Dt} = \frac{\partial}{\partial x}\left(\kappa_x \frac{\partial c}{\partial x}\right) + \frac{\partial}{\partial y}\left(\kappa_y \frac{\partial c}{\partial y}\right) + \frac{\partial}{\partial z}\left(\kappa_z \frac{\partial c}{\partial z}\right).$$

There is hardly ever any need for spherical geometry in marine science. However, modelling the spreading from a source on the sea surface could be a case for using cylindrical geometry, albeit only in two dimensions. One practical feature is that if pollution is spreading from an axisymmetric source in a river or estuary, and the coasts are far enough away not to have affected the behaviour of the contaminant, then the spreading does tend to be Gaussian. Assuming a variance σ^2 and a diffusion coefficient κ, it is possible to define κ by

$$\kappa = \frac{1}{2}\frac{d}{dt}(\sigma^2),$$

from which, on integration,

$$\sigma^2 = 2\kappa t.$$

This is consistent with the theoretical (one-dimensional) Gaussian distribution cited earlier. If advection is added, but in the simplest possible manner, then we arrive at an equation of the form

$$\frac{\partial c}{\partial t} + U\frac{\partial c}{\partial x} = \kappa\frac{\partial^2 c}{\partial x^2},$$

where U is the speed of the stream. This stream is constant and in the direction of the spread of the pollutant. This simulates the spread of pollutant in a flowing river. The solution is

$$c = \frac{A}{\sqrt{t}}e^{-(x-Ut)^2/4\kappa t}$$

and is not surprising. It represents a Gaussian profile being transported (advected) downstream with speed U. More realistic models involve representing the shear that is present in most estuaries and coastal seas. As a mechanism for diffusing a pollutant which is not yet uniformly distributed entirely across an estuary or throughout the depth of a sea, shear tends to dominate where it exists. We return to it later in this chapter, but meanwhile let us look at practical studies.

One of the most cited papers on diffusion was written by Akira Okubo, (Okubo, 1971). He, along with several other authors, recognised that oceanic diffusion occurs through many different mechanisms: turbulent eddies, shears of several origins, as well as biological agents. Obviously, it is not possible for a single model to simulate every diffusive effect. Okubo took on the task of examining many different measurements. In order to make some sense of all these data, it is useful (one could say essential) to provide a benchmark on which to measure and compare. Most experiments, certainly all those considered by Okubo, involve observing the spreading of a patch of dye. A patch of dye has a centre of mass and a distribution of mass about this centre. All distributions have a variance which measures how spread out they are. As time progresses, the patch of dye increases in size, spreading out. It seems sensible to ask, therefore, what relationship there is between this measure of spreading and time. The straightforward Fickian diffusion, which is diffusion by turbulent eddies in the absence of shears, can be represented exactly by the diffusion equation, as seen above and solved. The result of this solution is that variance is proportional to time. Okubo's results give that variance is proportional, not to time t, but to time raised to the power 2.3, $t^{2.3}$. The graph from Okubo's paper is given in Figure 7.3.

It can be seen from this graph that much of the data are scattered, and the line which corresponds to variance being proportional to $t^{2.3}$ is in reality a line of best fit or regression line. It will also be noticed that both axes are logarithmic, which enable a power law to become a straight line. However, here we are more concerned with the mechanisms of diffusion and trying to model them. Figure 7.3 exhibits large scatter, even larger if one considers the logarithmic axes. However, the same axes show a vast range on length and time scales over which the power law is approximately valid. The validity of this power law for small scales is very open to question. Okubo shows no data for scales of 10 m or less; even those that are shown are very scattered around the 50 m length scale. It is therefore reasonable

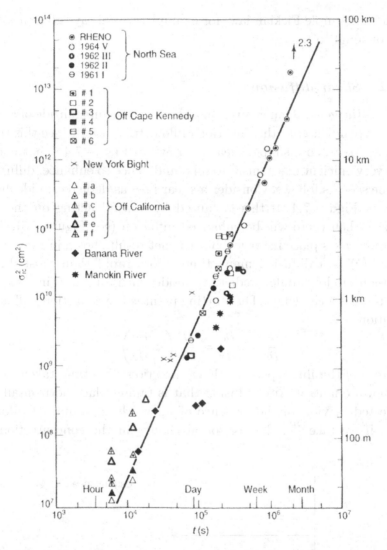

Fig. 7.3 Diffusion diagram for variance versus diffusion time. From A. Okubo (1971). Reproduced with kind permission from Elsevier Sciences Ltd., The Boulevard, Langford Lane, Kidlington OX5 1GB, UK.

to deduce that the diffusive mechanisms are most distinct at these very different length scales, and Okubo's power law, although very seductive, is too much of a simplification of reality. We look again at Okubo's model after the introduction of more theoretically based models. A simple Fickian law, for example, will always give variance proportional to t.

7.2.2   *Shear diffusion*

Shear is the general name given to a change in the magnitude of a current perpendicular to the direction of flow. It is easier to see this than to describe it, so a simple shear is drawn in Figure 7.4. That it can be a very efficient mechanism to mix, and hence to enhance, diffusion can be seen as follows. Consider a shear flow as depicted in idealised form in Figure 7.4. If this is caused by a wind acting on the sea surface, then there will be enhanced diffusion (see Figure 7.5). The diffusion takes place in stages, but the net result shown in Figure 7.5 (a) to (c) is a diffusion many times more rapid than most direct diffusion models would predict. To model such a process in a simple way is quite a challenge. The starting point is the advection–diffusion equation

$$\frac{\partial c}{\partial t} + u\frac{\partial c}{\partial x} = \frac{\partial}{\partial z}\left(\kappa_y \frac{\partial c}{\partial z}\right),$$

where u is the fluid speed in the x direction, this time taken as an unknown quantity. Any diffusion that is taking place horizontally is neglected in view of the presence of u, which is assumed to dominate. If we take $\frac{\partial c}{\partial x}$, the horizontal change in the concentration of

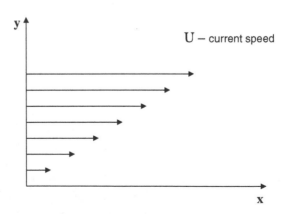

Fig. 7.4   A shear flow.

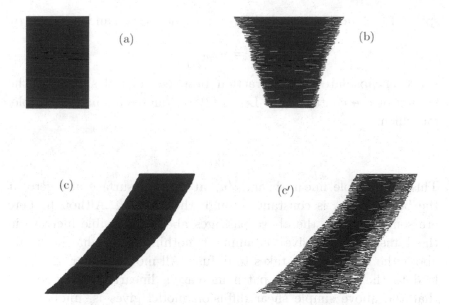

Fig. 7.5 (a) Vertically distributed pollutant, (b) pollutant diffused by a shear flow, (c) pollutant as at (a) but diffused by a shear, then (c') this sheared flow further diffuses both vertically and horizontally. This shows the enhanced {(a) to (c) then to (c')} mixing process.

contaminant, to be independent of depth, then it is permissible to integrate vertically between sea surface and sea bed. The right-hand side diffusive term integrates to zero provided there is no sea bed or sea surface input, which is certainly generally true. The result is therefore

$$\frac{\partial \langle c \rangle}{\partial t} + \langle u \rangle \frac{\partial c}{\partial x} = 0,$$

where $\langle c \rangle$ and $\langle u \rangle$ are depth mean values. If c and u are split into depth mean values, with the addition of variations from the mean, viz

$$c = \langle c \rangle + c',$$
$$u = \langle u \rangle + u',$$

where u' and c' are quantities representing the departures of u and c, respectively, from their depth mean values. Analysis can be performed on these equations (see Lewis, 1997). In particular, in the spirit of Fick, an "effective" diffusion coefficient κ_e can be defined by

$$\overline{u'c'} = -\kappa_e \frac{\partial c}{\partial x},$$

which encapsulates the interaction between vertical shear and the mixing of the contaminant. Lewis (1997) quotes a simple example, for which

$$\kappa_e = \frac{\langle u \rangle^2 h^2}{30\kappa_z}.$$

This is a simple linear shear ($2\langle u \rangle$ at the sea surface and zero at the bed) and κ_z is constant through the depth h. Although there are some limits in the above passages about a possible increase in the diffusion of a passive contaminant, nothing further has been said about the time a patch takes to diffuse. All models so far discussed lead to the variance of a patch increasing linearly with time. All that the above simple shear diffusion model gives is, under some circumstances, an enhanced diffusion coefficient. Suppose, however, we assume a constant value for the vertical shear $\frac{\partial u}{\partial z} = \chi_z$ say. This is consistent with the known character of flow close to the sea bed. With constant diffusion coefficients too, the advection–diffusion equation takes the form

$$\frac{\partial c}{\partial t} - \chi_z \frac{\partial c}{\partial x} = \kappa_x \frac{\partial^2 c}{\partial x^2} + \kappa_y \frac{\partial^2 c}{\partial y^2} + \kappa_z \frac{\partial^2 c}{\partial z^2}.$$

Under certain assumptions to do with partitioning of energy between large scale eddies that produce the shear and small scale eddies that merely contribute to diffusion, the variances of the solution to this advection–diffusion equation are

$$\sigma_x^2 = 2\kappa_x t + \frac{1}{6}\chi_z^2 t^3,$$

$$\sigma_y^2 = 2\kappa_y t.$$

The first terms on the right are to be expected, but the term $\frac{1}{6}\chi_z^2 t^3$ is due to the shear in the x (longitudinal) direction. The t^3 dependence here certainly indicates enhanced diffusion. The diffusion coefficient κ_e (an effective diffusion coefficient due to shear) defined earlier can be calculated to be proportional to t^2 under some circumstances.

7.2.3   *Homogeneous diffusion*

Let us go back a stage. Now that we have seen the kinds of models that are available, it is useful to return to a more fundamental approach and look again at the process of diffusion from first principles. It was Batchelor (1953) who first brought the ideas of the notable Russian mathematician and statistician A. N. Kolmogorov (1903–1987) to the general public. In particular, his work first published in 1941. The central idea is that in a field of turbulence locally all eddy characteristics are the same, independent of the direction considered. The name homogeneous, or isotropic, turbulence was coined to describe this view. Physically, energy is transmitted down from large scales to smaller scales until dissipation through viscosity and heat occurs. There is a well known parody due to Lewis Fry Richardson of Jonathan Swift's sonnet concerning fleas: "Big whorls have little whorls that feed on their velocity, and little whorls have lesser whorls and so on to viscosity" which admirably and succinctly describes the process. Turbulence can therefore be considered as consisting of a continuous range of eddy sizes. These eddies are fed by many sources. Tides are a particularly good source in estuaries and coastal waters, but there are others such as wind and solar energy. Energy will be input at various length scales. What is evident is that most of the important effects of turbulence occur well after the source that input the original energy has ceased to have any direct influence. Turbulence is a subject worthy of a substantial textbook by itself. The one by Hinze (1975), which is a classic, and the more recent Lesieur (1997) are particularly recommended. Let us proceed here by discussing turbulence in terms of variance, if only to allow easy comparison to the previous Fickian and shear models. If x is the separation of two particles of contaminant at time t, and u is the relative speed then, from the definition of variance, the following relationship holds between the variance of x at time t and its value at time $t = 0$

$$\overline{x^2} = \overline{x_0^2} + 2\int_0^T \int_0^{t'} \overline{\delta u(t)\delta u(t+\tau)}\,d\tau\,dt'.$$

The integral is a measure of agreement called the *autocovariance* of a signal with itself and is the same as the autocorrelation that we met when discussing Kalman filters in Chapter 2. The only difference between autocovariance and the autocorrelation is that the latter is normalised to take the value one when $\tau = 0$. As was mentioned in Chapter 2, it is zero for white noise where there is no relationship between a given value of the signal and any other, no matter how adjacent the second might be. It is constant for a deterministic function. This way of looking at diffusion is due to L. F. Richardson and the function that is a locally based variance is sometimes called a Richardson distance neighbour function.

If the diffusion time T is very small, the integrals can be dispensed with and, very nearly,

$$\overline{x^2} = \overline{x_0^2} + \overline{(\delta u)^2} T^2.$$

This corresponds to the particle's trajectories being straight lines. For very large times, the particles would be so far apart that they would behave individually. The following bit of reasonably simple mathematics sorts out what will be the result in this case. Let

$$\delta u = u_2' - u_1'$$

so that

$$\overline{\delta u(t)\delta u(t + \tau)} = \overline{(u_2'(t) - u_1'(t))(u_2'(t + \tau) - u_1'(t + \tau))},$$

from which the right-hand side becomes

$$\overline{u_2'(t)u_2'(t + \tau) + u_1'(t)u_1'(t + \tau) - u_1'(t)u_2'(t + \tau) - u_2'(t)u_1'(t + \tau)}.$$

In this expression, the last two terms are zero because the particles are so far apart that their speeds can be assumed to be uncorrelated. The first two terms are the same as there can be no distinction between these autocovariances. This is because by the very nature of autocovariance and from its definition there can be no distinction as over a long time they will, statistically speaking, have the same history. Hence, we have derived:

$$\overline{\delta u(t)\delta u(t+\tau)} = 2\overline{u'(t)u'(t+\tau)},$$

whence

$$\overline{x^2} = \overline{x_0^2} + 4 \int_0^T \int_0^{t'} \overline{u'(t)u'(t+\tau)}d\tau dt'.$$

Dimensional analysis has already been introduced. In particular, the Buckingham Pi theorem gave a method by which to infer functional relationships between variables. This can be usefully employed here. Let us assume that the time rate of change of x^2 suitably time averaged depends solely on the initial separation x_0, the diffusion time T, and the rate of transfer of turbulent energy ϵ. Having hijacked the usual dimensional symbol for time (T), for this section we will use T_1. The three variables x_0, ϵ and T only involve length L and time T_1 and not mass, so there is one dimensionless combination which, since

$$\epsilon \sim L^2 T_1^{-3}, \quad x_0 \sim L; \quad \text{and} \quad T \sim T_1$$

is

$$\epsilon x_0^{-2} T^3$$

or put another way, the reciprocal square root of this $x_0 \epsilon^{-1/2} T^{-3/2}$ is dimensionless. The view taken here is that of Kolmogorov in that the time rate of change of x^2 suitably time averaged does not depend on u, the separation speed, in the interesting range. This intermediate range, which is between the initial linear range and the range when every particle has a very large separation, is called the inertial sub-range. The deduction from dimensional analysis is that

$$\frac{d}{dt}\overline{(x^2)} = \epsilon T^2 F(x_0 \epsilon^{-1/2} T^{-3/2}),$$

where the function F is general. Some of you might need reminding that the time average is taken over a time scale appropriate to turbulent eddies whereas the outer time derivative refers to longer times; precisely this kind of averaging was done in Section 3.8 when we had to model turbulence without losing the time dependence at longer scales. It is like a moving average that slowly varies with time. Some special cases deserve individual consideration and are illuminating. If T is small, we return to a linear separation rate and so we must have

$$\frac{d}{dt}\overline{(x^2)} \sim \epsilon T^2 (\epsilon x_0^{-2} T^3)^{-1/3} = T(\epsilon x_0)^{2/3}.$$

Direct integration shows that for small T

$$\overline{x^2} \propto T^2.$$

For intermediate values of T we can assert that there will be no dependence on the initial separation x_0 and dimensional analysis thus forces the function F to be a constant from which

$$\frac{d}{dt}\overline{(x^2)} = \epsilon T^2$$

and so for intermediate T we have

$$\overline{x^2} \propto T^3.$$

For large values we return to a linear relationship

$$\overline{x^2} \propto T.$$

To summarise, therefore, we have a power law for the variance of a diffusing patch of contaminant

$$\overline{\sigma^2} \propto T^p.$$

If T is large, $p = 1$ and we have the standard Fickian diffusion. For intermediate values of T, $p = 3$ which, in the light of the previous section on shear diffusion, is consistent with the interaction between "shear" and the concentrate (pollutant) producing enhanced faster diffusion rates. For very small values of T, $p = 2$ and particles are simply moving away from each other in straight lines. Interestingly, since

$$\kappa = \frac{1}{2}\frac{d}{dt}(\sigma^2)$$

it can be shown that from the case $p = 3$ (intermediate T; most rapid diffusion) we can derive $\kappa = \sigma^{4/3}$, known as "Richardson's four-thirds law". In fact, Richardson proposed independently (Lewis Fry Richardson (1881–1953) was a *very* independent original thinker) that it was not really possible to distinguish between the flow and turbulent fluctuations. Richardson also deduced that the time over which quantities are averaged was crucial. The conclusion drawn was that the number of particles in a unit length, n, obeyed an equation

$$\frac{\partial n}{\partial t} = \frac{\partial}{\partial l}\left[K(l)\frac{\partial n}{\partial l}\right],$$

where l is the standard deviation of the particles from their mean position. κ is a Fickian diffusion, but dependent on l. It is only after considering a large range of length scales that Richardson deduced the following functional form for K

$$K = 0.2l^{4/3}.$$

This is consistent with Kolmogorov isotropic turbulence theory for intermediate times and hence reinforces both theories. Some further dimensional analysis is useful here. The turbulent kinetic energy spectrum $E(k,t)$ tells us how much turbulent kinetic energy is present at any given wave number and has dimension L^3T^{-2}. At these equilibrium ranges, the spectrum will only depend on the wave number k and the rate of dissipation of energy ϵ. Accepting this, we can use dimensional analysis given k has dimension L^{-1} and ϵ has dimensions L^2T^{-3}. As only two variables are involved, the elementary form of dimensional analysis can be used, that is the relating of powers without grouping variables into dimensionless clusters. Let

$$E \propto k^\alpha \epsilon^\beta,$$

in dimensional terms only, of course. As only L and T are involved, this is

$$L^3T^{-2} = L^{-\alpha}(L^2T^{-3})\beta$$

$$\text{giving} \quad 3 = -\alpha + 2\beta$$
$$\text{and} \quad -2 = -3\beta.$$

The solution of this is

$$\beta = \frac{2}{3} \quad \text{and} \quad \alpha = -\frac{5}{3}.$$

The result is the turbulence law

$$E = C_k \epsilon^{2/3} k^{-5/3},$$

which is known universally as Kolmogorov $-5/3$rds law. This law, together with the Richardson $4/3$rds law and the t^3 dependence of variance are all consistent and give an accepted picture of turbulence at intermediate scales (the inertial sub-range). There are many observations that back up these laws, see Okubo (1974) for one example and the text by Lewis (1997).

This model due to Kolmogorov is called similarity theory. This is not to be confused with the use of *similarity variables*, which is a technique used to obtain the exact solution to certain kinds of partial differential equation, amongst them the diffusion equation indeed. Similarity variables are used extensively in aerodynamics, but not here. Here, the word similarity arises from assuming that diffusion is homogeneous in all directions. We have indicated that similarity theory gives variance proportional to t^3. Okubo reasons that the diffusion in the oceans is close to homogeneous, but constraints such as the ocean surface and perhaps the thermocline restrict the spreading. In fact, some of the data he presents do locally fit a t^3 power law. Okubo's $t^{2.3}$ power law is still accepted today as the best simple law of oceanic diffusion. Okubo's actual relation is

$$\sigma^2 = 1.08 \times 10^{-6} t^{2.34}.$$

Since the diffusion coefficient κ is related to σ^2 through the relationship

$$\kappa = \frac{\sigma^2}{4t} = 2.05 \times 10^{-7} t^{1.34},$$

we have a power law for the diffusion coefficient too. Interestingly, drawing a regression line through the original data, but this time plotting κ against l, gives the relation

$$\kappa = 2.05 \times 10^{-4} l^{1.15}.$$

This can be compared to the theoretical law due to Richardson, the $l^{4/3}$ law.

Let us turn next to a different method of modelling diffusion that theoretically is quite old but practically has only been of much use in the last 15 or so years since the advent of high speed computers that are freely available and which produce good graphical output. These models also make use of some of the ideas introduced in this section but have a very distinct *modus operandi*.

7.2.4 *Particle tracking*

In the 1970s, with the political crises in the Middle East and a general focus on saving energy, the world's attention on oil transportation by tankers intensified. Coincident with this was an increase in public awareness of matters environmental. The final piece of the jigsaw was the unfortunate incidence of several oil spillages. Around the UK, the worst of these were the wreck of the *Torrey Canyon* in 1967 and, 11 years later, the break up of the *Amoco Cadiz*. There were many lesser spillages between these dates. The early 1970s also witnessed the beginnings of the computer revolution, making available a vast increase in computing power to modellers. The pressure to produce models of oil slick behaviour was therefore great.

A good method to model the behaviour of oil slicks, or any water-borne pollutant for that matter, is to hold the pollutant as a number of marked particles. The position of each of these particles is individually held in the computer, and the power of the computer is such that all these positions are held at each time step, being updated by the model. In this way, the evolution of a patch of pollutant can be tracked. Commercially produced software began in the mid-1970s; for example, one called SLIKTRAK was produced by one of the major oil companies. Since then more and more sophisticated oil spillage tracking models have been produced. There is a tendency to call these models "random walk models" and, to a large extent, the name is a good one. In the simplest of models, the marked particles are assumed to behave exactly like tiny bits of fluid. Hence, the equations obeyed by the sea (the conservation of mass and momentum) are imposed as normal on all the fluid, including the marked particles. In addition, however, these particles are also subject to local turbulence and so

306 *Modelling Coastal and Marine Processes*

are pushed and pulled around by the eddies. If it is assumed that this turbulence has the characteristics implied by Fickian diffusion, then this can be simulated by assigning a random number by which to move a given particle off the line dictated by the local fluid velocity. If the number of particles is large enough, the result will be that the marked particles are distributed about the mean according to a Gaussian distribution. The variance of this distribution is dictated by the user and is related to the diffusion coefficient. Specifically, the mean square deviation of the particles $\overline{x^2}$ is related to the value of the diffusion coefficient κ_x and the time elapsed since the process began T, which must be assumed large through the formula

$$\overline{x^2} = 2\kappa_x T.$$

Of course, the advection that took a great deal of attention to capture theoretically in the last two sections is there automatically simply as a result of the equations obeyed by the fluid (sometimes generally called the *primitive equations*, although this name should be preserved for constant density formulations only). So the net velocity of any given particle will be the vector sum of the local fluid velocity u and the displacement assigned by the random number divided by the time step of the numerical procedure used to implement the random walk. If this time step is Δt, then using the above equation, the displacement of a fluid particle is $\alpha\sqrt{2\kappa_x\Delta t}$, where α is the random number. Various modifications of this model can be used to simulate, for example, three dimensional diffusion. In this particular case, a three dimensional grid is defined and the movement of a marked particle within this box dictated by three random numbers corresponding to the three co-ordinate directions. A continuous or intermittent discharge can similarly be simulated by allowing new marked particles to enter at specified times and places, the method looks to be very powerful indeed the only drawback being that a random number method can only hope to produce a spread whereby the variance of the patch grows linearly with time. Having said this, of course, enhanced diffusion usually results from shear present in the numerical solutions of the primitive equations to which the particles are being subjected. In the final analysis, all that is missing is the interaction between eddies and flow that Richardson, in particular, insisted characterised turbulent diffusion in the sub-inertial range. We will later look at the implementation of one of the first random walk models.

A relatively straightforward particle tracking model — one of the first practical ones — was produced by Hunter (1980), and it is this model that is outlined as a case study later. It is worth pointing out that the seductive idea of being able to move particles around in a manner that seems so closely to mirror reality has been explored by several other researchers. Dyke and Robertson (1985) proposed a theoretical turbulence which was composed of randomly distributed and randomly sized eddies. They showed that the variance of a patch of oil in such a flow grows as time to the power of between 1.8 and 3. This is consistent with Okubo's result, but problems of initialising and calibrating such a model make it rather impractical for most purposes. Jenkins (personal communication) uses a model based on waves. The waves are random in direction with correlation times differing by factors of two. Jenkins reproduced the Kolmogorov t^3 law for the increase in variance. It is true, therefore, that there remains much scope in models of this kind. Perhaps, one day, such a model will give a true picture of turbulent diffusion. A cautionary note, however: Complicated models that involve many parameters can, chameleon like, be made to simulate any set of data without improvement of understanding.

7.3   Box Models

One of the earliest ways of modelling diffusion in estuaries and bays is to use box models. The basic method is to divide the estuary or bay into boxes, usually rectangular or square regions. Normally, attention was restricted to two-dimensional models, mostly because two dimensional and one-dimensional models were all there were back then. These two-dimensional box models were either depth integrated area models over the horizontal or a vertical section along the centre line of an estuary. However, in principle there is no reason why the box model concept could not be used for three-dimensional modelling.

Once the region is segmented into boxes, then in any given box all variables are taken as constant. This is equivalent to averaging quantities over each box. The box model itself is built up using continuity of variables. What goes into a box through adjacent boxes must equal what emerges from other adjacent boxes. This could be done for salinity, current, temperature and other chemical and biological variables. The network of boxes are a means of "modelling" given data and there is really no modelling in the numerical sense, nor some would say, in any other sense either. Certainly, it is only extremely crude. However, there is a parameter which is still quite a favourite and sought after amongst coastal engineers working with estuarial pollution and that is the "flushing time". This is the time taken for any given foreign matter present in the estuary or other semi enclosed environment to exit into the open sea. Flushing time was commonly estimated using box models by estimating the time fluid particles take to cross each box and then adding these times together. The worst case is the longest possible time and this is an upper bound for the flushing time. However, box models have largely been superseded by numerical models. In fact, numerical models based on finite difference grids are very fine scale computationally based box models, except that the variables at the grid points are based on the equations, not on measurements (except perhaps at the open boundaries).

7.4   Case Studies in Diffusion

Some of the earlier parts of this chapter could be thought of as over theoretical. It is certainly true that the physics of diffusion has fascinated physicists and has attracted the attention of many excellent theoretical researchers over the years. However, it is also true that their ideas have fed into practical modelling, particularly since the computer has enabled complicated calculation. In this section, we take the opportunity to introduce real diffusion models in various contexts.

7.4.1   *A particle tracking model*

As we have already stated, a good method to model the behaviour of oil slicks, or any waterborne pollutant for that matter, is to hold the pollutant as a number of marked particles. The position of each of the particles is individually held in the computer, and the power of the computer is such that all these positions are held at each time step, being updated by the model. A relatively straightforward model, produced by Hunter (1980), is now outlined to give the general flavour of these particle tracking models.

Hunter's aim was to produce a computer model that could sim-ulate the behaviour of an oil slick lying on the surface of the sea. Hunter also wished to develop computer software that could be used by an unskilled operator. To this end, the software would have an easy to use menu-driven front end so that anyone who had an interest in oil slick movement, but who was not perhaps familiar with the details of modelling, could input data and interpret output. This is now quite a common feature, but it was not back in 1980.

When creating a model like this, one needs to decide what effects to include and what to exclude. This model predicted the move-ment of surface oil slicks, so the following factors were assumed to be important:

(1) the tidal or other water motion in the underlying seawater;
(2) a wind-driven motion localised at the air–sea interface and caused by such mechanisms as surface water velocity, surface gravity waves and the direct action of the wind stress on the oil slick;
(3) Other processes — gravitational and surface tension effects, sur-face wave activity and horizontal turbulence.

The underlying water movement, encapsulated in (1) and (2) above, causes the sea currents. These currents cause the patch of oil to move around as a whole, as well as to distort. However, the principal effect is to move the centre of mass of the oil slick. It is the spreading and mixing processes (3) that cause the slick to diffuse, i.e. to increase in size.

The tidal and other water movements are obtained, not by the model, but by interpreting observations. Other, later oil slick models have been entirely model based, using as their database the currents predicted from a primitive model of the type outlined before. Hunter's model, however, simply uses a file of current data from Admiralty charts and other sources. In order to supply the wind-driven data (2), an empirical formula relating wind speed and direction is used. If a more sophisticated wind-driven current were to be introduced, some account would have to be taken of the effects of Coriolis acceleration. Indeed, one of the suggested improvements by Hunter himself is to include some kind of Ekman effect. In this simple model, however, all such effects were omitted in favour of a direct drift that solely moves the centre of mass of an element in the direction of the wind but at some fraction of the speed. The name *element* here refers to one of the triangles into which the domain of the slick is divided to facilitate numerical computation. It is a novel treatment of (3) that deserves most attention here. In order to simulate spreading by diffusion, Hunter incorporates a so-called Monte Carlo technique. This is a technique whereby a particular particle is moved randomly in a direction with a particular speed. Thus, if one imagines a speeded-up film of a bird's-eye view of an oil slick composed of such particles, a kind of Brownian motion would be observed with individual particles dashing madly about, but with the whole patch growing steadily in size. This has a comforting feel of realism about it, but unfortunately only Fickian diffusion can be modelled by a random walk. It can be shown that, under a random walk, the variance of the distribution of oil in an oil slick has to vary directly with time t and not, for example, with $t^{2.3}$, as Okubo's data indicate might be the case. In his model, Hunter divided the domain into triangular elements (in the manner of a conventional FEM), as shown in Figure 7.6. The particles of oil are tracked across the triangular mesh using an implicit finite difference scheme. The position of each particle is thus continually updated. The underlying velocity due mostly to tides, is taken as a constant throughout the element, as is the wind drift effect. Omitting the computational details, an analysis of the diffusive spreading process is given. The example of a

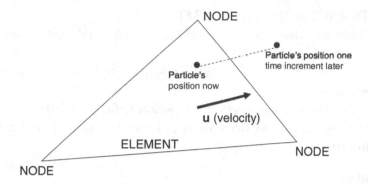

Fig. 7.6 A typical element.

radially symmetric patch after n applications of the finite difference scheme gives a variance of

$$\frac{(\text{speed} \times \text{time increment})^2 \times n}{2}.$$

Putting $t = n \times$ time increment and writing s as the distance travelled by a particle in one step gives variance $= ust/2$, as required by theory (and a Fickian model). Hunter's model also catered for the addition and removal of particles. It could not, however, cope with coastlines or with oil that sank or differentiated into several different types of oil with different characteristics. Hunter's model was distinctive in that it was the first to incorporate the user explicitly, and that was its principal value. It is recognised that the model itself was too simple to be widely applicable but it was a prototype. Here is a summary of model inputs and outputs:

Inputs

(1) Wind drift factor (proportion and direction).
(2) Diffusivity in SI units.
(3) Decay constant.
(4) Number of particles to be released initially (typical values would be 10 to 100).
(5) Initial patch size, for instantaneous release.
(6) Release position in latitude and longitude.
(7) Time increment for computation (typically 900 s).

(8) Time of initial release in GMT.
(9) Whether the particle checking is required within critical elements.
(10) Elapsed time for next model output in hours, minutes and seconds.
(11) Wind speed and direction for period defined in (10).
(12) Release rate of particles from point source over period defined in (10).

Outputs

(1) Number of particles.
(2) Co-ordinates of the centre of gravity of the patch.
(3) The direction of the principal axis of the patch.
(4) The standard deviations parallel to and normal to the principal axis of the patch.

Initially, the patch is an ellipse of particles which is inputted under input (5). An additional map output with representative coast-lines was also possible. A flowchart indicating the process is given in Figure 7.7.

7.4.2   *Modelling diffusion in the North Sea*

The motivation for the study of diffusion in the North Sea is tracking and monitoring of pollutants. In the late-1980s, a collaborative European study (the North Sea Programme) brought together researchers from countries that border the North Sea with the object of furthering our understanding of the underlying processes governing the behaviour of the sea itself, the sediments and the life within the sea. It was widely recognised that, within this ambitious programme, one of the most important processes to get to grips with was diffusion.

The overall circulation pattern in the North Sea is reasonably well known and is free of controversy. There is an anticlockwise flow; southwards down the east coast of the UK, eastwards along the northern coasts of Belgium, The Netherlands and Germany, and then northwards along the Danish coast, and finally along the west Norwegian coast to exit the North Sea (Figure 7.8 is a schematic version). There is some 'leakage' across the North Sea in the form of the

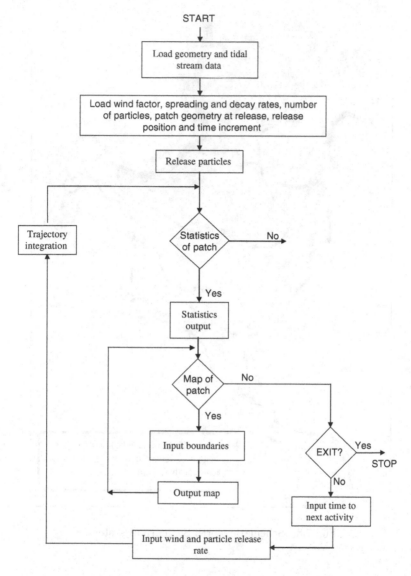

Fig. 7.7 Flow chart for the Hunter oil slick model.

Dooley current, also shown in this schematic map. It is an intermittent feature flowing from the Scottish–Northumbrian coast, across the North Sea, towards the Skagerrak. The model itself is based on a finite difference discretisation of the type described before. On top of the numerical model of water movements need to be added the

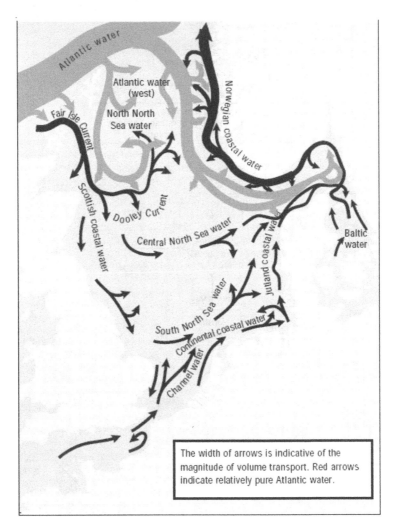

Fig. 7.8   Schematic North Sea current system.

inputs from the Firth of Forth, the rivers Tyne, Humber and Thames in the UK, and from the rivers Rhine/Meuse, Elbe, Schelde, Seine and Ems from Continental Europe. These rivers often contain foreign material in the form of discharges (industrial waste) and runoff from farmlands containing fertiliser, which are carried into the North Sea

as pollution. The North Sea, being a semi-enclosed basin, is particularly vulnerable to environmental stress (or environmental impact
as it is known nowadays).

These discharges are assumed to be in the guise of particles. The
model is based on solving the momentum equations but in a form
that uses particle tracking (Lagrangian) techniques, in order that
the pollution can be tracked explicitly. The horizontal resolution, by
which we mean the spacing between the grid points, is about 20 km,
and the time step is three hours. Two models are in fact used: a
two dimensional model with the stated resolution and time step,
in which all quantities are integrated through the vertical (depth
averaged), and a more sophisticated model which has ten levels in
the vertical and a much smaller time step of 12 minutes. The results
from the two models, are not significantly different. In both models,
the time-dependent tidal current, not the mean, is used as the basis
for computing the diffusion (spreading) of the particles, and in order
to facilitate the handling of what would otherwise be too much data,
the spreading simulations are vertically integrated. Spreading itself
takes place within the model in two ways. The currents themselves
exhibit turbulent fluctuations which, by virtue of small eddies, can
spread passive contaminants (as represented by the 100 particles)
that lie within the sea. In addition, there are mixing co-efficients in
the model that purport to represent directly the turbulent diffusion
process. Other diffusion processes that are not directly hydrodynamic
in origin, such as those due to biological or chemical agents, are not
present in this model. There is an important distinction between
the diffusion processes modelled here: *passive*, in which particles are
carried around by ambient flow (or perhaps at some fraction of the
ambient flow); and biological and chemical processes which, although
they may also be largely diffusive, are also *active* in that internal
biology and chemistry can take place. Biological organisms can also
propel themselves, of course. Let us now follow with a more detailed
description of a case study; this one is an attempt to model the Gulf
oil spill that was the result of the first Gulf War of 1991. This is the
largest known oil spill; may its like never be seen again.

7.4.3   *Modelling the motion of spilt oil*

The spill model outlined here follows Proctor, Flather and Elliott (1994). This is the primitive equation model that is used to predict the underlying current. The model for tides and storm surges used the depth averaged equations — good enough in these circumstances. The difficulties of three dimensionality, stratification and turbulence closure are avoided. Here are the equations:

$$\frac{\partial \eta}{\partial t} + \boldsymbol{\nabla} \cdot (H\mathbf{u}) = 0,$$

$$\frac{\partial \mathbf{u}}{\partial t} + \mathbf{u} \cdot \boldsymbol{\nabla}\mathbf{u} + f\mathbf{k} \times \mathbf{u} = -g\boldsymbol{\nabla}\eta - \frac{1}{\rho}\boldsymbol{\nabla}p_a + \frac{1}{\rho H}(\boldsymbol{\tau}_s - \boldsymbol{\tau}_b) + \nu_H \nabla^2 \mathbf{u}.$$

The usual notation has been adopted, with the addition that H is the total water depth, p_a is atmospheric pressure — an important feature to include for storm surge modelling, ν_H is the horizontal diffusion of momentum (eddy viscosity), and $\boldsymbol{\tau}$ is the stress at the surface (s) and the bed (b). At the surface, a quadratic law connects the stress to the wind vector $\mathbf{W}$

$$\boldsymbol{\tau}_s = C_D \rho_a \mathbf{W}|\mathbf{W}|.$$

Here, ρ_a is the density of air and the drag coefficient C_D is usually weakly related to the wind speed W. At the sea bed a similar quadratic law is assumed to hold, viz.

$$\boldsymbol{\tau}_b = \kappa \rho \mathbf{u}|\mathbf{u}|.$$

This time the drag coefficient $\kappa = 0.0015$, a constant value. An explicit finite difference scheme on a spherical but regular longitude–latitude grid was used. Obviously the authors used the tried and tested North Sea basic storm surge model refined by Roger Flather, the second of the authors of this Gulf oil spill paper. The open boundary condition used was

$$u_n = \hat{u}_n + \sqrt{\frac{g}{h}}(\eta - \hat{\eta}),$$

where $\hat{\eta}$ and $\hat{u}_n$ are specified functions of space and time including contributions from both tide and surge. This enabled waves to propagate out of the sea area in a natural manner. The solid boundary conditions were the usual zero normal flow and the grid itself was $5'$ by $5'$ or approximately 9 km square. An additional complication was that the model required ten tidal constituents to get the tides sufficiently accurate. Meteorological data from a global model were used for wind and air pressure input and the storm surge plus tide model validated at two ports. This then gave a sound basis upon which to add the spill model itself. The spill model used in this case is based on random walk modelling and not the conventional advection–diffusion equation. The particles of oil are pushed and pulled around by current shear in the form of advection, turbulence, which is manifest through diffusion (particle separation), and buoyancy, which depends on size of droplet and oil density. There is no independent movement of the oil in fluid dynamics terms; as a fluid oil too obeys the laws of fluid dynamics, but this is not accounted for here. The model also includes evaporation and decay through the water column (or should that be "water" column). The actual random walk formula used enables both horizontal and vertical diffusion modelling to take place. If the diffusion co-efficients are labelled D_H (horizontal) and D_V (vertical) then

$$D_H = R(12E_H\Delta t)^{1/2} \qquad \text{in the direction} \qquad \theta = 2\pi R$$

and

$$D_V = (2R - 1)(6E_V\Delta t)^{1/2},$$

where R is a random number chosen between 0 and 1, E_H and E_V are horizontal and vertical diffusion co-efficients and, of course, Δt is the time step. For small droplets (less than a critical diameter d_c obtained by matching droplet motion and considering Reynolds number) it is assumed that the rise velocity w can be described by

$$w = \frac{gd^2(1 - \rho_0/\rho)}{18\nu},$$

whereas for large droplets it is assumed that

$$w = \left[\frac{8}{3}gd(1 - \rho_0/\rho)\right]^{1/2}.$$

In these formulae, besides the droplet size d, we have the viscosity of seawater ν and ρ_0 is the density of oil. The probability p of removal of a droplet at each time step is given by

$$p = 1 - e^{-(\lambda \Delta t)}.$$

In practice, for each droplet a random number R is generated at each time step, and if $R \leq p$ the droplet is removed. The decay constant λ is related to the time scales of evaporation and degradation. Obviously, only oil droplets close to the surface experience evaporative decay but oil throughout the water column experiences degradation. Using standard properties of hydrocarbons, typically in less than ten days most of the oil that is going to evaporate (25%) has done so and about 5% has degraded. This allows a decay or "e-folding" time, which is the reciprocal of λ, to be determined. The droplets are, of course, also moved horizontally and if they are either beached (that is hit a solid boundary) or move out from the open boundary then they take no further part in the simulation. The paper by Proctor, Flather and Elliott (1994) then goes on to discuss the specifics of a particularly large oil spill in the Gulf in January 1991. The data used were $\rho_0 = 870$ kg m^{-3}, $E_H = 10$ m^{2}s^{-1}, $E_V = 0.005$ m^{2}s^{-1} and Δt is one hour. The droplets of oil are between 60 and 120 μm. Simulated and observed results are reproduced in Figure 7.9. The result of the simulation was broad agreement between model and observation, which was gratifying. It is unusual for such a comparison to be possible. Deliberately causing a slick then monitoring it to validate a model is not an option. All of us would rather the Gulf oil slick did not happen, but taking a small grain of comfort from such an ecological disaster, it is good to test a sophisticated spill model and have it vindicated. More details, including a further account of shortcomings, are given in the original paper. Interestingly, it is thought that, even given the disastrous short term pollution caused by the vast spill, the long term pollution would be significantly decreased by the decrease in tanker traffic during the Gulf War itself.

Finally, the paper by Elliott (1991) outlines the model EUROSPILL, which includes the kind of droplet behaviour indicated above together with an accurate numerical tide and surge model. Wind shear is included through a parameter which represents the thickness of the wind shear layer. An additional mechanism not before mentioned is the Stokes drift due to the nonlinear

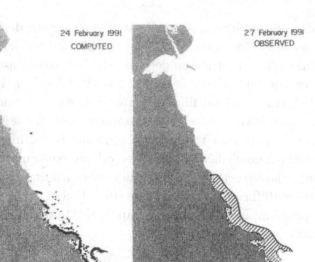

Fig. 7.9   Oil slick prediction and observed position.

interaction of surface waves. These give rise to a drift that can be significant, but detailed computation of Stokes drift is delayed until the next chapter. Diffusion is, of course, present as is buoyancy. The really novel aspect of EUROSPILL is the inclusion of a user-friendly menu and associated graphics which enable a novice user to drive the model. The usual warning applies; the glamorous output is only as good as the modelling behind it. The modelling behind this one looks pretty good though. In more recent times, there have been models that have modelled the actual hydrodynamics of the oil itself; it is worth looking at one of these next.

7.4.4   *Multi-phase oil spill modelling*

In these models, it is assumed that there is enough spilt oil to form a layer on the top of the sea. Even if there is enough of it to form a continuous layer, it is still distinctive and insubstantial enough not to be directly acted upon by the wind and tide; instead any influence of wind and tide will come via the sea upon which the oil rests, and this is done through a multi-phase fluid dynamics approach. Models of this type have an oil layer of thickness h_{oil} and the fluid dynamical equation obeyed by h_{oil} is usually averaged over the depth of the slick. It is assumed that the slick migrates with bulk velocity dictated by the fluid velocity of the underlying sea but that there is a friction between the oil and the sea. The model of Tkalich, Huda and Gin (2003) includes an oil film — water friction co-efficient that gives a value of $0.03\bar{W}$, where $\bar{W}$ is the magnitude of the wind. Another distinctive feature of this particular model is the inclusion of three phases of oil: emulsified oil, dissolved oil and particulate oil, with each having separate settling or buoyancy rates leading to three separate advection–diffusion equations. The spreading of the oil takes the form of a spreading function. In this paper, this spreading function takes the form

$$\frac{gh_{oil}^2(\rho - \rho_0)\rho_0}{\rho f_r},$$

where f_r is the oil–water friction coefficient, ρ_0 is the density of the oil, ρ is the water density and g acceleration due to gravity. There are thus four equations, one hydrodynamic one for the bulk flow of the oil, which is subject to spreading and chemical kinetics as well as interaction with the underlying sea, and three advection–diffusion equations for the three phases of the oil. The actual tide and wind-driven flow is an input to the hydrodynamic equation of the oil slick. It is therefore possible to embed this kind of model into a more conventional hydrodynamic sea model. In the paper, a great deal of attention is given to the modelling of the interaction of the oil with sediments, with the inclusion of many measured mass exchange rates for the different interactions, and the results of the numerical model with these co-efficients are verified by comparison with test cases. The resulting model is a flexible tool that can be used in many different environments; it is expected that future commercial software will have such models as optional add-ons.

7.4.5   *Modelling plumes*

The civil engineering industry requires plume models that are valid for rivers and estuaries. This is because many factories are sited beside rivers and estuaries so that the adjacent water can be used, perhaps for cooling and also so that there can be a waste discharge. Since the 1970s, the legislation governing such discharges has become much more severe. Europe has now also got in on the act. There are long lists of chemicals that are either banned or are such that only a very low level concentration is permitted. Turning to what actually is done, typically there is a discharge pipe that ends part way into a river or estuary and through which the material flows. Usually, this is not a pipe but has holes strategically placed along the sides to maximise the dilution of any substance that might be present in the outflow material. This device is termed a "diffuser". Once the discharge water is in the river or estuary, then it is important to model its behaviour as accurately as possible in order to decide whether or not legislation is breached. The usual method is to consider the plume as a buoyant jet which diffuses in accordance with a Gaussian law. The diffusing jet is then subject to tide, wind, river flow — all of which under normal circumstances help the dilution. The case of most interest is always the worst case scenario where all dilution mechanisms are least. This is simple to model, of course as the case of zero flow due to river, wind or tide does the trick. The other environmental question to ask is whether any unwanted material is likely to find its way on to a nearby sensitive part of the river or estuary — an Site of Special Scientific Interest (SSSI) perhaps, a holiday beach or yachting marina. In all these questions modelling plays a pivotal role. However, let us turn to a larger model for a detailed case study: the Rhine river plume as it exits into the North Sea.

The Rhine is the largest river discharge into the North Sea and, therefore, there is good reason to model it. The model presented here is a summary of the paper by Ruddick *et al.* (1994). The initials Region of Freshwater Influence (ROFI) join the plethora of acronyms, this set being employed to indicate the waters around the mouth of the estuary that are stratified and otherwise influenced by the river discharge. The model used is not particularly simplified but employs three-dimensional hydrodynamics with stratification incorporated via buoyancy and a conservation of salinity. The momentum equations themselves use only the constant reference density in the pressure term; this is called the Boussinesq approximation. The momentum equations employ an eddy viscosity, and the conservation of salinity a diffusion co-efficient. Some of the symbols used are different to those introduced here, so to give the equations of the paper verbatim would confuse — suffice it to say that the horizontal momentum conservation equations contain nonlinear terms, Coriolis terms and eddy viscosity terms. Instead, only certain differences will be highlighted. The buoyancy b is given by

$$b = \beta_S g(S_0 - S),$$

where g is the acceleration due to gravity, β_S is the coefficient of haline expansively and S_0 is a reference salinity. A reduced pressure is defined by

$$q = \frac{P - P_{\text{atm}}}{\rho_0} + gz$$

and the buoyancy is then

$$b = \frac{\partial q}{\partial z}.$$

The conservation of salinity is the standard advection–diffusion equation

$$\frac{\partial S}{\partial t} + \frac{\partial}{\partial x}(uS) + \frac{\partial}{\partial y}(vS) + \frac{\partial}{\partial z}(wS) = \frac{\partial}{\partial z}\left(\lambda_S \frac{\partial S}{\partial z}\right),$$

this equation and the momentum equations not quoted here are written in what is called "flux" form. We met this in Chapter 3 and shall do so again in Section 8.5.3. Briefly, simply take the continuity equation

$$\frac{\partial u}{\partial x} + \frac{\partial v}{\partial y} + \frac{\partial w}{\partial z} = 0,$$

multiply it by S and add to the standard form of the advection–diffusion equation. The combinations

$$u\frac{\partial S}{\partial x} + S\frac{\partial u}{\partial x} \qquad \text{etc.}$$

then appear, which are combined to give the fluxes uS (etc.) differentiated. This is what is seen above.

The Arakawa C grid is used and the variable depth domain transformed via the sigma co-ordinate into lying between two constant values. Let us look at features that pertain to the river plume modelling. Boundary conditions were dealt with in the last chapter, and in particular, what boundary conditions are appropriate to impose at the open boundaries, and there are three of them here. Figure 7.10 shows the domain and the three open boundaries. The model has to allow for tide, wind stress and river as forcing mechanisms, therefore the tide must be allowed to enter and leave the model area. Here are

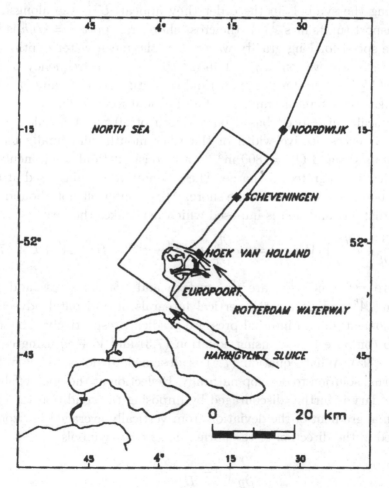

Fig. 7.10   The Rhine plume model study area.

the boundary conditions imposed. They will be explained once they have been stated.

$$U + c\eta = 2c\eta_0 \sin\omega_0 t, \tag{7.1}$$

$$U - c\eta = 0, \tag{7.2}$$

$$V - c\eta = c\eta_0 \sin\left(\frac{\omega_0 x}{\bar{c}} - \omega_0 t\right), \tag{7.3}$$

$$V + c\eta = \frac{2}{A}(Q_0 - Q_1 \sin\omega_0 t). \tag{7.4}$$

Taking the symbols in the order they appear, U is the alongshore transport (units m^2s^{-1}), V the cross shore transport, $c = \sqrt{gh}$ is the wave speed for long gravity waves, h is the mean water depth, η is the surface elevation, $\omega_0 = 1.405 \times 10^{-4}$ s^{-1} the frequency of the tide (M_2 — semi-diurnal lunar tidal constituent). The value $\eta = 0.9$ is taken as a typical amplitude for the tidal forcing, and $\bar{c} = \sqrt{20g}$ a typical wave speed based on a depth of 20 m. A is taken as 1 km, a representative width of the river mouth, and finally $Q_0 = 1200$ m^3s^{-1} and $Q_1 = 1800$ m^3s^{-1} are constant tidal components of vertically integrated discharge. These conditions are imposed at the two boundaries that cross the shore. At the outer offshore boundary, a radiation condition is imposed which here takes the form

$$\left(\frac{\partial}{\partial t} + c\frac{\partial}{\partial y}\right)(V + c\eta) = -c\frac{\partial U}{\partial x} + \tau^{\text{surf}} - \tau^{\text{bed}} - fU - \hat{\mathcal{A}}^h + \hat{\mathcal{L}}_d, \tag{7.5}$$

where τ^{surf} and τ^{bed} are the surface and bottom stress and the terms $\hat{\mathcal{A}}^h$ and $\hat{\mathcal{L}}_d$ are the vertical integrals of horizontal advection of momentum and internal pressure gradient, respectively. The idea is to compute $V - c\eta$ using equation (7.3) and $V + c\eta$ using equation (7.5). What equation (7.5) is designed to do is to allow the internal solution to develop naturally. Reflection at the open offshore boundary is further discouraged by imposing the condition that the normal gradient of the deviation from vertically averaged horizontal speed in the direction of the normal is zero. In symbols

$$\frac{\partial}{\partial y}\left(v - \frac{V}{H}\right) = 0.$$

Salinity needs careful treatment too; it is carefully allowed to leave and enter at open boundaries by using upwind differences and by supposing that incoming fluxes correspond to advecting the reference salinity S_0. At the river discharge, a two layer situation (5 m upper layer, 25 m lower layer) is applied. At the river, the flux enters the domain. Given the uncertainty of the modelling assumptions that surround the open boundaries, validation and verification studies are essential. The authors were encouraged to apply the model to real situations thanks to an accurate Kelvin wave simulation (for more about Kelvin waves, see Chapter 6). Figure 7.11 gives one output graph from the model. The next picture, Figure 7.12, gives a

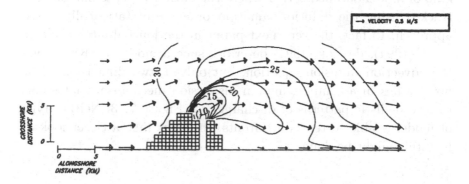

Fig. 7.11   The Rhine plume: Surface currents and salinity at high tide.

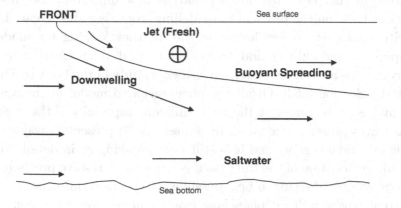

Fig. 7.12   A schematic alongshore transect through the plume.

schematic alongshore transect. The freshwater jet is indicated spreading into the offshore tidal stream. It remains identifiable as a buoyant jet and stays close to the surface of the sea. This is a complex area to model, and this model of Ruddick *et al.* (1994) is certainly not the final word on the subject. Sharp fronts occur in reality and cannot be resolved using this model. Arguments about the use of diffusion co-efficients for plume dynamics must be listened to. Unrealistically, large diffusion would render the model useless for water quality purposes. Variations in river flow have also been ignored and these are patently a very important aspect of the plume. It is interesting to compare the kind of results given in Ruddick *et al.* (1994) with the kind of model outlined earlier where the outfall water is marked and tracked using the diffusion equation or a more statistically based approach. In fact, the very next paper in the journal after Ruddick *et al.* (1994), de Kok (1994) describes such a model. de Kok solves the advection–diffusion equation, but only in two dimensions. The novelty lies in a complex uneven dispersion mechanism that seems to give realistic patterns of discharge, see Figure 7.13. Neither type of model is able to reproduce fronts, so let us take a closer look at how this might be done.

7.5   Modelling Fronts

Although the paper by James (1996) is now quite old, the way it tackles head on the problem of modelling fronts is still current. This writer has spent a few hours on the internet looking for modern papers on the subject and has failed to find them. A reasonable account has already been written as a case study in the book by Dyke (2001), Section 3.5.3. Of all the phenomena to model in the sea, a front has to be amongst the most difficult, especially if the regime has waves present. The paper by James (1996) presents a way to do this in a tidal regime, and it is still worth looking at in detail. First of all, the location of the front itself is not easy to resolve precisely — the moving grid seems to be a possible way forward but this remains current research. To explain how fronts can be preserved, consider

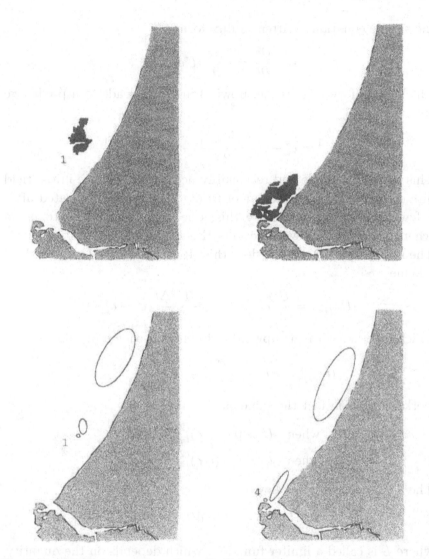

Fig. 7.13   Dispersion patterns, from de Kok (1994), reprinted with permission from Munksgaard International Publishers Ltd., 35 Nørre Søgade, PO Box 2148, DK-1016, Copenhagen K, Denmark. Originally published in *Tellus*, **46A**, page 167 in 1994.

the simple equation, written in flux form

$$\frac{\partial c}{\partial t} = -\frac{\partial}{\partial x}(Uc),$$

which using forward in time, upwind in space, leads to a predictive equation

$$c_i^{s+1} = c_i^s - U\frac{c_i^s + c_{i+1}^s}{2} - |U|\frac{c_i^s - c_{i+1}^s}{2}.$$

This scheme is stable and reasonably accurate but if the initial field of c contains a sharp gradient of front, then this is dissipated after a few time steps. To remedy this, one answer would be to use a scheme that in some way prevents the spread of an initial gradient. The following TVD scheme does this. It is based on a Lax–Wendroff scheme

$$(Uc)_{i+1}^{(1)} = \frac{U}{2}(c_i^s + c_{i+1}^s) + \frac{U^2\Delta t}{2\Delta x}(c_i^s - c_{i+1}^s),$$

which together with the upwind differencing scheme (as above)

$$(Uc)_{i+1}^{(2)} = U\frac{c_i^s + c_{i+1}^s}{2} - |U|\frac{c_i^s - c_{i+1}^s}{2}$$

works. This is in fact the same as

$$\text{when} \quad U > 0 \quad (Uc)_{i+1}^{(2)} = Uc_i^s,$$
$$\text{when} \quad U < 0 \quad (Uc)_{i+1}^{(2)} = Uc_{i+1}^s.$$

The TVD scheme is to allow

$$(Uc)_{i+\frac{1}{2}} = (Uc)_{i+\frac{1}{2}}^{(2)} + L(\alpha)[(Uc)_{i+\frac{1}{2}}^{(1)} - (Uc)_{i+\frac{1}{2}}^{(2)}],$$

where L is called a limiter function, which depends on the quantity

$$\alpha = \frac{(Uc)_{i+\frac{1}{2}-n}^{(1)} - (Uc)_{i+\frac{1}{2}-n}^{(2)}}{(Uc)_{i+\frac{1}{2}}^{(1)} - (Uc)_{i+\frac{1}{2}}^{(2)}},$$

where n is either 1 or -1 defined by

$$n = \text{sign}(U).$$

Notice that the superscript s is absent here as all variables are at the current time step. The choice of L is wide and several were tried with varying degrees of success: they are listed in James (1996) — a good one turns out to be

$$L(\alpha) = \frac{\alpha + |\alpha|}{1 + \alpha}.$$

The kind of scheme outlined briefly here is good at preserving fronts. They do so, in fact, by creating (locally) negative viscosity that re-sharpens fronts. The complexity of such schemes makes their adoption over large domains impractical. Indeed, there is no evidence that they would continue to work and one is once more looking at time varying grids to develop more practical schemes that preserve fronts. The result of applying this scheme to a tidal front that moves over a step in a continental shelf environment is shown in Figure 7.14.

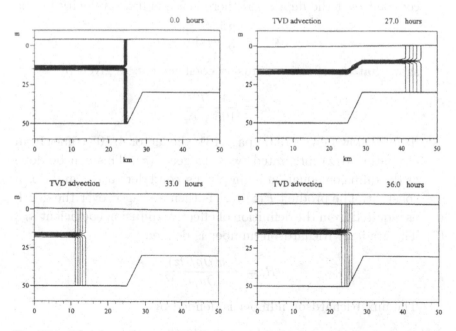

Fig. 7.14   The sharp front, although blurred a little, is more or less preserved as the model is run for 36 hours.

7.6   Exercises

(1) Two waters of salinity $S_1 = 35$ ppt and $S_2 = 32$ ppt are mixing and producing water of salinity $S = 34.5$ ppt. Find the proportions of S_1 and S_2 water in the mix.

(2) Two separate contaminants with concentrations c_1 and c_2 are present in an estuary. Use a method similar to that of section 7.2.1 to analyse the diffusion of these two pollutants and confirm that the equations satisfied by c_1 and c_2 are still advection — diffusion equations. How is this changed if the two pollutants interact chemically?

(3) Give a qualitative explanation for Okubo's $t^{2.3}$ law for the spreading of pollutants in the open sea.

(4) Describe how the process of shear diffusion works. Suppose that a current $u(z)$ depends on the depth z and that $u = \langle u \rangle \alpha(z)$ so the function $\alpha(z)$ gives the deviation of u from its depth mean value $\langle u \rangle$. If the vertical diffusion of the contaminant is assumed constant over the depth and there is a quadratic velocity profile

$$\alpha(z) = \frac{3}{2}\langle u \rangle (1 - z^2),$$

show that the effective diffusion coefficient κ_e is given by

$$\kappa_e = \frac{4}{105}\frac{\langle u \rangle^2 h^2}{\kappa_z}.$$

[Hint: Follow Lewis (1997) page 131; the factor $4/105$ comes from evaluating $\alpha(z)$ integrated twice to get $F(z)$. This can be done as the diffusion equation is simply a second derivative, and α is a constant. The product $F(z)\alpha(z)$ is then averaged over the depth as required from the definition of effective diffusion coefficient κ_e.]

(5) The gradient Richardson number is defined by

$$R_i = -\frac{g(\partial \rho / \partial z)}{\rho(\partial u / \partial z)^2}.$$

The flux Richardson number is defined by

$$R_f = -\frac{g\kappa_z(\partial \rho / \partial z)}{\rho \nu_z(\partial u / \partial z)^2}.$$

Discuss the inequalities

$$R_i < \frac{\nu_z}{\kappa_z}, \qquad R_f < 1.0$$

in the context of stability of the sea to disturbances.

(6) An estuary plume due to freshwater intrusion is modelled using a three dimensional primitive equation model of the sea environment together with an advection–diffusion equation for the salinity and a buoyancy term that links the two. Discuss the problems of implementing such a model, in particular the imposition of boundary conditions. How useful is such a model to an engineering company who wishes to site a factory on the estuary?

Chapter 8

Shoreline Management

8.1 Introduction

There has been a long history of man's interaction with the sea. Most of our towns and cities are where they are due to the proximity of water, either rivers, an estuary or the coast. Since the industrial revolution, this has become reinforced as coastal and estuarial ports were used for the transport of heavy material such as metal ores imported coal and the like, first by sea and thence by canal and river. The coast has thus been altered through civil engineering structures such as harbours and jetties as well as protected through the building of breakwaters, groynes and sea walls. In more recent times, the coast has become a popular place of recreation and for wildlife reserves. Now, more than ever, man is very keen on protecting the coast. Since the 1970s this interaction has extended to offshore and shallow sea regions following the advances in engineering activities linked to the exploration of oil under the sea. Thus civil engineers who developed wave prediction schemes in the 1950s principally for coastal use now applied them to forecasting the wave climate around offshore production platforms and the like. The original wave prediction schemes were developed for war-time activities, not least the Normandy landings and attacks on the Japanese, nevertheless they found peacetime application to shipping and maritime transport and leisure.

Changes in the shape of coastlines have always taken place to a greater or lesser extent. There are places in the world that are particularly vulnerable. In the United Kingdom, the coast of Norfolk has been changing continuously — so much so that the entire village of Dunwich has disappeared into the North Sea over the last few thousand years. Normally it is a storm that causes a sudden land slip and over the years the cliff edge gets inexorably closer to vulnerable buildings, which are vacated and lost or, in some circumstances, demolished carefully brick by brick and resurrected in a safer place. Such was the fate of an historic lighthouse on the Sussex coast. Of course, in other places the opposite is happening. In South Yorkshire, just north of the River Humber, the coast north of Spurn Head is being built up and once coastal towns and villages now find themselves inland. The processes that cause such changes are not completely understood, but there are models that incorporate the essential features and it is useful to summarise these features before describing models. The wave climate is crucial, especially its extreme manifestation under storm conditions. Local tides and other currents need to be modelled as best as possible. Finally, the help of geologists is required to gauge the composition of the coast. Sand and silt beaches, and cliffs composed of easily erodible rocks such as limestone or soft sedimentary rocks are vulnerable. In East Anglia for example, the annual budget of sand being eroded from the north Norfolk cliffs at Overstrand is about 400 m^3 per year. This material mostly ends up replenishing sandy holiday beaches further south. Coasts are very variable places in terms of look, composition and use by man. There are cliffs, beaches of various sorts, coastal marshes, dunes etc. The photographs in the introduction to the excellent text by Reeve, Chadwick and Fleming (2004) give a flavour of the variety. The many and varied coasts are protected from the ravages of the sea in different ways with differing success. Sea walls are still quite popular, but a storm will subject this to large forces and most will need renewal processes eventually. Offshore structures such as man-made offshore breakwater and the natural tombolo are perhaps more successful and also more environmentally acceptable. However, a tombolo only forms under certain circumstances; circumstances where the wave and current climate gives rise to this kind of deposition. This climate is not the same as an eroding climate, so the tombolo is useful only in helping us to understand the sediment drift that forms it. As sediment in shallow seas is largely a consequence of wave climate, it certainly means that we have to understand the wave climate as fully as possible in order to design successful coastal defences.

As the climate becomes warmer and more energetic, such processes will speed up so it becomes ever more important to understand and model them. Only then might something be done to control, or at least minimise, the effects of costly erosion. Another consequence of climate change (global warming) is sea level rise, and this is certainly extremely unhelpful in the constant war against coastal erosion.

This chapter, then, has the following aims. First, to run through wave prediction, which has moved on apace since the advent of the computer of high speed and vast memory. Second, to look briefly at models of a different kind of water wave, the edge wave, and thirdly, to look at models of coastal erosion, including the all important modelling linked to the prevention of it. The starting point to all of these is the wave climate. Let us study this first; and to do this we need to start where we left off with linear wave theory in Chapter 3.

8.2 Applied Linear Wave Theory

In Chapter 3, the dispersion relation

$$c^2 = \frac{g}{k} \tanh kh$$

was derived. To recap, c is the wave speed (or celerity), and k is the wave number and the theory upon which this relationship is based is linear. An important quantity in waves is their energy, and so this is calculated for these linear waves. Waves have both potential and kinetic energy. They also have energy due to surface tension, but this can be safely ignored as it is very small for sea waves. The potential and kinetic energies of a wave per unit area of sea are obtained by adding up the respective energies of all the water particles that comprise the wave over one wavelength, then dividing by the wavelength. We are dealing with waves in one dimension, and so they will have unit width already. It turns out that potential and kinetic energies are the same for linear sinusoidal waves. That is, a linear wave partitions energy equally between potential and kinetic. This

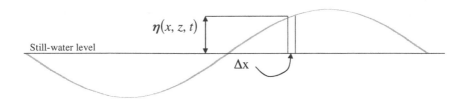

Fig. 8.1   The sinusoidal linear wave.

is interesting in itself, but the real reason for actually calculating this energy lies in the future application to real sea waves, where the energy plays an important role. So, some time will be spent deriving this result from first principles. First of all, the linear sinusoidal wave is represented in Figure 8.1. The mass of water of a small strip shown schematically in Figure 8.1 (assumed to be of unit width out of the page) is volume times density, i.e. $\rho\eta\Delta x$. This strip will have potential energy $\frac{1}{2}\rho g\eta^2\Delta x$ as the strip has a centre of mass "half way up", i.e. at coordinate $\frac{1}{2}\eta$. The total potential energy is the integral of this quantity over the wave, which is

$$\frac{1}{2}\rho g a^2 \int_0^\lambda \cos^2(kx - \omega t)dx = \frac{1}{4}\rho g a^2 \lambda,$$

where we have denoted the amplitude of the wave by a and its wavelength by λ. So the (potential) energy per unit wavelength is

$$\frac{1}{4}\rho g a^2 = \frac{1}{16}\rho g H^2$$

in terms of total wave height ($H = 2a$ here, but it will not always be). Most of you who know about integration can perform the calculation here. The calculation of the kinetic energy is a little more tricky. The kinetic energy of a fluid particle under the wave (or anywhere else in the fluid for that matter) is $\frac{1}{2}\rho(u^2 + w^2)\Delta x\Delta z$. The expressions for u and w were derived in Chapter 3 as

$$u = C\cosh k(z + h)\cos(kx - \omega t)$$

and

$$w = C\sinh k(z + h)\sin(kx - \omega t),$$

where C is a constant. This constant is determined through relating these velocity components to the sea surface elevation. The mathematical details are routine for those who are familiar with this kind of thing. Here they are; they can be skipped if preferred. First of all, ϕ the velocity potential introduced in Chapter 3 by

$$u = \frac{\partial \phi}{\partial x}, \quad \text{and} \quad w = \frac{\partial \phi}{\partial z}$$

is given by

$$\phi = \frac{C}{k} \cosh k(z + h) \sin(kx - \omega t).$$

Now, since we have the boundary condition

$$\eta = -\frac{1}{g} \frac{\partial \phi}{\partial t}\bigg|_{z=0},$$

the constant C can be found. You will remember that this condition is the dynamic surface condition and tells us that the free surface is a surface of constant pressure. The formula is a linearised Bernoulli equation. So we must have

$$\eta = \frac{C\omega}{gk} \cosh kh \cos(kx - \omega t) = a \cos(kx - \omega t),$$

which gives

$$C = \frac{agk}{\omega \cosh kh}.$$

The kinetic energy $\frac{1}{2}\rho(u^2 + w^2)\Delta x \Delta z$ of a fluid particle is thus

$$\frac{\rho a^2 g^2 k^2}{2\omega^2 \cosh^2(kh)}[\cosh^2 k(z+h)\cos^2(kx - \omega t)$$

$$+ \sinh^2 k(z+h)\sin^2(kx - \omega t)],$$

and this has to be integrated over the wave. The integration is again not too difficult. One useful result is

$$\int_{-h}^{0} \cosh^2 k(z + h)dz = \frac{1}{2}\left(h + \frac{\sinh 2kh}{2k}\right)$$

and the equivalent result for $\sinh^2 k(z + h)$ is obtained by using

$$\cosh^2 \theta - \sinh^2 \theta = 1.$$

The solution, once the algebra has been done and division by the wavelength has taken place, is

$$\frac{1}{4}\rho g a^2 \quad \text{or} \quad \frac{1}{16}\rho g H^2,$$

the same as the potential energy. These integrals are called "elementary" by mathematicians, but if they should give trouble, software like MAPLE or MATHEMATICA eat them for breakfast. So, the energy is equally partitioned between kinetic and potential. The question is now how fast does the energy in a wave move? In order to answer this question, first we revisit the expression derived in Chapter 3 as equation (3.15), namely

$$\frac{DE}{Dt} = -\int_{\text{wavelength}} \int_{-h}^{0} \rho c \left(\frac{\partial \phi}{\partial x}\right)^2 dz dx.$$

We evaluate this with

$$\phi(x, z, t) = \frac{ag}{\omega \cosh kh} \cosh k(z+h) \sin(kx - \omega t).$$

Differentiating and squaring this gives

$$\left(\frac{\partial \phi}{\partial x}\right)^2 = \frac{a^2 g^2 k^2}{\omega^2 \cosh^2 kh} \cosh^2 k(z+h) \cos^2(kx - \omega t),$$

and evaluating the integrals with respect to x and z gives

$$\frac{DE}{Dt} = \rho c \frac{a^2 g^2 k^2}{4\omega^2 \cosh^2 kh} \left(h + \frac{\sinh 2kh}{2k}\right)$$

(the integral with respect to x over a "wavelength" is the average value of $\sin^2(kx - \omega t)$ over any wavelength, which must be $1/2$). If the negative sign in the formula troubles you, don't let it. This is one of those occasions when the sign is not that important; in fact it denotes that our progressive waves happen to be travelling in the other direction than the conventional normal that points outward from surfaces. We take out a factor of $\sinh(2kh)/2k$ to get

$$\frac{DE}{Dt} = \rho c \frac{a^2 g^2 k^2}{4\omega^2 \cosh^2 kh} \cdot \frac{\sinh 2kh}{2k} \left(\frac{2kh}{\sinh 2kh} + 1\right)$$

and using

$$\sinh 2\theta = 2 \sinh \theta \cosh \theta$$

gives

$$\frac{DE}{Dt} = \rho c \frac{a^2 g^2 k^2}{4\omega^2 \cosh^2 kh} \cdot \frac{\sinh kh \cosh kh}{k} \left(\frac{2kh}{\sinh 2kh} + 1\right).$$

Finally, cancelling a cosh kh and using the dispersion relationship

$$\omega^2 = gk \tanh kh$$

after recalling that

$$\tanh \theta = \frac{\sinh \theta}{\cosh \theta}$$

gives the result

$$\frac{DE}{Dt} = \frac{\rho a^2 g}{2} \cdot \frac{c}{2} \left(\frac{2kh}{\sinh 2kh} + 1 \right).$$

Try not to be overawed by all the manipulation because, in fact, this expression gives us an important result. It has already been derived above that the average energy in the wave is

$$E_{av} = \frac{1}{2} \rho g a^2.$$

It is the value that counts here, not that it is equally partitioned between kinetic and potential. If the rate of change of energy is E_{av} multiplied by some quantity, then this quantity has to be the speed at which the energy is travelling. So, we have

$$\frac{DE}{Dt} = E_{av} c_g,$$

where

$$c_g = \frac{c}{2} \left(\frac{2kh}{\sinh 2kh} + 1 \right).$$

So, energy does not travel at the speed of the wave but at the velocity c_g, which turns out to be the group velocity of the wave, but we have not shown this yet. This fact was hinted at in Chapter 3 where the group velocity of a couple of waves close in wavelength and frequency was discussed. The derivation of c_g here needs to be broader because in the next section we deal with real wave records and these have a continuum of wavelengths and wave frequencies. Now, the mathematics of trigonometric functions such as sine and cosine do not permit such generalisations, more's the pity. So, though so far we have managed to describe waves with a cosine, this can no longer be so, and instead we shall have to use the exponential form $ae^{i(kx - \omega t)}$, where $i = \sqrt{-1}$. It can still be one dimensional though, so it only involves x and t. Suppose that instead of having a single wave or a pair of waves we have a continuum of them. This is represented by the integral

$$\phi(x,t) = \int \bar{\phi}(k)e^{i(kx-\omega t)}\,dk.$$

Here, $\bar{\phi}(k)$ is called a wave number spectrum and indicates a range of continuous wave numbers. The actual range would be specified by the limits of the integral. If these limits are close, then there is a narrow wave number spectrum. This is the case here, as we want to focus on a group of waves that stay close and this can only be the case if their wave numbers are close too. Specifically, suppose this wave "package" has wave numbers that are all clustered around some value k_0, and let ω_0 be the frequency that corresponds to this through the dispersion relation. A more precise formulation of the wave packet would then be

$$\phi(x,t) = \int_{k_0-\Delta k}^{k_0+\Delta k} \bar{\phi}(k)e^{i(kx-\omega t)}\,dk.$$

Some simple manipulation (adding and subtracting k_0 and ω_0) gives

$$\phi(x,t) = \int_{k_0-\Delta k}^{k_0+\Delta k} \bar{\phi}(k)[e^{i(k-k_0)x-(\omega-\omega_0)t}e^{i(k_0x-\omega_0t)}]\,dk.$$

Now, the exponent $e^{i(k_0x-\omega_0t)}$ does not depend on k and so only modifies the amplitude. This integral will be constant if the exponent $e^{i(k-k_0)x-(\omega-\omega_0)t}$ changes little, and this will be the case provided

$$\frac{x}{t} = \text{wave speed} \approx \frac{\omega-\omega_0}{k-k_0}.$$

In the integrand (this means the thing being integrated), although k and therefore ω can assume a continuum of values, they are all around k_0 (look at the range of integration), therefore we can say that the right-hand side of the integral for ϕ is least changing at the wave speed given by

$$\frac{\omega - \omega_0}{k - k_0}$$

and this value approaches

$$\frac{d\omega}{dk} = c_g$$

in the limit. This can be shown by using Taylor expansion about k_0 and ω_0, see the book by Mei (1989) for more details. We call this the *group velocity* of the waves, and use the notation c_g. Since

$$\omega^2 = gk \tanh kh$$

c_g can be calculated explicitly. Here are the details for those who like to know them. Starting with the dispersion relation we have

$$\omega^2 = gk \tanh kh.$$

Differentiating gives

$$\frac{d}{dk}(\omega^2) = \frac{d}{dk}(gk \tanh kh)$$

and, using the product rule on the right and implicit differentiation on the left, we find that

$$2\omega\frac{d\omega}{dk} = g \tanh kh + ghk\text{sech}^2 kh.$$

We now divide through by ω and then manipulate the right-hand side to get

$$2\frac{d\omega}{dk} = \frac{g}{\omega} + \frac{\omega ghk}{\omega^2}\text{sech}^2 kh$$

and finally

$$2c_g = c + \frac{\omega ghk}{gk}\frac{\text{sech}^2 kh}{\tanh kh},$$

where the dispersion relation has been used again. The factor gk is cancelled from the second term on the right and a little manipulation occurs through knowledge of hyperbolic functions as follows

$$\omega h \frac{1/\cosh^2 kh}{\sinh kh/\cosh kh} = \frac{2\omega h}{\sinh 2kh} = \frac{2ckh}{\sinh 2kh}.$$

Thus, we get the result

$$c_g = \frac{1}{2}c\left(1 + \frac{2kh}{\sinh 2kh}\right),$$

which matches the result for the speed of propagation of energy. For those who cannot follow this kind of manipulation, the computer algebra packages are worth trying. However, they often give correct results, but not the ones wanted especially with trigonometric and hyperbolic identities where the choice is so rich. Anyway, it is the result that is important not the mathematical details of its derivation. It can be seen that c_g is always less than the wave celerity c, and if kh is large $c_g \sim \frac{1}{2}c$. This is consistent with the result obtained in Chapter 3 for pure sinusoidal waves (equivalent to $h \sim \infty$ here). The other end of the range is more interesting. If kh is very small, then $c_g \sim c$. This is true for waves that have a very large wavelength compared to the depth of the sea. There are two very different species of wave that fall into this category, tides and tsunamis. Tides have a wavelength typically several hundreds of kilometres, so k is of order 10^{-4} and over the continental shelf the depth is ~ 200 m. Thus kh is about 10^{-2}. Tides therefore transmit a lot of energy and this is attempted to be harnessed by tidal power schemes. This energy pales into insignificance when compared to that of a tsunami (see Chapter 6), the wave caused by a subterranean earthquake. These were dealt with in some detail in Chapter 6, but to recap briefly, although in the deep ocean the depth can be several thousand metres (with an average of 3700 m) the wavelength of these waves is more than ten times this so the energy propagates as if it is a shallow water wave. The celerity is a mind bogglingly large 500 plus metres per second (it is given by $\sqrt{gh}$ here). Armed with these numbers, the devastation caused by tsunamis is less surprising. Fortunately, they are rare phenomenon.

Before going on to discuss real sea waves, there is another kind of wave found nearshore — this is a wave that travels not towards the shoreline, but parallel to it. It is called an edge wave. The study of them follows on nicely from a study of water waves, but they are also waves that are trapped against the coast so, in some respects, belong in Chapter 12. Nevertheless, the opportunity is taken to introduce them here.

8.3 Edge Waves

To model edge waves, we need waves that propagate along the shore. A wave will be assumed sinusoidal, but it is essential that its amplitude decays with offshore distance otherwise there will be problems. Infinite energy for one. So, to firm ideas, let us have the y-axis as the shoreline, and assume it is straight. The x-axis is pointing directly out to sea. Then an edge wave will take the form

$$\eta(x, y, t) = F(x)\cos(ly - \omega t),$$

where complex exponentials have been avoided; they are not really necessary for this treatment. These waves will have a wavelength that is small enough for them not to be influenced by the rotation of the Earth, which is why they are being considered here and not in Chapter 12. Confirming this assumption, the frequency of these edge waves is much greater than the local value of the Coriolis parameter; observations indicate a typical edge wave period of about 13 minutes (Yanuma and Tsuji, 1995), which is much less than $2\pi/f$, close to 24 hours. In terms of frequency, these measured edge waves have frequency of about $2\pi/(13 \times 60) = 8 \times 10^{-3} s^{-1} \gg f \approx 10^{-4} s^{-1}$ so, ignoring Coriolis effects, is confirmed as a reasonable assumption. Also, let us assume that they are linear in much the same way as surface waves are assumed to be. The equations can thus be a simple set as follows

$$\frac{\partial u}{\partial t} = -g\frac{\partial \eta}{\partial x},$$

$$\frac{\partial v}{\partial t} = -g\frac{\partial \eta}{\partial y},$$

$$\frac{\partial \eta}{\partial t} + \frac{\partial}{\partial x}(uh) + \frac{\partial}{\partial y}(vh) = 0.$$

These are the LTE without rotation and seem a good place to start. As edge waves do not simply pop out as a natural solution once boundary conditions are set, we need to come up with a suitable form that might generate them. For the best full mathematical treatment, students are guided towards the books by Leblond and Mysak (1978), Johnson (1997) and Mei (1989). To progress we try our wave solution and see where it gets us. So, let us put the above surface wave η together with associated velocity

$$(u, v) = (U(x) \sin(ly - \omega t), V(x) \cos(ly - \omega t))$$

into the above three simplified LTE. It will be noticed that the choice of sine for u and cosine for v works; the trigonometric terms cancel from each equation — this is not luck. The phase of u is always 90° ahead of the phase of v; older books used to say that they are "in quadrature", a term this author always found unhelpful. Substituting these expressions into the first two equations gives

$$\omega U(x) = gF'(x) \quad \text{and} \quad \omega V(x) = lgF(x).$$

The third equation then gives

$$(hF')' + \left(\frac{\omega^2}{g} - hl^2\right) F = 0,$$

where the dash denotes differentiation with respect to x, and the depth $h(x)$ also is x dependent. This is a differential equation for the amplitude of the surface wave. It is second order and, therefore, will need two boundary conditions for solution. One of these must be the decay to zero offshore, so $F(x) \to 0$ as $x \to \infty$. No flow through the coast would be $F'(0) = 0$, physically replaced by no mass flux through the coast $h(x)F'(x) = 0$, where $x = 0$. If there is zero depth at the coast, this is replaced by $U(0)$ being well-defined so $F(x)$ has to be differentiable at $x = 0$. At this stage, of course, one could input a real offshore depth profile for $h(x)$ and solve numerically, however, having come this far analytically, choosing either an exponential depth profile or a simple constantly sloping bed is worth doing to see the structure of the edge wave. With a depth given by

$$h(x) = \alpha x,$$

the equation for $F(x)$ becomes

$$(xF')' + \left(\frac{\omega^2}{g\alpha} - l^2\right) F = 0.$$

If one makes the substitution

$$F(x) = e^{-lx} L(2lx),$$

then one can show that (see Exercise 8.7.6)

$$XL'' + (1 - X)L' + \gamma L = 0,$$

where $X = 2lx$,

$$\gamma = \frac{1}{2}\left(\frac{\omega^2}{g\alpha l} - 1\right)$$

and $L(X)$ are Laguerre functions satisfying Laguerre's differential equation. The key to transforming our original version to a "well-known" equation is that its properties are documented. In particular, with $\gamma = n = 0, 1, 2, \ldots$ that there are polynomial solutions given by the generation function

$$L_n(X) = e^X \frac{d^n}{dX^n}(X^n e^{-X}), \quad n = 0, 1, 2, \ldots.$$

The first few are

$$n = 0: \quad \omega = \sqrt{\alpha g l}, \quad L_0 = 1,$$
$$n = 1: \quad \omega = \sqrt{3\alpha g l}, \quad L_1 = 1 - X,$$
$$n = 2: \quad \omega = \sqrt{5\alpha g l}, \quad L_2 = 2 - 4X + X^2.$$

The lowest mode ($n = 0$) gives the edge wave

$$\eta = C e^{-lx} \cos(ly - \omega t),$$

which obviously decays to zero offshore. Higher modes have more interesting offshore structure with nodal lines parallel to the coast.

Also, the higher the mode, the higher the frequency. For example, the second mode ($n = 1$) is the edge wave

$$\eta = Ce^{-lx}(1 - 2lx)\cos(ly - \omega t),$$

which is obviously zero at the line $x = 1/2l$ as well as a frequency that is $\sqrt{3}$ times the frequency of the lowest mode. The dispersion relation for these edge waves is given by the integer value of γ, rearranged as an equation for ω, viz.

$$\omega^2 = (2n + 1)gl\alpha.$$

This is quite different from previous dispersion relations. The frequency of the wave is critically dependent on α, that is the slope of the offshore bed at right angles to the direction of travel of the wave. So, when the slope of the bed offshore disappears, $\alpha = 0$, the wave vanishes too. This is a truly trapped wave.

Other sea bed profiles are of course possible, in particular the exponential profile

$$h(x) = h_0(1 - e^{-\beta x})$$

that represents a zero depth at the coast tending to the constant depth h_0 far offshore, but analytical progress is difficult for such cases. Practically, if there are edge waves then, because there is a wave travelling parallel to the coast, the actual line of zero h has to be wavy in the direction of y. Crescentic bars are features of edge waves that are not modelled by the previous simple model. Another feature, of perhaps more practical importance, is rip currents. Bathers everywhere have to beware of these on many seaside beaches and they are, very unfortunately, the cause of deaths by drowning every year. Now, crescentic bars and rip currents are not directly caused by edge waves, but edge waves give rise to longshore currents. Longshore currents move sand, and these can form offshore bars. The gaps in these bars are where there can be rip currents. Figure 8.2 shows a cartoon of these features.

Trapped waves are considered further in Chapter 12. Let us now move to consider practical issues of modelling sea waves.

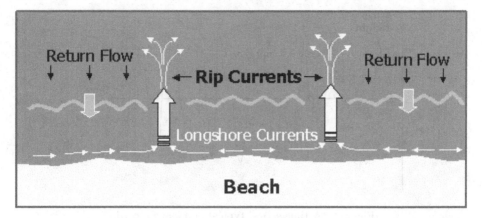

Fig. 8.2 A cartoon showing some coastal features relevant to edge waves and longshore currents.

8.4 Real Sea Waves

In Chapter 3, linear wave theory was introduced and it is this that has been used so far in this chapter. The advantage of linear wave theory lies mainly in the tractable mathematics. The equations of fluid mechanics are (more or less) satisfied, mass continuity is obeyed and we obtain a very convenient formula (the dispersion relation) connecting allowable wavelengths and allowable frequencies, or alternatively allowable wavelengths and allowable wave speeds. We can, as above, also calculate energy and group velocity. These waves are pure sinusoids with no net drift; the water particles move in closed orbits (ellipses). Of course, in a real sea there are very few pure sinusoidal waves. The waves that are visible on the surface of the sea certainly exhibit the to-and-fro motion one thinks of as an essential characteristic, but their form is usually complicated, particularly in stormy conditions. A single record of a sea wave will have an appearance like Figure 8.3. There are a number of different parameters that can be useful in the analysis of such a record or bunch of such records. Here are some of them. If you count the number of times the graph crosses the time axis from under, these are called "zero up-crossings" — there are 11 in Figure 8.3 — then this gives information on how variable the waves are, called wave climate variability. The mean of this period

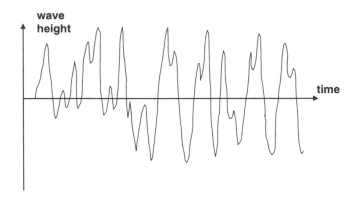

Fig. 8.3   A time series, typical of a wave record.

indicates, for example, the "average" period of the waves and hence gives an indication of the peak frequency, which is 2π divided by this value. An alternative clue to this value would be to measure the mean interval between successive peaks. If these two periods are very different, then this tells us something about the character of the waves; in particular that they are far from sinusoidal. The spectrum of the waves is a term used to indicate the spread of energy in different frequencies, and the broader the spectrum, the less sinusoidal the wave climate. If the peaks themselves are noted, and the mean of the highest one-third of these measured, then this is a useful guide to wave height. It used to be standard practice to assume that a perfectly sinusoidal sea of this height (remember that wave height is twice the wave amplitude) is a good proxy for the real sea in terms of calculating forces on structures and that kind of thing, but this is not a good assumption in general. Other measurements from wave records are the maximum distance between adjacent crest and trough ($H_{\max}$) and the root mean square value of the wave height (H_{rms}). If there are many wave records such as the one given, they of course will all be different in detail. If the values of all the different kind of statistical parameters that indicate period and wave height remain the same, then the process being measured by the time series is called *stationary*. If these parameters are also unaltered as the process is measured at different *places*, as may be the case if waves are being measured by an array of devices (wave spars or accelerometers on buoys) or deduced from information via a succession of satellite images, then the process is also *ergodic*. Most civil engineering calculations on waves have to assume stationarity and ergodicity to a certain extent, but time and length scales must be defined over which both assumptions hold. When we consider extreme waves later in this chapter, stationarity will only be assumed over very long time scales indeed. For mathematicians, the term ergodic often includes stationary — ergodic encompasses lack of variability in statistical parameters over both time and space. Be that as it may, we will go on to discuss the waves of the type represented by the time series of Figure 8.3 in a little more detail. First, however, something on the history of wave forecasting; this will give a useful perspective. Scientific attempts to forecast waves really started in the Second World War with the need to know in advance the sea state so as to judge whether or not certain military operations, such as the Normandy landings, were going to be possible. Two methods emerged: the Pierson–Neumann–James (PNJ) method and the Sverdrup–Munk method, soon to become the Sverdrup–Munk–Bretschneider (SMB) method. The methods have similarities, but the PNJ method was seen to be the more practical method; in this method, each wind velocity produces a certain range of wave periods with a well-defined maximum, with the total range of periods increasing with the wind velocity along with the energy within the total spectrum. Charts were used and maximum wave heights read off using the complete spectrum of the waves. This method was often thought a black art and has not survived the computer age. It has been superseded by the less empirical significant wave method upon which the SMB method is based. This method starts with a theory of wind generated waves and deduces a spectrum to describe, in a natural sea, the amount of energy one can expect at each wave frequency.

The distribution of wave heights within such a time series is often assumed to obey the following rule: the probability that a particular wave height h_w exceeds a given prescribed wave height H_w is

$$e^{-2(H_w/H_s)^2},$$

where H_s is the significant wave height (the average height of the highest one third of the waves, the definition that was mentioned above). This expression is the cumulative distribution, which is the integral of the corresponding probability density function

$$p(h) = \frac{2h}{H_{\text{rms}}^2} e^{-(h/H_{\text{rms}})^2},$$

where H_{rms} is the root mean square wave height of all the waves. This is the Rayleigh distribution. Practically, there will only be a finite number of waves and the graph of p against h (Figure 8.4) is a "line of best fit" through a wave record analysis. The total number of waves in a particular record is N_w, say, and, assuming a Rayleigh distribution, it is possible to find estimates of various parameters. For example, the maximum wave in the record is estimated by

$$H_{\max} = H_s \left(\frac{1}{2} \ln N_w \right)^{1/2} \sim 1.6 H_s$$

for a typical 20 minute wave record.

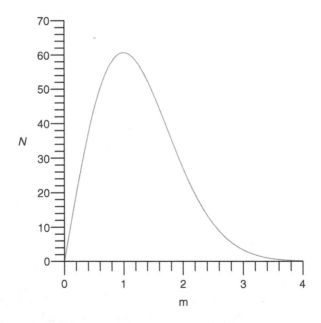

Fig. 8.4   The general shape of the Rayleigh distribution: Relative number of waves N against wave height h in metres m. p in N divided by the total number of waves measured.

In turn, we can also deduce that $H_s = H_{rms}\sqrt{2}$. Unfortunately, the Rayleigh distribution only really applies to a train of linear (sinusoidal) waves whose wavelengths do not vary too much, and although it can still be used successfully for swell waves and some wind waves of the variety that might cause long term erosion, it is not good for, say, storm waves. For example, measurements of real waves indicate that $H_s \sim 1.42 H_{rms}$ for swell waves, which is in accord with the Rayleigh distribution, but give $H_s \sim 1.48 H_{rms}$ for storm waves. Clearly, more sophisticated spectra are required. Before moving on, it is beneficial to spend a little time understanding this concept of a wave spectrum. The underlying assumption behind wave spectra is that a real sea with all its complexity can be expressed as a sum, perhaps an infinite sum, of sinusoidal waves. For those who know about Fourier series, it will come as little surprise that this is possible. For those who do not, the idea is that *any* periodic signal can be decomposed into a sum of sinusoidal signals of differing frequencies. An extension of this is that any signal whatsoever is composed of a spectrum of frequencies. This might seem preposterous, but one quite convincing analogy is that a complicated television picture can be generated through the careful combination of just three primary colours. For signals, however, there are no primary frequencies and usually many are required to generate a given signal. It is tempting to believe that the spectrum of a signal is merely a function which expresses for each frequency, the intensity of that sinusoid required successfully to generate the original signal. This, however, is not quite the case. The reason is because signals do not start or finish but are infinitely long and this gives convergence problems. A single wave, if perfectly sinusoidal, will have a form

$$\eta = a\cos(\omega t + \epsilon),$$

where a is the amplitude of the wave, ω its frequency and ϵ the phase. A whole train of these waves will take the form

$$\eta = \sum_{n=0}^{\infty} a\cos(\omega nt + \epsilon_n).$$

The infinity is, in reality, replaced by a very large integer. The right-hand side is almost a Fourier series and can be made precisely so with some mathematical tidying up. This is not done here. Fourier series can emulate any time series in theory, but in practice other considerations dominate. The energy of a sinusoidal wave is proportional to half of the square of the amplitude. For the above wave train, therefore, the energy is E, where

$$E = \frac{1}{2}\rho g \sum_{n=0}^{\infty} a_n^2$$

and will, in general, be infinite. The dimensions of E are in fact MT^{-2}, which represents energy *per unit area of sea surface*. Hydraulic and coastal engineers will be more familiar with the expression for the single sinusoidal wave of the form $\frac{1}{8}\rho g H^2$, where $H = 2a$. The wave train comprising infinitely many waves will, however it is expressed, remain stubbornly infinite in most cases. One of the most successful ways of overcoming this problem is to define a function called the autocovariance, which is derived directly from the signal. This function, which we have already met in Chapters 2, 3 and 7, measures agreement between two portions of a signal and decays to zero for large times (gaps between the portions). A version that is normalised to take the value one at the origin is called the autocorrelation. As it is formed by multiplying one part of the signal with another, it has a similar form to the energy and can be made to have the same dimensions if multiplied by ρg. Indeed, it is common parlance in electronic engineering to call the time integral of the square the "energy", so if $x(t)$ is a time series of the type depicted in Figure 8.3, then the integral

$$\int_{-\infty}^{\infty} [x(t)]^2 dt$$

is the energy. If $x(t)$ is a length, then this integral will have dimension L^2T, remembering the integration with respect to time. As true energy per unit area has dimension MT^{-2}, one needs to multiply this by ρg and divide by time, perhaps the duration of the record, in order to be dimensionally consistent. It is the Fourier decomposition of the autocovariance of the signal that forms what is called the spectrum. This is done by using Fourier transforms, and we briefly show this. If a signal (time series) $x(t)$ is a continuous function of time, then perform the integral

$$X(i\omega) = \int_{-\infty}^{\infty} x(t)e^{-i\omega t}dt.$$

This integral encapsulates the behaviour of the signal as a function of frequency, ω. The presence of $i = \sqrt{-1}$ should not trouble you, but the possible convergence problems associated with the infinite integrals are the subject of modifications that we return to later. The function X is the Fourier transform of x. It can be shown by double application of the Fourier transform that

$$\int_{-\infty}^{\infty} [x(t)]^2 dt = \frac{1}{2\pi} \int_{-\infty}^{\infty} |X(i\omega)|^2 d\omega.$$

The quantity $|X(i\omega)|^2$ is called the *energy spectral density* of the signal and the above relationship is Parseval's theorem. The appearance of 2π is unavoidable and is due to the period of sine and cosine functions. In fact, this factor can appear all over the place in Fourier transforms and cause confusion. Some books prefer dealing not with ω but with frequency in Hertz (f) where $\omega = 2\pi f$; in this way all (2π)s are in the exponent. In order to widen the scope of Fourier transform theory in the representation of sea waves (and waves in general), the power

$$\lim_{T\to\infty} \frac{1}{T} \int_{-T}^{T} [x(t)]^2 dt,$$

rather than the energy, is considered. The spectrum of this function is termed the "power" spectral density, or just the spectral density, and it is this that is usually dealt with by coastal and offshore engineers. This is as much detail as it is appropriate to delve into here. The subject of signal analysis and signal processing is a complex one and forms significant parts of electronics courses. The mathematical prerequisites are also beyond what this text has assumed.

Coastal engineers and marine physicists have expended considerable effort over the years to find the spectra that describe sea waves. After several oversimplified spectra, two have emerged that, in general, seem to fit observations. These are the Pierson–Moskowitz or

PM spectrum and the JONSWAP (JOint North Sea WAve Project) spectrum. These are the most widely used spectra that are in closed form; however, the WAM (Wave Model) is a so-called third generation model and is of a complexity apart from the PM and JONSWAP spectra. However, it applies to deep water waves and so is only of peripheral interest here. The waves on the surface of the sea are caused by the action of the wind. Precisely how this occurs remains one of the great unsolved mysteries, although there are several very well thought out contenders; nevertheless, the consensus is that this is the case. If the wind has blown for long enough at a requisite speed and has enough sea area to act upon, then the waves are neither duration limited nor fetch limited. Such a sea is termed fully developed and it is the spectrum of such a sea that is well described by the PM spectrum. The form of this spectrum is

$$S_{PM} = 5 \left(\frac{H_s}{4} \right)^2 \frac{\omega_p^4}{\omega^5} \exp \left\{ -\frac{5}{4} \left(\frac{\omega}{\omega_p} \right)^{-4} \right\},$$

where ω is the frequency, ω_p is the frequency at the peak of the spectrum and H_s is the significant wave height (the average height of the highest one third of the waves). The shape of the spectrum $S_{PM}(\omega)$ is shown in Figure 8.5. If the sea is not fully developed, then the spectrum is a lot harder to determine. However, the importance of such a determination was enough for an expensive observational programme to be instigated in the early-1970s. The JONSWAP spectrum was the eventual outcome of these observations. The functional form of the JONSWAP spectrum is as follows

$$S_J = 3.29 \left(\frac{H_s}{4} \right)^2 \frac{\omega_p^4}{\omega^5} \exp \left\{ -\frac{5}{4} \left(\frac{\omega}{\omega_p} \right)^{-4} \right\} (3.3)^{\phi(\omega/\omega_p)},$$

where

$$\phi(x) = \exp \left\{ -\frac{1}{2\beta^2} (x-1)^2 \right\},$$

$$\beta = 0.07, \quad \omega \le \omega_p \quad \text{and} \quad \beta = 0.09, \quad \omega > \omega_p.$$

The JONSWAP spectrum is not so widely accepted as the PM spectrum, especially outside Europe, in particular the "3.3" can be anything between 1 and 7; 3.3 is simply an average value. The JONSWAP

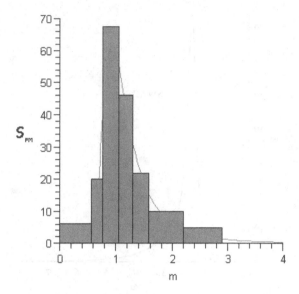

Fig. 8.5   The PM spectrum: The histogram denotes data to which the PM spectrum is fitted.

spectrum is applicable to seas of limited fetch and duration; these are called "not fully developed" seas. In practice, this means that the spectrum is a lot more peaked and the power of 3.3 that multiplies what is essentially a PM spectrum is there to represent this peakiness. The factor in the exponent of this term is large and negative almost everywhere, which means that the term itself is very close to unity outside a small range around ω_p, the peak frequency. Hence, the general shape of the PM spectrum is preserved outside a small range. This is shown schematically in Figure 8.6.

The "tail" of the PM and JONSWAP spectra both obey the inverse fifth power "law",

$$S(\omega) \propto \frac{g^2}{\omega^5},$$

first proposed by Owen Phillips, the pioneer of the study of ocean waves, in 1958. He reasoned this from considering that the waves in the high frequency part of a wave spectrum will be controlled by gravity. By looking at the definition of S in terms of the integral of the square of displacement ($x(t)$ — wave amplitude) we found that

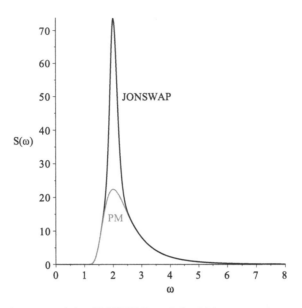

Fig. 8.6   A schematic of the JONSWAP and the PM spectra shown on a spectral density S vs. frequency ω graph; the peak frequency is at $\omega = \omega_p = 2$.

it has dimensions L^2T. Following Phillips, we can use dimensional analysis as follows and assume that $S(\omega)$ will, at the high frequency end of the spectrum, only depend on gravity and the frequency. Thus, we can put

$$S \propto g^A \omega^B,$$

which gives the dimensional equation

$$L^2T = (LT^{-2})^A(T^{-1})^B,$$

leading to the two equations

$$A = 2 \quad \text{and} \quad B = -2A - 1 = -5,$$

as stated. Once this is accepted, then the analytical form of the PM spectrum follows from the assumption that the low frequency part is exponential; the PM spectrum glues the two together. The JONSWAP spectrum preserves these properties whilst also coping with the aforementioned peakiness. The extra complications of the JONSWAP spectrum are to be expected given its applicability to fetch and duration limited seas; the two values of β indicate a (slight) asymmetry and its shape really due to the fetch or duration limited sea having less time to disperse than the fully developed sea, which will have had more time to spread energy through its spectrum via various interactions. The shape of the JONSWAP spectrum is shown in Figure 8.6.

One aspect of real sea surface waves yet to be mentioned is directionality. One only has to glance at a real sea surface to realise that, in almost all cases, the waves have a direction of propagation. The PM and JONSWAP spectra refer only to one dimensional waves, and this could be the main direction of propagation; however, in reality sea surface waves do have a component at right angles to this main direction, so that they are definitely two dimensional. In offshore engineering applications, such as the forces due to waves on a floating or fixed structure, this directionality is very important and a great deal of research has taken place to model it. At the coast, it is less important as refraction tends to line waves to be parallel to the shore. If waves are taken as two dimensional, then particular angular distribution functions have been proposed. A common choice is $|\cos \frac{1}{2}\theta|^s$, where θ runs from 0 to 2π and denotes the direction and s is a parameter dependent upon the frequency controlling the directional distribution of the wave energy. The expressions

$$s = s_p \left(\frac{\omega}{\omega_p}\right)^{-2.5}, \quad \omega \geq \omega_p,$$

$$s = s_p \left(\frac{\omega}{\omega_p}\right)^{5}, \quad \omega \leq \omega_p,$$

can be shown to fit a JONSWAP type of spectrum offshore. Here, ω_p remains the peak frequency and $s_p = 11.5(\omega_p)^{-2.5}$. More information about directional spectra can be found in the paper Mitsuyasu *et al.* (1975). Fortunately, we are principally concerned with coastal processes where Snell's law of wave refraction means that the waves align themselves parallel to the coast. In fact it gets better, in applications to modelling coastal erosion, although knowledge of the PM and JONSWAP spectra can be useful, as indeed is some knowledge of directional spectra, it turns out that whichever spectrum is used the predictions are similar. In attempting to formulate models to predict the movement of bed material, a commonly used technique is to find an "equivalent monochromatic wave" which turns out to be $\sqrt{2}U_{\mathrm{rms}}$, where U_{rms} is the root mean square value of the spectrum (Soulsby, 1997). This is not true for very small values of mean speed, but these small speeds are of no concern here. The reason that U_{rms} features is that this quantity is the standard deviation of all the speeds that emerge from superposing all the orbits in a JONSWAP or PM spectrum. It is perhaps correct to be suspicious of linear theory and models based on linear theory when this is being applied to storm waves eroding a coast. There are alternatives, but each has its problems. The technique whereby linear theory is modified through expansion of variables in a suitable small dimensionless parameter (the wave slope is a popular choice) is called a perturbation technique, and this has been successfully employed for over 100 years. Military and civil defence shore protection codes are formulated using Stokes 2nd–5th-order solutions. This means that the variables in the governing nonlinear equations have been expanded up to 5th power in the small parameter and like powers equated. This is used for water deeper than $0.01gT^2$ where T is the wave period. The wave has steep crests and shallow troughs when compared to sine waves (linear theory). For very shallow water (between $0.003gT^2$ to $0.016gT^2$) cnoidal theories have been used. Cnoidal wave theory is based on nonlinear equations and the surface wave profile is expressed in terms of Jacobi elliptic functions, which were never exactly common knowledge and are certainly rather obscure these days. Theories based on streamfunctions (similar range to cnoidal waves) together with other theories, can be found in specialist books such as Sleath (1984), but none of them are without problems. The second order Stokes theory gives the following maximum speed of current under a crest due to a monochromatic wave U_c

$$U_c = U_W \left[1 + \frac{3kh}{8\sinh^3(kh)} \frac{H}{h} \right],$$

whereas the maximum speed of current under a trough is U_{tr}, where

$$U_{tr} = U_W \left[1 - \frac{3kh}{8\sinh^3(kh)} \frac{H}{h} \right].$$

The asymmetry is important and tends to drive sediment onshore (Soulsby, 1997). In these expressions, the quantity U_W is the wave orbital speed as dictated by linear wave theory

$$U_W = \frac{\pi H}{T \sinh(kh)},$$

h is the water depth, H is the wave height and $k = 2\pi/L$ is the wave number (L is the wavelength).

8.4.1 *Extreme events*

As good as the JONSWAP and PM spectra are at wave prediction, they are at their worst when the situation is at all unusual. In offshore and coastal engineering, there has long been interest in being able to predict the "100 year wave". This is commonly thought to indicate the wave that only occurs every 100 years. This is wrong. The 100 year wave is that wave that has a low probability of returning with a frequency of less than 100 years. Precisely what level of probability to give depends on the statistical distribution assumed, and these statistical distributions are special to the statistics of extreme events. Exceptional storms, their associated floods and extremely high winds are the kind of phenomenon that environmental protection agencies are interested in predicting. They turn to the statistics of extremes to help them. Of course, a dynamic model based securely on hydrodynamics that, in the natural course of running, will predict the largest high water, the extreme current, is ideal, but this approach has only really been successful in predicting storm surges. The flooding associated with storms, rivers spilling over their banks and the like still unfortunately surprises us. The major problem in predicting extreme events is the lack of data upon which to base predictions. Commonly, the kind of statistical distribution assumed contains a double exponential, that is the exponential of an exponential. Taking logarithms twice is necessary in order to render the line of best fit a straight line. Unfortunately, this bunches up almost all the data and renders extrapolation to home in on the extreme event subject to large uncertainty. In the theory of the statistics of extremes, there are many distributions from which to choose and the normal way of proceeding is to estimate parameters and allow these estimates to govern the choice of distribution. It turns out that the estimated parameters lead to the choice of something called the Gumbel distribution (named after Emil J. Gumbel (1891–1966), an outspoken and courageous Jewish but German statistician and pacifist, exiled eventually to the USA). Here are some details for the Gumbel distribution

$$p(x) = \frac{1}{\alpha} \exp \left\{ \frac{x-k}{\alpha} - e^{[(x-k)/\alpha]} \right\},$$

where α is a scale parameter of the distribution to be estimated. The second parameter k is a location parameter, which again is estimated from data. Writing

$$F(x) = \exp(-e^{[(x-k)/\alpha]})$$

and taking logarithms twice gives

$$\frac{x-k}{\alpha} = y = \ln\{-\ln[F(x)]\}.$$

Plotting y against x thus gives a straight line, but the nested logarithm bunches up most data one would want to use to estimate α and k. Special "Gumbel" paper is produced to help with practical use. Another distribution, perhaps more widely used in the civil engineering community, is the Weibull distribution

$$p(x) = \begin{cases} abx^{b-1} \exp\{-ax^b\}, & x > 0, \\ 0, & x \leq 0. \end{cases}$$

This distribution is named after Waloddi Weibull (1887–1979) an unusual Swedish engineer and scientist who developed the distribution in 1914 in response to the analysis of explosions. Weibull was also an inventor, amongst whose inventions are the ball bearing and electric hammer. The Gumbel and Weibull distributions come under the general class of general extreme value (GEV) distributions, but they all have the same problem: Very long records are required for accurate estimation. It is the case that, to predict with any certainty an event that has a return period of N years, a record of length at least $N/2$ is required. In how many sensitive locations are there even 25 years worth of wave data?

It is useful to do some practical examples that demonstrate the use of extreme wave prediction. As a reminder, the significant wave height is the mean of the highest one third of the waves and is given the symbol H_s. Let T_s be the period associated with this wave. The following empirically derived formulae relate these to the speed of the wind at 10 m above the sea

$$H_s \sim 0.025 U_{10}^2,$$
$$T_s \sim 0.79 U_{10},$$

for the PM spectrum, and

$$H_s \sim 5.1 \times 10^{-4} U_{10} F^{0.5},$$
$$T_s \sim 0.059 \{U_{10} F\}^{0.33},$$

for the JONSWAP spectrum. In the latter, F is the fetch as it will be remembered that this spectrum is for use with non-fully developed seas. Although the above relationships are empirical, they have been arrived at by engineers and are the best that can be done under present knowledge. Engineers have codes of practice that govern how they can design and predict from a legal standpoint. The above formulae are part of this code. Let us calculate the significant wave height and period for a wind of 20 m s^{-1} at a height of 10 m above the sea if the sea is fully developed. The PM formulae are used, which gives answers $H_s = 10$ m, and $T_s = 15.8$ s. Suppose now that the fetch is limited to the width of the North Sea, which is about 450 km. We now find the new values of H_s and T_s. This time, the JONSWAP spectrum is appropriate and these formulae give $H_s = 6.8$ m and $T_s = 12.3$ s. Note that these values are significantly less than the fetch unlimited values. Most may think that the North Sea must be wide enough to "wind up" the sea surface; it is not. The two spectra only give the same values if a fetch of close to 1000 km is assumed. Of course, this is very simplified. How much fetch is required depends

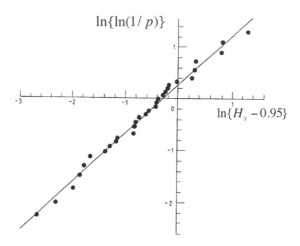

Fig. 8.7   A wave height extrapolation plot: The line has equation $\ln\{\ln(1/p)\} = 0.945\ln\{H_s - 0.95\} + 0.278$.

crucially on the value of U_{10}, which in this example has been assumed to be constant. In reality, the variability of the wind will play an important role, but a constant fierce wind is a worst case scenario.

Figure 8.7 shows a straight line arising from many measurements and forms part of BS6349 (Maritime works.General.Code of practice for geotechnical design). We will try to use the straight line to predict one year, 50 year and 100 year significant wave heights. This graph is actually based on the Weibull distribution. The factor p denotes what is called the "exceedence probability". It is tempting to think that the probability is linked to the return period through the formula

$$p = \frac{1}{T}.$$

The larger the return period, the smaller the probability that the large wave will occur. However, upon examination, this formula only works if T is very large. (If $T < 1$ we get nonsense as probability has to lie between 0 and 1, and if T is close to 1 but larger we still get what amounts to nonsense.) A workable formula is

$$p = \frac{1}{T+1},$$

where the extra 1 in the denominator actually stems from using the formula

$$T = \frac{1-p}{p},$$

which states that the return period is the ratio of the probability of having no large waves to the probability of having such a wave. For large T, (or small p),

$$p \approx \frac{1}{T}$$

is regained. Engineers have to be able to predict long term waves with incomplete data sets. The methods used are based on sound enough statistical techniques, but the margins of error, hardly ever referred to, are high. Nevertheless, here is what is done. What wave records there are available are analysed and all waves that exceed a threshold value are subtracted out. For this example, suppose that in a year long record there are 39 "storms", that is 39 times that the threshold value has been exceeded. The exceedance probability is then given by

$$p_1 = \frac{1}{39+1} = \frac{1}{40}$$

for these data. The exceedance probability for a 50 year wave is then simply one fiftieth of the exceedance probability for a one year wave

$$p_{50} = \frac{1}{50}\frac{1}{40} = \frac{1}{2000}$$

and the exceedance probability for the 100 year wave is half this

$$p_{100} = \frac{1}{4000}.$$

More about the theory behind these formulae can be found in statistics texts that include the practical use of probability. Texts on hydrology and flooding are particularly good as they tend to be written using the right language. Assuming the values of p given above, now estimate the wave heights using Figure 8.7. The answers are

$$1 \text{ year wave} = 3.9 \text{ m},$$
$$50 \text{ year wave} = 7.3 \text{ m},$$
$$100 \text{ year wave} = 7.9 \text{ m},$$

Now, if global warming provides a stormier environment, there is a good case for seeing what happens to these predictions when there are more storms. Suppose the rate increases from 39 to 49 storms each year, we calculate the new wave heights for one year, 50 year and 100 year return periods to be

$$1 \text{ year wave} = 4.1 \text{ m},$$

$$50 \text{ year wave} = 7.5 \text{ m},$$

$$100 \text{ year wave} = 8.1 \text{ m},$$

which, given that the energy is proportional to the square of the wave height, could give designers of offshore platforms and the like headaches. Even more bizarre is the attempted forecast of highest waves. If extreme value statistics are used to predict three hour maximum waves, the Gumbel distribution comes up with waves of height $H_{\max} \sim 30$ m for the northern part of the North Sea. Of course, this is based on records that are too short (three years) and so the findings are suspect. No waves of this height have been seen there and no structure yet built there could survive such a wave. Does it need to? There are more sophisticated approaches that rely on MLE (see Chapter 1). However, this is a technique that uses data to estimate the parameters in a chosen distribution, and since good data are precisely what we do not have, the extra complexity of the MLE technique is hardly worth going through. Confidence intervals can be estimated using "bootstrapping", which is essentially a resampling method that enables one to improve the estimation of error bounds. You may still be wondering why more straightforward techniques cannot be used; perhaps this example will help. Suppose that, there is a time series record where wave heights have been sampled every hour. Suppose that, in this record, only three times has the wave height exceeded a prescribed value (say H_0). Is there any way that a return period for waves of this height can be estimated? The answer is only if the record is long enough. So, if there are $876,000$ records (which corresponds to 100 years of one hourly records), then the probability of exceedance is, directly from the definition of probability, $3/876000$. The return period is thus the reciprocal of this, which translated into years, gives $33\frac{1}{3}$ years. The flaw is now obvious. If the record was only 10 years long, the return period would be $3\frac{1}{3}$ years. The shorter the record, one would expect such large waves would be rare to encounter. Perhaps there is only one, or none at all. No waves means that the calculation of a return period is impossible, one wave gives a return period of 10 years, and two a return period of five years. Our suspicions are aroused. We then see that the three waves all occurred in one stormy dark February (or whatever) and this is why we should question the validity of such "predictions"; the statistics are neither stationary nor ergodic. The book by Reeve, Chadwick and Fleming (2004) has wise words on this subject and delves more into the use of maximum likelihood methods. There is no getting away from the truth however: The longer the record, the better any predictions.

On a practical level, engineers do what they can to design sea defences and the like that can withstand the largest waves and strongest current likely to hit in, say, 100 years. The fact that the kind of stormy weather likely to give these extreme events can cause two "100 year waves" to arise in the same season is unfortunate and drives the non-technical to curse the modellers. The correct reaction is to call for more research; it has already been stated that all of the statistical techniques mentioned in this section, both the extreme statistics and the standard spectra, are based upon the wave field being statistically stationary. This means it is assumed that the overall statistics over a very long time are unchanging. Global warming, is therefore, precisely what cannot be taken into account. All we can do is to suggest some over designing and hope that the extra cost is money well spent.

8.4.2   *Interactions and wider issues*

Here, the effects of currents and waves have been examined separately. A good question to ask is whether there is interaction and, if so, what are their effects? Any interaction will be due to non-linearity, but this can take many forms. The interaction of a wave with itself produces currents and this is called the Stokes drift. Nonlinear interaction of waves occurs because the linear formulation, necessary to perform the mathematical calculations earlier in this chapter, and still a very good approximation, nevertheless are an approximation. The ignored nonlinear term

$$(\mathbf{u} \cdot \nabla)\mathbf{u}$$

means that the orbits of the fluid particles under the wave do not precisely close, and there is a steady drift (which is in the direction of propagation of the wave for a progressive wave). The exact form that this Stokes drift takes depends on the nature of the approximation, but it is useful to approach this via reconciliation of Lagrangian and Eulerian viewpoints, as discussed in Chapter 7 (see Mei (1989) for the full story, but page 462 in this book for Stokes drift). Here is a non-vectorial treatment. Suppose that the wave is one dimensional and has a surface wave $\eta(x,t)$ and associated current $u(x,z,t)$. Earlier in this chapter we derived

$$\eta(x,t) = a\cos(kx - \omega t)$$

and assuming a linear relation (to first order), then we deduced the x wise current under this wave as

$$u(x,z,t) = -\frac{agk}{\omega \cosh(kh)} \cosh k(z+h) \cos(kx - \omega t).$$

However, it is simpler for the moment not to use these formulae explicitly and to keep to a general $\eta(x,t)$ and, in particular, a general $u(x,z,t)$.

Now we take up the ideas of Lagrangian and Eulerian descriptions. The variable z will be taken as a constant in these calculations and so ceases to feature explicitly. The Lagrangian description refers all variables back to the position once occupied by the particular fluid particle, whereas the Eulerian description refers to an origin that is fixed in space. At a particular time and location, $t = t_0$ and $x = x_0 = x(t_0)$. The Lagrangian description u_L is given by

$$u_L(x_0, t) = u(x,t).$$

Note that the actual *value* of the current is the same, it is that they differ in how they are referred. The fluid particle that is at the location x_0 at the time t_0 is the one that is now at the general position (x, t). From the definition of u_L, we have

$$\frac{dx}{dt} = u_L(x_0, t)$$

so that

$$dx = u_L(x_0, t)dt$$

and by integrating,

$$x = x_0 + \int_{t_0}^{t} dt' u_L(x_0, t').$$

What this expression does is relate the starting position to the speed and time taken to get from this start position to the general position. We can use this to determine the Lagrangian current as a function of the Eulerian current at x_0

$$u_L(x_0, t) = u\left(x_0 + \int_{t_0}^{t} dt' u_L(x_0, t'), t\right).$$

The technique now is to use Taylor's theorem (Taylor's series) to expand the right-hand side, then average it over a wave to deduce that part of this flow that is not periodic. We have

$$u_L(x_0, t) = u(x_0 + \epsilon, t),$$

where

$$\epsilon = \int_{t_0}^{t} dt' u_L(x_0, t').$$

Taylor's series is

$$u(x_0 + \epsilon, t) = u(x_0, t) + \epsilon \frac{\partial u}{\partial x}\bigg|_{x=x_0} + \cdots$$

and inserting the expression for ϵ we get

$$u(x_0 + \epsilon, t) = u(x_0, t) + \int_{t_0}^{t} dt' u_L(x_0, t') \frac{\partial u}{\partial x}\bigg|_{x=x_0} + \cdots .$$

Now, notice that on the right-hand side there is a term $u_L(x_0, t')$, but this is multiplied by $\frac{\partial u}{\partial x}$. Hence, *to lowest order* (this phrase is explained in a little while) this can be replaced by $u(x_0, t')$. Secondly, the time average of this needs to be taken over a wave period, and let us denote this averaging by an overbar. So

$$\overline{u_L(x_0, t)} = \overline{u(x_0 + \epsilon, t)} = \overline{u(x_0, t)} + \overline{\int_{t_0}^{t} dt' u(x_0, t') \frac{\partial u}{\partial x}\Big|_{x=x_0}} + \cdots .$$

The left-hand side is termed the *mass transport velocity* (actually it is the *mass transport speed* here in one dimension, but the concept generalises to two and three dimensions where direction is more important). This last expression is worth pondering a little. Suppose $u(x, t)$ is represented by a power series of the type

$$u(x, t) = \delta u_0(x, t) + \delta^2 u_1(x, t) + \delta^3 u_2(x, t) + \cdots ,$$

where $u_0(x, t)$ is the linear speed, $u_1(x, t)$ the first correction etc. and δ is a small parameter, usually the wave slope. This kind of representation is called a perturbation expansion and was all the rage before numerical methods dominated. It is still useful for precisely this kind of modelling as it is a good way of representing a process that is largely linear but with small nonlinear effects. The lowest order is called "first order", written $O(\delta)$, and this would be the linear sinusoidal system with

$$u_0(x, z, t) = -\frac{agk}{\omega \cosh(kh)} \cosh k(z + h) \cos(kx - \omega t).$$

The expression for the mass transport velocity is definitely second order $O(\delta^2)$. The term $\overline{u(x_0, t)}$ is termed the Eulerian mass transport, and tends to be zero for inviscid sinusoidal waves. The third term, which is an interaction, is called the Stokes drift. The equation

$$\overline{u_L(x_0, t)} \approx \overline{u(x_0 + \epsilon, t)} \approx \overline{u(x_0, t)} + \overline{\int_{t_0}^{t} dt' u(x_0, t') \frac{\partial u}{\partial x}\Big|_{x=x_0}}$$

therefore relates the Lagrangian and Eulerian mass transports and can be written in words:

Lagrangian mass transport $\approx$ Eulerian mass transport + Stokes drift.

If the flow is inviscid, the Stokes drift can be a reasonable first approximation to the Lagrangian mass transport. For the sinusoidal wave given, we can do the calculus and calculate the Stokes drift to be

$$u_S(z) = \frac{a^2 g^2 k^3}{2\omega^3 \cosh^2(kh)} \cosh^2 k(z + h).$$

It is tempting to think that if friction is included near the sea bed or sea surface, then away from the boundary layers at the surface or bed, this value of Stokes drift still prevails. Sadly, this is not the case. By including friction in the form of eddy viscosity, the governing equations and boundary conditions change significantly and a different mass transport current is induced in the interior. The (rather complicated) theory for this was done for general oscillatory flows in 1953 by M. S. Longuet-Higgins, but the book by Mei (1989) contains a useful summary. These further calculations will not be pursued here.

The zeroth order current due to a purely sinusoidal wave has an elliptic orbit, and if there is a current (for example a wind-driven flow) in addition to this, then the interaction can enhance the net current to be greater than a simple summation of them would indicate. Of course, the interaction also causes Langmuir circulation (see Chapter 11), which is a special helical current with distinctive important characteristics. Longshore currents can also be generated. These are principally caused by the angle between the crests and the line of the beach. In particular, if the waves break, then there can be sediment transport along the beach, which is very important to quantify in the study of coastal changes (erosion and accretion). Although no detail is given here of wave refraction and defraction, if the waves break, there is a longshore stress term that takes the form $S_{xy} = Ec_g \sin\alpha \cos\alpha/c$, where E is the wave energy density, α is the angle that the wave crests make with the shoreline and c_g and c are the group and phase speeds of the wave, respectively. Now it can be shown that, because of Snell's law, if the depth contours are parallel to the coast and if the bottom friction is negligible, then $S_{xy} = \text{constant}$ and the wave energy is (virtually) all expended when it breaks on the beach. Of course, in this case there is no thrust to cause longshore current. The generation of thrust is done by the onshore gradient of S_{xy} which, using linear wave theory on a sloping beach, is given by

$$\frac{\partial S_{xy}}{\partial x} = \frac{10h^2}{8h^2 + 3a^2}\rho g h \frac{dh}{dx} \sin\alpha\cos\alpha.$$

Because both water depth h and the angle α reduce (the latter due to refraction) as the waves approach the shore, this gradient of S_{xy} also approaches zero from its maximum at the breaker zone. For practical purposes, a formula such as

$$v_l = A\sqrt{gh_b}\frac{dh}{dx}\sin\alpha_b\cos\alpha_b$$

is used to estimate the longshore current speed v_l. Here, A is an empirical constant arising from the combination of the various numerical factors and the drag coefficient and the suffix b denotes the breaker position.

Another, quite separate, interaction is the modification of wavelength and wave celerity by the presence of a current. This modification can lead to wave breaking and, hence, lead to increased erosion; also, a longer wave will attenuate less with depth and, hence, disturb the sediment more as well as possibly producing greater wave generated currents.

The all important interaction, of course, is that between the sea and the coast. Just exactly how does a storm induced wave and current tear into a vulnerable cliff? Of course, we do not *exactly* know, but what is certain is that if the waves are prevented from slamming into the cliff or sea wall then the erosion is reduced. There is a considerable literature on the design of groynes and other coastal protection methods using physical barriers, and for further information the reader is referred to this, which is clearly in the field of coastal engineering. (See Reeve, Chadwick and Fleming (2004), and Masselink, Hughes and Knight (2011).)

As far as managing is concerned, the plan has to be as follows. First of all, there is a pooling of knowledge supplemented by observation exercises where there are gaps. This knowledge must be recent, although it is important to know the historical development of particular coastlines. From this knowledge comes a recognition of those areas which are at risk from flooding or erosion, and cliffs that are unstable can be identified. All this takes place in a routine way, but these days the effects of global warming need to be taken into account. It is expected that there will be increased storm activity, which means that there will be larger mean winds. These winds will, in turn, give rise to stronger wind driven currents as well as a more energetic sea surface. The JONSWAP spectrum will, it is expected, remain reasonably valid, however the parameters within it will change to reflect this enhanced energy. The models of waves and currents given in this chapter and earlier can still therefore be used.

As there is more flooding due to increased rainfall as well as sea level rise, so the characteristics of groundwater are likely to change. This is a very important subject for land and river management but lies outside the scope of this text. As there is still some uncertainty about exactly how severe global warming will be, it would be very expensive to guard against a "worst-case" scenario, which is the natural instinct of all civil engineers. The bolstering of coastal flood defences alone would run into billions of dollars.

Another advance in wave forecasting has taken place, but mainly for applications to waves in the open sea. This is the bringing together of spectral methods with finite difference modelling. The lead here is taken from weather forecasting. In principle, if the winds are forecast with some accuracy, then these winds can be used in wave generation models and wave heights can be predicted. The first generation of such wave prediction models used only the linear terms and contained components that suddenly stopped growing as soon as they reached a universal saturation level (the f^{-5} spectrum as proposed by Owen Phillips in 1958 and justified by dimensional analysis), the next generation used the nonlinear terms but not fully. These models suffered from several limitations and the spectrum shape, far from evolving naturally, had virtually to be prescribed as the nonlinear interactions were insufficiently resolved. In the late 1980s, a concerted international effort led to the publication in 1994 of the WAM model. This is a model of the physics of wave generation and interaction according to present knowledge. It includes wave–wave interaction, dissipation due to whitecapping and sea bed processes and incorporates the ability to cater for rapid change, which previous methods did not. Although simpler methods based on the JONSWAP spectrum, for example, yield similar results most of the time, the advantage of the WAM method lies in its ability to cater for extreme circumstances. There is also satisfaction in having a model that contains better physics. The WAM model also contains a good advection scheme capable of correctly forecasting the propagation of long period swell. The book by Komen *et al.* (1994) contains all the details, and it can be seen there that a lot of effort is given to the correct modelling of the propagation of sea waves on a global scale using spherical co-ordinates. There is interplay between the shape of the spectrum and the source terms, which demonstrates that the nonlinear interaction is both important and modelled realistically. Validation of the WAM model is possible using satellite data and Synthetic Aperture Radar (SAR), though there are technical difficulties converting these data into the right form to do with sampling correctly from such a dense data set. The impact of the WAM approach to wave modelling has not impacted on nearshore and coastal management yet, but it is surely the next step.

8.5   Coastline Change

So far in this text, not very much has been said about the modelling of sediment transport. The study of sediment movement has long been a prominent aspect of coastal engineering and its importance is probably greater now than ever. In Chapter 5, there was a section on modelling the bottom boundary condition using turbulence closure modelling. The modelling of sediment transport follows these lines. Sediment, if it stays on the sea bed, tends not to move to any great extent and so does not hold a lot of interest to the modeller. Movement only happens if the sediment becomes suspended, and this occurs if the sediment is fine enough and if the currents are strong enough. The book by Sleath (1984) gives many empirically derived formulae for the initial flow of sediment over a flat bed. Complications due to assuming laminar or turbulent flow, and whether the bed is approximately flat, is wavy or irregular with certain statistical properties can be incorporated at a later stage. The theory behind some of the formulae is due to Shields (1936) and can be summarised as follows. If there is flow over a bed that consists of loose material then there must be a critical stress that has to be attained if the sediment is to move at all. The value this takes is wrapped up with whether the flow is laminar, turbulent or in transition between the two. Additionally, the physics of the pick up of sediment may be rather different for steady flow and oscillatory flow, such as one finds under surface waves or in a tidal regime. There has been a lot of research examining the relationship between this threshold stress and the diameter and density of the loose grains on the bed, but it is not appropriate to delve into great detail here, although some more will be said in the next subsection. In this book, we are really only interested in sediment transport as it might affect the coastline in terms of erosion and deposition.

8.5.1 *Modelling erosion*

The title of this section is a shorthand for the wearing away of the coast. Of course, the coast can take variety of forms. There are all the different types of beach: muddy, sandy, shingle etc. Then there are rocky shores, cliffs; and then there are places where the forest meets the shore and finally there are man-made shorelines — these are usually permanent structures — where "permanent" means lasting 50 plus years. Erosion is a name usually preserved for the loss of beach or cliff; of course, the problem with man-made permanent structures is that, because nature is prevented from eroding a particular stretch of coast, there is a danger that a point somewhere else on the coast, where material would have been deposited, ceases to be a deposition point. Man has to be very careful about such non-local effects, and modelling can help prevent mistakes being made. Tied up with the erosion process is sediment transport which, as has already been said, is complex. In the past ten or so years, there have been large co-operative modelling and linked observation exercises that have taken place based around specific physical locations, for example the Humber estuary in the UK. These have helped in our understanding, but they have also confirmed how complex the processes are.

In order to make any headway, let us first describe some of the processes that transport sediment. Here, we are fully in the realm of coastal engineering, so the reader is guided to books by authors fully involved in that subject for more detail. The book by Reeve, Chadwick and Fleming (2004) is a good modern example, and the descriptions here have mainly been distilled from there. Most of the material that lies along the coast in the form of sand and silt has been washed there from adjacent sea or washed down from rivers. The origin of the material is more than likely the last ice age, but recent storms can also contribute. Civil engineers calibrate as follows: sand is very fine (0.00625–0.037 mm); fine (0.037–0.25 mm); medium (0.25–0.5 mm); coarse (0.5–1.0 mm) and very coarse (1–2 mm). After that the sediment is classified not as sand but as gravel (sometimes called shingle in the UK), and this is subdivided into the following classes: granular (2–4 mm); pebble (4–64 mm); cobble (64–256 mm) and sediment bigger than 256 mm are called boulders. The astute amongst you will see that this scale contains powers of two, which means that there is an underlying power law classification. It is not only the size that is important of course, but the density and perhaps also the shape. Between the larger pieces of gravel there will be spaces (called voids) which can be filled with either air or water. Water, in particular, can aid the rolling or slipping of gravel or sand and facilitate transportation. If we just consider beaches for a moment, any particular beach will contain a variety of sizes of sediment, but not as much as you might think. Many beaches are largely of one particular size range and this is one of the factors that has governed the way in which the grain size has been classified. The slope of the beach also is highly dependent on the grain size: the larger it is the greater the slope can be. Now let us address the way sediment is transported. This is done principally in two distinct ways: first of all the sediment can roll or slide along the bed over other sediment, or it can move in suspension. The suspension is called the "suspended load" by civil engineers, whilst the sediment that moves on the bed is called the "bedload". Bedload is the principal mechanism for sediment transport, suspended load only becoming important for fine sand or very high speed flows. An important criterion is at what current speed does sediment start to move. This brings us to think of the two mechanisms that make the fluid particles that form the sea move: currents and waves. If the current is dominant, then all sediment transport has to be bedload as there is no mechanism for suspending the sediment. On the other hand, under large waves, the bedload can be transformed into suspended load as the finer particles get caught up in the fluid, which away from the bed itself might have a significant (non-hydrostatic) vertical speed. A common way to relate the mean current speed to the bed shear stress is to use the quadratic law, derived by dimensional analysis in Chapter 3

$$\tau_0 = \rho C_D \bar{u}^2.$$

Again using dimensional analysis, we can derive a quantity called the shear speed that characterises the bed stress and the fluid density. From the above expression (or again from first principles, the choice is yours), the shear speed will be dependent on $\sqrt{\tau_0/\rho}$, so we define the shear speed as

$$u_* = \sqrt{\frac{\tau_0}{\rho}}.$$

There are many empirically derived expressions that have been proposed that are used to determine currents generated under specific sediment and wave environments. The main difficulty with modelling erosion looks insurmountable. The important mechanisms that control the erosion process lie precisely in the interface between a violently moving sea and a soft moving land. This is notoriously difficult to model. Indeed, even assuming a quiet sea near the bed, very small length scales are required to capture the physics; this formed part of the discussion in Chapter 3 when formulating models of sea bed friction. These small scales are at odds with the kind of larger scale finite difference models adopted in coastal sea modelling. There is thus something of an impasse. When the sea is anything but quiet, which is of interest here, there is not much hope of an accurate model. Nevertheless, some attempts have to be made to model the effects of a violent nonlinear sea with its foaming breaking waves crashing against and undercutting sea cliffs. The secret lies in averaging over time in order to forecast long term trends, remembering that this is the long term goal.

8.5.2   *Currents*

One common assumption for currents is that close to the sea bed they have an approximately logarithmic profile. This was alluded to in Chapter 5. The horizontal current u is a function of height above the sea bed z, an indicative roughness length z_0 and a "friction velocity" u_*, which is approximately $\sqrt{\tau_0/\rho}$, where τ_0 is the bed shear stress. This profile is given by

$$u(z) = \frac{u_*}{\kappa} \ln\left(\frac{z}{z_0}\right) = \frac{1}{\kappa}\sqrt{\frac{\tau_0}{\rho}} \ln\left(\frac{z}{z_0}\right),$$

where κ is von Karman's constant, usually attributed to be 0.4. Estimates of z_0 usually rely on classical experiments done in the 1930s by J. Nikuradse. However, z_0 has been related to the viscosity of seawater ν (not eddy viscosity; we are firmly in a laminar boundary layer this close to the bed). The constant k_s is called the Nikuradse roughness, which is directly related to the grain size. Experimental fit gives

$$z_0 = \frac{k_s}{30}\left[1 - \exp\left(-\frac{u*k_s}{27\nu}\right)\right] + \frac{\nu}{9u_*}.$$

The expression $\left(\frac{u_* k_s}{\nu}\right)$ is a Reynolds number (ratio of inertia to viscous terms), which is traditionally the important dimensionless ratio in a fluid lacking waves or the influence of Coriolis effects. Some simplification of this formula for certain ranges of Reynolds number is possible, and the more specialist text of Soulsby (1997) is recommended for further details.

Power laws are prevalent in the coastal engineering literature. For example, the formula

$$u(z) = \left(\frac{z}{0.32h}\right)^{1/7}\bar{U}, \quad 0 < z < 0.5h$$

has been successfully employed to model a tidal current in the bottom half of a flow where the depth of water is h. In the top half, the constant value $1.07\bar{U}$ suffices. (Here, $\bar{U}$ is the depth mean tidal flow.) The one seventh power law is only justified by fitting a power law through a considerable amount of data taken from a variety of coastal seas and estuaries. Fitting a straight line to a log–log plot inevitably produces a power law, and the exponent 1/7 has a long pedigree from mechanical engineering (Schlichting, 1975, p. 600). The logarithmic law can be derived from dimensional analysis in that the velocity gradient can be deduced to be proportional to a representative velocity (u_*) divided by the vertical co-ordinate z, which is a (the only) representative length scale. Integration of this yields the logarithmic law. However, it is freely admitted that this is not an area where the science is exact; remind yourself of what we are trying to model here.

Currents are, of course, important, but so are waves and, in order to assess environmental impact, both need to be considered together. There is modern software that does this, and some of this software is examined in the next section.

8.5.3 *Software for erosion modelling*

The first, and perhaps simplest, model to mention uses the acronym GENESIS, which stands for the GENEralised model for the SImulation of Shoreline change. It is a model and contains limited physics. On the other hand, it is reasonably accessible to students. The program requires wave characteristics such as its period, height and direction. It also requires a shoreline position. The grid is set up with the x axis along the shore and the y-axis pointing seaward. The software is based on finite difference methods, and because of its one dimensional nature, there are a number of simplifying assumptions that have been made. The shoreline is assumed to move seaward or shoreward with no change of profile shape. The sediment is assumed to sit on the bed rock as a lump, something coastal engineers call a "berm". The height and width of this berm dictates how much sediment there is to move. In the original specification that incorporates the movement of breakwaters as well as shorelines, there are 14 parameters that specify the processes. Typically for a coast, the berm lies in front and over shoreline rocks and approximates a beach, and as it moves so does the shoreline. GENESIS assumes that this berm is moved by the waves and currents in a simple way, given the above parameters. There are no hydrodynamics as such and the waves and currents directly move the sand in a way dictated by the input data, particularly the extent of the berm and the source (or sink) of sediment.

It turns out that most beaches maintain an average shape that depends on the general shape of the coast. Although there are seasonal changes, these are cyclical and only over a very long time does the coast change appreciably. The changes in coastline tend to be parallel to themselves, that is propagating shoreward or seaward without overall change in shape. In keeping with this, sand are thus transported between two extremes. Since the actual dynamics of the transported sand are difficult to model accurately, there are empirical formulae and these tend to be governed by breaking waves. The assumption that the shoreline always moves parallel to itself is, of course, violated near structures such as groynes and jetties. In reality, there is also longshore transportation, which is incorporated, albeit crudely, by GENESIS. The equations inside GENESIS are not difficult to derive conceptually and they are a useful guide to the form of the model, so this derivation is repeated here. If D_B is the elevation of the berm and h the water depth from chart datum (mean sea level usually), then the change in volume is simply $\delta x \delta y (D_B + h)$. This change comes about due to sand entering and exiting its four sides. If there is a difference in longshore transport rate, call it δQ, then

$$\delta Q \delta t = \frac{\partial Q}{\partial x} \delta x \delta t$$

gives the net volume change due to this mechanism. Additionally, there will be the shoreline change itself that can be represented by the rate q of sand that produces a volume change $q \delta x \delta t$. The total balance is thus

$$\delta x \delta y (D_B + h) = \frac{\partial Q}{\partial x} \delta x \delta t + q \delta x \delta t.$$

Applying the usual calculus limit taking infinitesimal quantities to zero gives the equation

$$\frac{\partial y}{\partial t} - \frac{1}{D_B + h} \left(\frac{\partial Q}{\partial x} + q \right) = 0.$$

Just looking at this one equation, already there are four parameters, Q, q, D_B and h, that have to be prescribed. The quantity Q is given by the formula

$$Q = (H^2 c_g)_b \left(a_1 \sin(2\theta_{bs}) - a_2 \cos(\theta_{bs}) \frac{\partial H}{\partial x} \right)_b,$$

where H is wave height, c_g the group velocity of the waves, b is a sub-script denoting breaking conditions and θ_{bs} is the angle of the waves to the shoreline. The constants a_1 and a_2 are given by empirical formulae that are calibrated according to the property of the sand

$$a_1 = \frac{K_1}{16(\rho_s/\rho - 1)(1 - p)(1.416)^{5/2}}$$

and

$$a_2 = \frac{K_2}{8(\rho_s/\rho - 1)(1 - p) \tan \beta (1.416)^{7/2}},$$

where K_1, K_2 are calibration constants, ρ_s the density of the sand $(2.65 \times 10^3 \text{ kgm}^{-3})$, ρ the density of water $(1.03 \times 10^3$ for seawater), p the porosity of the sand, usually taken as 0.4, and $\tan \beta$ is the average bottom slope from the shoreline to the depth of the active shoreline transport. The values 1.416 that appear in these formulae are to convert from significant wave height (mean height of the third highest waves) to the root mean square (rms) height that is required by GENESIS. The average profile shape is given by the simple formula $h = Ay^{2/3}$, where A depends on beach grain size. The average near-shore slope is related to the width of the littoral zone y_{LT} through

$$\tan \beta = A(y_{LT})^{-1/3}.$$

If D_{LT} is the depth of this zone, then $y_{LT} = (D_{LT}/A)^{3/2}$, whence

$$\tan \beta = \left(\frac{A^3}{D_{LT}} \right)^{1/2}.$$

The last parameter to discuss here is the depth of closure. This is a theoretical depth and represents the seaward depth beyond which the depth profile is constant. It is the very definition of what is meant by "the shore". Common sense needs to be applied to home in on a figure appropriate to particular coastlines, however this does bring home the empirical nature of GENESIS and, although it is a very useful tool, this empiricism is a limitation.

The next model discussed in some detail here is called MIKE
21. The name derives from Mike Abbott, who led the initial developments in the UK, The Netherlands, Denmark, Belgium and indeed all
over the world. There are now a whole suite of sophisticated models.
At the time of writing, the website boasts the suites called MIKE 3,
MIKE 11, MIKE SHE as well as MIKE 21. They all have different roles, including river modelling and ecosystems; we shall focus
on MIKE 21, which is "the industry standard for advanced modelling of coastal hydrodynamics, environmental conditions, waves and
sediment transport" according to the website based at the Danish
Hydraulics Institute: http://www.dhigroup.com/. There is a whole
industry based on this modelling suite, including large projects and
short courses. Let us run through the details of MIKE 21, chosen
because of its sophistication and because it happens to be available. (It is still too common for the modelling details of proprietary software to be hidden inside a black box so one is buying a
pig in a poke. This is not the case with MIKE 21.) The version
explained here is dated 2004, and add-ons and updates do occur
from time to time. Unlike the GENESIS software mentioned previously, MIKE 21 is unashamedly based on hydrodynamics and finite
difference approximations. Perhaps if Chapters 3 and 4 were a real
problem to understand, the technical aspects of this next part might
be skipped over.

There are many differences between the approach of MIKE 21 and
that given in Chapters 4 and 5, but perhaps the biggest is the use of
the flux form of the equations to the exclusion of the standard form.
The flux form was met briefly in Chapter 4, Chapter 5 and again in
Chapter 7. Here, it is examined more closely and derived carefully. It
can be most easily demonstrated through a simple example. Take the
x wise equation of momentum without Coriolis and without friction,
which are irrelevant for the moment

$$\frac{\partial u}{\partial t} + u\frac{\partial u}{\partial x} + v\frac{\partial u}{\partial y} + w\frac{\partial u}{\partial z} = -\frac{1}{\rho}\frac{\partial p}{\partial x}.$$

Now look at the continuity equation in its original form

$$\frac{\partial u}{\partial x} + \frac{\partial v}{\partial y} + \frac{\partial w}{\partial z} = 0.$$

Multiply it by u and add the resulting equation to the x wise momentum equation. This results in

$$\frac{\partial u}{\partial t} + 2u\frac{\partial u}{\partial x} + \left(v\frac{\partial u}{\partial y} + u\frac{\partial v}{\partial y}\right) + \left(w\frac{\partial u}{\partial z} + u\frac{\partial w}{\partial z}\right) = -\frac{1}{\rho}\frac{\partial p}{\partial x}.$$

The second term simplifies, and each term in parentheses combines, to give

$$\frac{\partial u}{\partial t} + \frac{\partial}{\partial x}(u^2) + \frac{\partial}{\partial y}(uv) + \frac{\partial}{\partial z}(uw) = -\frac{1}{\rho}\frac{\partial p}{\partial x}.$$

This does not seem much of an advance, which is why it did not feature in Chapter 3. However, the product terms can be more easily related to energy and momentum and so does have advantages, particularly in trying to get to grips with the physics of a difficult environment like the impact of waves on the coast. It also seems to offer numerical advantages. More to the point here, instead of using u, v the horizontal components of current, MIKE 21 uses p, q, which are $p = uh$ and $q = vh$ and represent the components of the fluxes in each horizontal direction. Again, a seemingly trivial change but one that helps with the physics. Here, h denotes the total depth (depth from mean sea level plus surface elevation).

The (flux form) equations used by MIKE 21 do include the Coriolis terms, quadratic friction terms and terms containing the input from wind. The pressure gradient is incorporated through gradients in sea surface elevation by assuming hydrostatic balance. Although there are additional shear stress gradient terms in the stated equations, they seem not to feature in the subsequent numerical application. Nevertheless, this is a sophisticated two dimensional homogeneous treatment of the hydrodynamics. Each term of the balances is subjected to individual numerical algorithms tailored to minimise truncation error, eliminate instability and to speed up computation without compromising physical principles. To give a flavour of this, let us run through some of the numerical methods used.

The square grid is the standard one used for all the Arakawa grids, but the method of computing the elevation (η), the fluxes p and q is an ADI scheme, see Chapter 5. Figure 8.8 shows the procedure diagrammatically. Once all the equations are put in finite difference

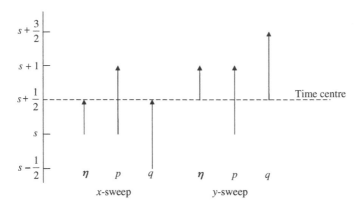

Fig. 8.8   The MIKE 21 software uses an ADI method; the time steps are on the left $(s - \frac{1}{2}$ to $s + \frac{3}{2})$ and the arrows indicate how each variable, elevation (η), x wise flux (p) and y wise flux (q), is calculated.

form, they are solved by one-dimensional sweeps. In the x sweep, the continuity and x wise momentum equations are solved with the elevation η being taken from step s to step $s + \frac{1}{2}$, the x wise flux p being taken from step s to step $s + 1$ and the y wise flux from step $s - \frac{1}{2}$ to step $s + \frac{1}{2}$. The y sweep solves continuity once more and the y wise momentum equations. The value of elevation η at time s that has just been evaluated is used to progress η to step $s + 1$, and the y-wise momentum is progressed from step $s + \frac{1}{2}$ to step $s + \frac{3}{2}$. It will be noticed that the y sweep for x wise momentum p is the same, however the combination of the two sweeps results in the centering of the time at the step $s + \frac{1}{2}$. This is not perfect due to the nonlinear terms in the equations, and the optimal way of doing this is far from obvious as the way to be most accurate often compromises stability. MIKE 21 uses upwind differences. This means that (using the usual convention of x pointing east and y pointing north), for a southerly sweep, the y derivative uses values at $t = (s + 1)\Delta t$ and $y = (j + 1)\Delta y$ on the northerly side and values at $t = s\Delta t$ and $y = j\Delta y$ on the southerly side. For a northerly sweep, this same derivative will use $t = s\Delta t$ and $y = j\Delta y$ on the southerly side and $t = (s+1)\Delta t$ and $y = (j+1)\Delta y$ on the northerly side, in other words precisely the reverse. In this way, the computation zigzags across the domain, but the average time of $t = (s + \frac{1}{2})\Delta t$ is achieved. According to the MIKE 21 literature, the possibility of oscillations (instability) is not completely eliminated. To centre the computations in space looks to be complicated, but it is not in principle. As an example, the continuity equation (in flux form of course), which is exactly

$$\frac{\partial \eta}{\partial t} + \frac{\partial p}{\partial x} + \frac{\partial q}{\partial y} = \frac{\partial h}{\partial t}$$

in finite difference form using centred differences is

$$2\left(\frac{\eta^{s+\frac{1}{2}} - \eta^s}{\Delta t}\right)_{i,j} + \frac{1}{2}\left\{\left(\frac{p_i - p_{i-1}}{\Delta x}\right)^{s+1} + \left(\frac{p_i - p_{i-1}}{\Delta x}\right)^{s}\right\}_{j}$$

$$+ \frac{1}{2}\left\{\left(\frac{q_j - q_{j-1}}{\Delta y}\right)^{s+\frac{1}{2}} + \left(\frac{q_j - q_{j-1}}{\Delta y}\right)^{s-\frac{1}{2}}\right\}_{i}$$

$$= 2\left(\frac{h^{s+\frac{1}{2}} - h^s}{\Delta t}\right)_{i,j},$$

where the extra 2s in front of the first and last terms stem from using half time steps. The software is written with great attention to detail. For example, the finite difference form of each term in turn is worked out carefully, in particular the nonlinear terms are discretised so as to minimise both any chance of instability and truncation error. Although the MIKE 21 literature may be considered to be short on analysis, it makes up for this by drawing on a great deal of experience of using the code and modifying it in a wide variety of different and challenging situations. As an example of this, let us take a look at the nonlinear terms. In flux form, the x wise momentum equation contains the terms

$$\frac{\partial p}{\partial t} + U\frac{\partial p}{\partial x} + V\frac{\partial p}{\partial y},$$

all evaluated at the $(i\Delta x, j\Delta y)$ point in space but at the half time step $(s+\frac{1}{2})$. In order to lessen (the MIKE 21 literature says minimise, which is not actually proved) the truncation error, five correction terms

$$-\frac{1}{2!}\left\{U^2\Delta t\frac{\partial^2 p}{\partial x^2} + 2UV\Delta t\frac{\partial^2 p}{\partial x\partial y} + V^2\Delta t\frac{\partial^2 p}{\partial y^2}\right.$$

$$\left.+U\Delta t\frac{\partial^2 p}{\partial x\partial t} + V\Delta t\frac{\partial^2 p}{\partial y\partial t}\right\}$$

are used, where these correction terms are also evaluated at the $(i\Delta x, j\Delta y)$ point but at the forward time step $(s+1)\Delta t$. In these expressions, (U, V) is the (horizontal) velocity averaged over the time interval $(s, s+1)$. The usual criticism is that these correction terms (well the first and third certainly) are exactly the form of eddy viscosity and so merely act to dissipate, and dissipation always helps numerical schemes to be stable. There is a certain truth in this, but as these terms are multiplied by the averaged speeds, they are applied far more selectively than would be eddy viscosity. They are virtually zero in areas of very low velocity and conversely are maximum in regions of high velocity, which accords with reality. Without going into a lot more detail, it is worth noting that the finite difference form of the momentum terms results in having to solve matrix equations (due to the semi-implicit nature of the scheme). Moreover, the matrix has five, not the usual three, bands down the diagonal because of the use of half steps. Local substitution is employed to reduce the algebraic overhead. It is also worth saying a little about the upwind nature of the scheme used in MIKE 21. The convective momentum term is approximated by

$$\frac{\partial}{\partial x}\left(\frac{p^2}{h}\right)_i \approx \frac{1}{\Delta x}\left[\left(\frac{p^2}{h}\right)_{i+\frac{1}{2}} - \left(\frac{p^2}{h}\right)_{i-\frac{1}{2}}\right],$$

where other terms have been ignored. When the flow is positive in the x direction, the upwinded form of this is approximated by

$$\frac{\partial}{\partial x}\left(\frac{p^2}{h}\right)_{i-\frac{1}{2}} \approx \frac{1}{\Delta x}\left[\left(\frac{p^2}{h}\right)_i - \left(\frac{p^2}{h}\right)_{i-1}\right].$$

It turns out that this kind of upwinding is equivalent to the original centred space term as far as truncation error is concerned, plus a correction term as follows

$$\frac{\partial}{\partial x}\left(\frac{p^2}{h}\right)_{i-\frac{1}{2}} \approx \frac{\partial}{\partial x}\left(\frac{p^2}{h}\right)_{i} - \frac{\Delta x}{2}\frac{\partial^2}{\partial x^2}\left(\frac{p^2}{h}\right)_{i}.$$

This upwinding is only included selectively, notably where the local Froude number (denoted by Fr) is large enough. Define a parameter α where

$$\alpha = \begin{cases} 0 & \text{Fr} \leq 0.25, \\ \frac{4}{3}(\text{Fr} - 0.25) & 0.25 < \text{Fr} < 1.0, \\ 1 & \text{Fr} \geq 1, \end{cases}$$

then this weighting factor is applied according to

$$\frac{\partial}{\partial x}\left(\frac{p^2}{h}\right)_{i} \approx (1-\alpha)\frac{\partial}{\partial x}\left(\frac{p^2}{h}\right)_{i} + \alpha\frac{\partial}{\partial x}\left(\frac{p^2}{h}\right)_{i-\frac{1}{2}}.$$

So, in this way, the motion of the fluid itself is intricately woven into the design of the numerical scheme, and the experience of the builders of software intelligently incorporated into improving it. MIKE 21 is continually being updated and (hopefully) improved.

8.6   Risk and Forecasting

In this section, more is said about the use of statistics to help in the assessment of risk and the accuracy of forecasting. For those not familiar with basic statistics, Chapter 1 provides a brief introduction. In that chapter, concepts such as mean, variance and probability were defined. Suppose that an unlikely event has the probability p of occurring. It is interesting to try and quantify "risk" in the sense of predicting the correct insurance premiums and the like. Now, it would not be appropriate to delve too much into the calculation of insurance premiums, but in the context of flood prediction, coastal damage and loss of property, risk is certainly worth some consideration. $1-p$ will be the probability that this unlikely event does not occur. Suppose we segregate what happens each year so that $1 - p$ is the probability that this rare event has not happened in a particular year. Over n

years, the probability of it not happening would be $(1-p)^n$, provided the events are independent. The risk would then be defined by

$$R = 1 - (1 - p)^n.$$

For very small p, this is approximately

$$R = np,$$

which some may recognise as resembling an expectation value. So, if p is the probability of a rare event (storm surge, wave exceeding some value, etc.) but $n = 100$, then the value taken by the risk R might still be small enough for a chance to be taken. It has been written that to add a metre to an offshore oil production platform costs \$1.5M, so to do so on the off-chance of some rare event is not a decision to be taken lightly, and any quantitative help would be gratefully received.

In Chapter 1, there was also an introduction to the use of MLE as well as to non-parametric statistics, correlation and regression, although these latter ideas will not be used in this section. Earlier on in this chapter, we used various distributions when discussing wave spectra, and extreme statistics were used in the determination of the return periods of large waves. There is a future in using MLE for this too, but at present there is not the data. To get an idea of how to use statistics when enough data are available, here is an example of using statistics for the forecasting of sediment movement.

This is taken from Reeve, Li and Thurston (2001) and concerns the application of EOFs to sandbank morphology. The sandbanks in question lie off the eastern coast of the United Kingdom, close to the Norfolk town of Great Yarmouth. They are chosen to analyse because there are enough data to do so successfully, additionally they are important to both tourism and navigation. The EOF method is particularly suited to the analysis of time series, and so, armed with many offshore surveys that measured the depth offshore, a detailed EOF analysis was possible. The first thing that needs to be done is to abstract profiles from historical data. Sixteen charts were used covering the years 1846, 1864, 1875, 1886, 1896, 1905, 1916, 1922, 1934, 1946, 1954, 1962, 1974, 1982, 1987 and 1992. This ensures a rich source spread over enough years to capture the time constants required for EOF analysis. Given this spread of data, any analysis is not going to predict detailed sediment transport patterns but will predict large overall movement in the banks. The records from these 16 surveys were digitised and rendered comparable by the use of a common projection and reduction to an Ordnance Datum. Each of the 16 surveys gave rise to an image that resembles a contour map that had been coloured, with darker parts being shallower. Looking at the 16 images (not reproduced here for copyright reasons, but available in the original paper), there are only slight differences in most of the macroscopic sea floor features, which is a good reason for the success of the EOF method. The sea bed is described by the function $h(x, y, t)$. The data are then discretised from the surveys and this depth is estimated from the survey maps. At any particular location, a time series of h versus t can be drawn. It is also possible to estimate quantities such as the movement of the volume of sand. A one dimensional EOF analysis can then be carried out as follows. Denote the discrete data by $g(\xi_l, t_k)$ where $1 \le l \le L$, $1 \le k \le K$ and ξ is a co-ordinate that measures the volume of sand. EOF analysis proposes that

$$g(\xi_l, t_k) = \sum_{p=1}^{L} c_p(t_k).e_p(\xi_l),$$

where e_p are the eigenfunctions of the square $L \times L$ covariance matrix with elements

$$a_{mn} = \frac{1}{L.K} \sum_{k=1}^{K} g(\xi_m, t_k)g(\xi_n, t_k),$$

with m and n running from $1, 2, 3, \ldots, L$. Some may be perplexed at use of the term *eigenfunction*; there is no need because eigenfunctions are the same as eigenvectors, the names are interchangeable — the name function is preferred for representatives of continuous variables like our ξ, whereas vectors are traditionally discrete and represented in column form. Remember from Chapter 1 that the volume co-ordinate ξ has to have mean zero in other words the mean has to be subtracted out from the data. When this computation is done for these data, the first six eigenvalues have values: 0.9735, 0.0085, 0.0033, 0.0029, 0.0023 and 0.0022. It is obvious, therefore, that the first EOF, corresponding to the first eigenvalue 0.9735, carries the bulk of the information (97% in fact). The bad news is that this corresponds only to information on the mean depth and is quite useful but not all that surprising. More interesting analysis is carried out in the original publication. For example, a detailed analysis of the second, third and fourth EOFs reveal the following. The second EOF shows a slow variation from negative to positive with a proposed period of 200 years. This needs validation, of course, as the time series is not long enough. The third EOF shows a sharp drop from positive to negative from 1840 to 1880 but relative stability thereafter. The fourth EOF shows a variation that is strongly suggestive of repetitive behaviour with a period of 100 to 120 years. The original paper gives a great deal more information, but long term trends and wave-like morphological disturbances are apparent from the analysis of the data and, moreover, are not out of line with other studies and important factors to take into account in long term planning. This gives some idea of how EOFs can be used to make sense of a lot of data. Global warming, with its increase in storm activity, will no doubt play a part in the evolution of such banks, not only those off the east coast of the UK but sandbanks throughout the world.

Before finishing, let us summarise some other consequences of global warming. There are other conservation issues, such as the protection of sensitive salt marshes which house many rare and endangered species. There is the whole question of fisheries. Some species will thrive, others will dwindle, and the consequences for the industry, which is particularly bedevilled with bureaucratic rules and regulations, need to be thought through fairly. Water quality in terms of pollution has been dealt with elsewhere (Chapter 7) but issues such as the management of the quality of drinking water and the management of waste water, although all very important, are not tackled here. A discussion of all the consequences of global warming would fill several textbooks. We have glanced at those consequences that affect the coast, and a more extensive look at global warming takes place in Chapter 10.

8.7 Exercises

(1) Carry out the integration that shows that the potential and kinetic energies of a linear water wave are the same.

(2) Use Taylor's series to establish that the group velocity of waves is given by c_g, where

$$c_g = \frac{d\omega}{dk}.$$

(3) Compare and contrast the SMB and PNJ wave prediction methods. Why have the SMB method and its developments survived, whereas the PNJ method has not?

(4) Explain succinctly why directional wave theories are less important for coastal engineers than they are for offshore engineers.

(5) Explain why the mass transport velocity due to water waves when calculated in the presence of viscosity does not revert to the inviscid value once viscosity is allowed to get smaller and smaller until it is zero.

(6) Carry out the calculation that leads to the edge waves due to a constant offshore gradient $h(x) = \alpha x$, given in section 8.3.

(7) State the main features of the JONSWAP spectrum compared with the PM spectrum.

(8) Using that, due to global warming, there are now 59 storms each year, calculate via the wave extrapolation given in Figure 8.7 the one, 50 and 100 year waves.

(9) Explain why power laws are so popular in relating near bed current to height above the bed.

(10) Contrast the methodology of GENESIS with that of MIKE 21 as models for sediment and coastline prediction.

Chapter 9

Ecosystems and Other Biological Modelling

9.1 Introduction

It is perhaps in the general area of biological and ecosystems modelling that the largest strides have been made in the past few years. Ecology has been a branch of biology for some time. It deals with the interaction between biology and the environment. Ecosystem modelling has been around for not quite as long, perhaps 40 years, but it is only within the last few that good progress has been made. However, there are over 1.5 million species on this planet, and all of them interact. Therefore, ecosystem modelling is not easy and has to be done systematically. There are obvious environmental partitions, such as the difference between a marine- and land-based ecosystem with their different and largely non-interacting species. There are also ways of grouping animals and plants, for example using "phytoplankton" for all marine plant life and "zooplankton" for all marine animal life. Later, we can split the zooplankton pool into two groups: those that feed on animals (carnivores) and the rest (zooplankton). It is in this way that we can begin to make sense of the complex interactions between species and their environment. On land, it is agriculture that provides impetus for ecosystems modelling. Pest control is extremely important and understanding the ecosystem central in preventing a cure being worse than the problem. Genetically modified (GM) crops are with us and although they remain controversial, particularly in the UK, they could provide key solutions to developing world food shortages. It is vital, however, to understand how GM crops fit into their particular ecosystem jigsaw — if the "pests" are still around, what are they feeding on if not their once favoured crop? If the "pests" have gone, what has happened to the birds that used to feed on them? In the sea, a major reason for ecosystem analysis in the past has been the fishing industry. Fish are part of a complex ecosystem, some playing the role of top predator, so an important question will be how does the ecosystem respond if one major carnivore disappears. More appropriately, how much of the stock of a particular fish can be taken without deleterious effects on the ecosystem in which it is embedded? Recently, there is the question of climate change and global warming. The balance of various ecosystems is acting as a sensitive signal to subtle changes in climate. We have all heard of birds that are wintering or breeding further north — the Little Egret is now a UK species, whereas it was not ten years ago; the winter thrushes (Redwing and Fieldfare) are less common. These could be manifestations of climate change. So the study of ecosystem dynamics continues to be very important and has become more tractable with the advent of faster computers with enhanced graphics. Moreover, these computers do not need to be programmed by experts.

Modelling the biology and behaviour of particular plants or animals is a bit different, though there are similarities. A simplified biology is taken to be a proxy for the entire animal. Incorporating this proxy animal into larger ecosystems models or even coupled eco–hydrodynamic models is a new venture and still the subject of research. It will be assumed here that the idea of ecosystem modelling is completely new. The first step, therefore, is to begin with the simplest of models.

9.2   Population Models

An ecosystem is an interconnected biological system involving animals, plants, nutrients and waste products. The simplest ecological model, is the single variable population model, which is governed by a single equation. Typically, a population model will be governed by an equation such as

$$\frac{dP}{dt} = rP\left(1 - \frac{P}{K}\right),\tag{9.1}$$

where r and K are constants and P is the population of the organism. The constant r is something to do with how P increases and the constant K is the carrying capacity and is the maximum value P can take. In actuality, this value is never attained. Inserting $P = K$ in the equation immediately gives zero growth rate, so it is equivalent to a steady state that is never reached. We shall return to this equation, called the logistic equation, later but for now let us turn to describing general features of population models. In a typical population model, the variable (number of animals) has to be an integer, therefore any model might be thought to be naturally discrete and in precisely the form for a finite difference equation, rather than a continuous differential equation. This is indeed true, and many books on modelling populations do start with difference formulations that resemble those introduced in Chapter 4 but are not numerical. For example,

$$P_{i+1} = P_i + b_i - d_i,$$

where P_i is the population at instant i, b_i are the births in that time and d_i are the deaths in that time. The time period can be a few minutes, a day or perhaps a year.

In order to ground the idea of a population model, let us do a simple example based on the logistic equation.

Example 9.1. This example concerns the prediction of the future population of the United States. At a time t, the population of the USA is given by a function $P(t)$. Let t denote the year past 1900, and the function $P(t)$ is assumed to obey the logistic equation

$$\frac{dP}{dt} = aP - bP^2.$$

Solution. We will use the forward difference

$$\frac{dP}{dt} \approx \frac{P(t+h) - P(t)}{h}.$$

Table 9.1   The population model results.

Year	t	$P(t)$ (approx)	$P(t)$ (measured)
1900	0	76.10	76.1
1910	10	89.00	92.4
1920	20	103.64	106.5
1930	30	120.97	123.1
1940	40	138.21	132.6
1950	50	158.32	152.3
1960	60	179.96	180.7
1970	70	203.00	204.9
1980	80	227.12	226.5
1990	90	251.9	259.6
2000	100	276.9	281.4
2010	110	301.6	307.8
2020	120	325.6	?

The discretised version of the logistic equation using the above forward difference takes the form

$$P(t + h) \approx P(t) + h(aP(t) - bP^2(t)).$$

With data $a = 0.02$, $b = 4 \times 10^{-5}$, $P(0) = 76.1$ and $h = 10$, the value of $P(10)$ is easily calculated. We obtain $P(10) = 89.0$. We continue with $t = 10$ to find $P(20)$, $t = 20$ to find $P(30)$ and so on to complete Table 9.1. Note how remarkably good the predictions are. You may think that one reason for this is the parameters a and b, which are both adjustable and can be chosen so that the fit between actual and predicted population is minimised. However, in this case, both of these parameters were fixed using data *prior* to 1900; in fact the parameters a and b are fixed using measured population (census) statistics for the years 1880 and 1890. So it is remarkable that the model works so well for such a long time. Let us examine what happens to the population of the USA. The First World War and the influenza epidemic that followed slowed the population growth rate. In the 1930s, the Depression had the same effect. If it were not for this, the predictions could have been widely out — vastly underpredicting reality. In the 1950s and 1960s, the population had time to catch up, but in the late 1960s and 1970s there came the birth control pill and another slowing down in population. By the end of the 20th century, the social climate was very different to 100 years previously and large families were culturally less popular and economically less viable. Then there are racial issues and the increase in the number of blacks and hispanics. Since the year 2000 we have 9/11, attitudes following the Iraq invasion and global terrorism, however the model undershoots the measured population by six million. Who knows what may happen in 2020, but whatever, the parameters a and b cannot have taken all these factors into account and the good fit is down to a large slice of luck. If any reader is around in a few years time to test the formula, they can see how good it is. The global recession might have helped to retire the model yet again; it certainly has been successful so far. Who knows?

Figure 9.1 gives the picture according to this model up to 400 years from 1900 (the year 2300). It will be seen that, theoretically,

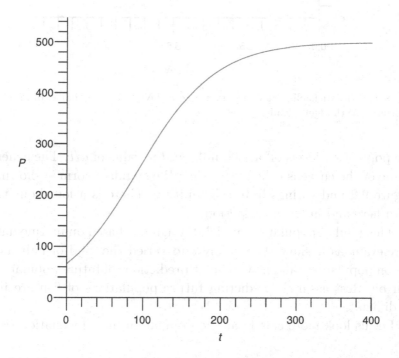

Fig. 9.1 The population of the USA according to the example; t is in years and P in millions.

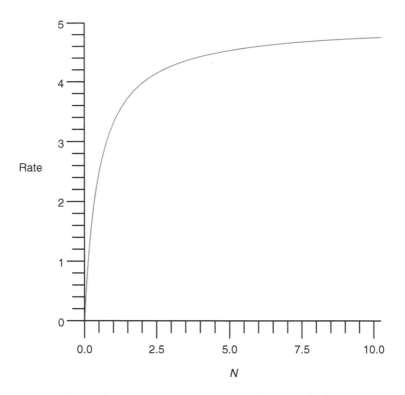

Fig. 9.2   The Michaelis–Menten curve: rate $= 5N/(0.5 + N)$ that expresses self limiting growth algebraically.

the population levels off at 500 million, the value of a/b. The general shape of the curve is called "logistic"; the standard form is shown in Figure 9.2 and springs from self limiting. There is a lot about this both here and in the next section.

This kind of population model, though not based on marine data, is relevant as it shows how many and varied the kind of influences are on populations and how difficult predictions of future populations can be. Rest assured, predicting future populations of fish are just as difficult.

Let us look more closely at the population model, equation (9.1)

$$\frac{dP}{dt} = rP\left(1 - \frac{P}{K}\right).$$

If K were very large, mathematically letting it tend to infinity, then the equation would be

$$\frac{dP}{dt} = rP$$

with a solution

$$P = P_0 e^{rt},$$

where P_0 is the value of the population at time $t = 0$. This represents unlimited growth with the constant r governing the rate of growth. Obviously this is unrealistic, but picking a model apart like this is useful. For it is now clear that the role of K is to limit the growth; in the USA population example above, where $r = a$ and $K = a/b$, the single parameter K stood for all the many and varied limitations on the growth of population in the United States during the 20th century, from war, disease, social deprivation through to cultural changes due to medical advances such as the oral contraceptive pill. Using a single parameter K for all these different effects is a little extreme but is what happens in ecosystem and biological modelling — it has to, as reality is just far too complex to model each mechanism individually. In mitigation for such a simplification, what we can say is that each mechanism certainly either inhibits or encourages growth in population so, although we as humans think each influence is very different, it isn't as far as the population is concerned. In this simple case, it is possible to solve the equation exactly. The solution to the logistic equation (9.1) is found straightforwardly using separation of variables as

$$P(t) = \frac{P_0}{\frac{P_0}{K} + e^{-rt}},$$

where P_0 is still the population at time $t = 0$. This gives the solution as displayed in Figure 9.1.

There are, of course, other kinds of population model. For example, a crude way of incorporating a time delay in the onset of limitation is by using an equation such as

$$\frac{dP}{dt} = rP\left(1 - \frac{P(t-T)}{K}\right),$$

where T is the time after which limitation effects kick in. For the time that $t < T$ the second term is positive, so although it is reasonable perhaps only to permit this equation for $t \geq T$, permitting it for all t is not too serious as all that happens is a rather different and nonlinear growth rate for $t < T$. Non-linearity in models is very important — an importance recognised only in the past 20 or so years when advances in computing enabled scientists to visualise non-linearity in the form of chaotic systems. (In fact, the great French mathematician Henri Poincaré (1854–1912) recognised what we now call chaos 100 years ago or more, he simply did not have the tools to express it the way we now can.) However, there are no "Newton's Laws" and so the rule to govern how the rate of change is related to various parameters and to the variable itself is only guided by observation and experiment. In the final analysis it is our choice. The logistic equation is not given to us on tablets of stone; there are other possibilities that also fit the data, for example

$$\frac{dP}{dt} = -rP + P(t-T)\left\{\alpha + \beta\left[1 - \left(\frac{P(t-T)}{K}\right)^{\gamma}\right]\right\}, \qquad (9.2)$$

where r is the decay rate and the constants α, β and γ are chosen to fit data. Although such models now have to be solved numerically, they all have the general shape of Figure 9.2, called the logistic curve. This behaviour can be summarised as follows: growth, initially indistinguishable from exponential and therefore not significantly limited by any factors, then tending to a limit (5 in Figure 9.2) that corresponds to the maximum allowable growth. In the population model this was K, but although the curve looks similar (compare Figures 9.1 and 9.2), the mathematics is different. Most importantly, the rate at which the Michaelis–Menten curve approaches its limit (5 in Figure 9.2) is a lot more gradual than the rate at which the population approaches K. The Michaelis–Menten equation is derived in the next section; it will be seen there that its origin is very distinct from population models. The population model, as represented by equation (9.2), is really just about the simplest single variable population model equation one can get away with, and it has its uses for modelling a single species of fish (for example, cod). The USA population example is only useful as an introduction. The plot displayed in Figure 9.1 shows that this example is the first (growth) part of the logistic curve, and in the first 100 years of such a model the limiting has not really had time to feature. Therefore, the goodness of fit is really down to luck and not model design; none of us really knows the self limiting features of the population of the USA in the next 300 years. Certainly, suggesting it levels off at 500 million is guesswork — from today's perspective, if mankind is still around then it seems most likely that the population will be a lot less than this.

Another way of predicting populations using models is to use discrete mathematics, that is difference equations. The simplest difference equation equivalent to the unlimited exponential growth given earlier takes the form:

$$P_{i+1} = rP_i,$$

where r remains the growth factor. In general terms, these difference equations can be written

$$P_{i+1} = F(P_i),$$

where $F(x)$ is some appropriate function. The form that F takes needs to take account of factors such as any equilibrium point for P where the population is stable and reasonably immune from outside influences and the return time, which is how quickly the population gets back to some kind of equilibrium value after a perturbation due to factors such as disease, war, etc.

These models are all deterministic in nature, but of course underlying all natural phenomena is uncertainty. Uncertainty might be thought to best be characterised by using stochastic (that is non-deterministic) modelling, but in fact this is not the case. One excuse might be that stochastic modelling must contain probabilistic notions and is therefore more difficult to do than deterministic modelling, but this excuse is unnecessary. All the modelling done so far in this section is not stochastic but deterministic, and there is no doubt that deterministic modelling bears dividends; this is because the underlying processes are growth and decay and these are not in themselves inherently stochastic. A little thought will tell you that the stochastic characteristics of a population model emerge from inputs rather than the actual processes being modelled. Thus stochastic equations are avoided. Let us see how a stochastic input might be implemented using the simplest difference model

$$P_{i+1} = r_i P_i,$$

but notice the addition of the subscript i on r_i. If r_i is a number dictated by some data input at the ith stage, then the model would be stochastic because it would reflect whatever effects were around to influence the population at this stage. For example, we could model a famine, an epidemic or other natural disaster that is normally well outside the kind of assumptions made in standard population models. It is indeed tempting to think that such ultra responsive models must be good as they seem to reflect reality. However, they are impossible to calibrate (but there is a statistical modelling technique called system identification theory that can be used for short term prediction — see papers by Huixin Chen, for example Chen and Dyke, 1998). A more important effect would be the overall density of the population; the larger the number, the smaller the growth. This can also be built into r, but a better way is to develop the kind of model that is self limiting through the parameters, and we have already met this kind of model: it is described through the differential equation

$$\frac{dP}{dt} = rP\left(1 - \frac{P}{K}\right),$$

which in the present discussion converts to the difference equation

$$P_{i+1} = P_i + r\Delta t P_i \left(1 - \frac{P_i}{K}\right),$$

where Δt is the time step. Nevertheless, "density dependent models" is still an expression used by ecosystem modellers for models that have some dependence on overall population — usually a limitation. Another possibility is extinction. If the value of P_i and r_i are such that P_{i+1} gets very small (less than one), then the population has died out. A model will predict this if r_i is small enough over successive values of i. This might be due to a number of factors such as non-availability of food, disease or excessive hunting.

If $P_{i+1} = 0$ for some $i+1$, then that is it. The value of P stays at zero and the population has become extinct. Nature is full of examples of extinction: all the dinosaurs, the dodo, the American passenger pigeon and the Tasmanian tiger, to name a few at random. Perhaps we know why these died out, but could their extinction be modelled accurately, even with hindsight? Certainly, the hunting of the passenger pigeon brought its population from the millions crashing down to zero in not many years, but the extinction of the dinosaurs is more of a modelling challenge. Mathematically, of course, once the value of P becomes zero it stays zero; this has to be written into the code of the model as a condition. Also, populations are negative despite this being mathematically allowable.

9.3 Michaelis–Menten Relationship

In this brief section, the Michaelis–Menten relationship is outlined. This is such a widely accepted relationship that it is hard not to call it a law. The "law" itself originates from enzyme kinetics, and a brief foray into this is worthwhile, if only to derive it. This derivation can be skipped if biochemistry is completely alien, but the relationship itself is central to ecosystem modelling. There are other names that are used for this relationship, for example Holling type II response or the Holling disc equation. Michaelis–Menten is, however, the most widely used.

Observations of enzyme kinetics set Leonor Michaelis (1875–1949) and Maud Menten (1879–1960) thinking about the underlying reasons why a curve plotting the initial reaction rate against molecular concentration should follow one particular shape and led them to derive an algebraic equation that now bears their names. In the experiment, the amount of enzyme is kept constant and the concentration of the substrate (particular compound or collection of molecules) is gradually increased; then it is found that the reaction velocity increases until it reaches a maximum. The result is the curve shown in Figure 9.3, which is very similar in form to Figure 9.2.

Deriving the Michaelis–Menten equation in terms of biochemistry proceeds as follows. Start with the generalised scheme for

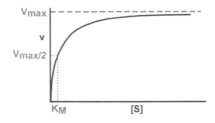

Fig. 9.3   The Michaelis–Menten relationship in its original form: the x axis is concentration of substrate, the y-axis is reaction rate (called "velocity" by biochemists).

enzyme-catalysed production of a product P from substrate S. The enzyme E does not magically convert S into P, it must first come into physical contact with it, i.e. E binds S to form an enzyme-substrate complex ES. Michaelis and Menten therefore set out the following scheme, whereby

$$E + S \underset{k_{-1}}{\overset{k_1}{\rightleftharpoons}} ES \overset{k_2}{\longrightarrow} E + P.$$

The terms k_1, k_{-1} and k_2 are rate constants for, respectively, the association of substrate and enzyme, the dissociation of unaltered substrate from the enzyme and the dissociation of product (= altered substrate) from the enzyme. The overall rate of the reaction (v) is limited by the step ES to $E + P$, and this will depend on two factors — the rate of that step (i.e. k_2) and the concentration of enzyme that has substrate bound, i.e. $[ES]$, where square brackets are a shorthand for "concentration of". This can be written as

$$v = k_2[ES].$$

At this point, it is important to draw attention to two assumptions that are made in this scheme. The first is the availability of a vast excess of substrate, so that $[S] \gg [E]$. Secondly, it is assumed that the system is in steady state, i.e. that the ES complex is being formed and broken down at the same rate, so that overall $[ES]$ is constant. The formation of ES will depend on the rate constant k_1 and the availability of enzyme and substrate, i.e. $[E]$ and $[S]$. The breakdown of $[ES]$ can occur in two ways, either the conversion of substrate to product or the non-reactive dissociation of substrate from the complex. In both instances the $[ES]$ will be significant. Thus, at steady state we can write

$$k_1[E][S] = k_{-1}[ES] + k_2[ES] = (k_{-1} + k_2)[ES]$$

so that

$$\frac{k_1[E][S]}{k_{-1} + k_2} = [ES].$$

Notice that the three rate constants are now on the same side of the equation. As the name implies, these terms are constants, so we can actually combine them into one term. This new constant is termed the Michaelis constant and is written k_m. Hence,

$$\frac{k_{-1} + k_2}{k_1} = k_m$$

so we substitute into the previous equation to get

$$\frac{[E][S]}{k_m} = [ES].$$

Now some stuff about enzyme dynamics. If E_0 is the total enzyme, then

$$[E_0] = [E] + [ES]$$

whence $[E] = [E_0] - [ES]$ and substituting for $[E]$ into the above equation gives

$$\frac{[E_0][S]}{[S] + k_m} = [ES].$$

Substituting for $[ES]$ into the above equation for v gives

$$v = k_2 \frac{[E_0][S]}{[S] + k_m}.$$

The maximum rate, which we can call v_m, would be achieved when all of the enzyme molecules have substrate that is bound. Under conditions when $[S]$ is much greater than $[E]$, it is fair to assume that all E will be in the form ES. Therefore $[E_0] = [ES]$. Substituting v_m for v and $[E_0]$ for $[ES]$ in $v = k_2[ES]$ then gives $v_m = k_2[E_0]$, whence eliminating $[E_0]$ gives, finally,

$$v = \frac{v_m[S]}{k_m + [S]},$$

and this is the original form of the Michaelis–Menten equation. Put $v = v_m/2$, exactly half the maximum rate, and the Michaelis–Menten equation gives

$$k_m = [S],$$

so the k_m of an enzyme is therefore the substrate concentration at which the reaction occurs at half of the maximum rate.

In terms of the enzyme reaction, k_m is an indicator of the affinity that an enzyme has for a given substrate, and hence the stability of the enzyme–substrate complex.

At low $[S]$, it is the availability of substrate that is the limiting factor. Therefore, as more substrate is added there is a rapid increase in the initial rate of the reaction — any substrate is rapidly mopped up and converted to product. At the $[S] = k_m$ rate, 50% of active sites have substrate bound. At higher $[S]$, a point is reached (at least theoretically) where all of the enzyme has substrate bound and is working flat out. Adding more substrate will not increase the rate of the reaction, hence the levelling out observed in the graph.

Let us now move away from biochemistry and translate the Michaelis–Menten relationship into a wider context. In terms of nutrient N, the Michaelis–Menten law can be written

$$\text{nutrient uptake} = \frac{\alpha N}{k + N},$$

that is the uptake of nutrient has a maximum value of α that can never be exceeded. This is because in the field, as more nutrient is absorbed, the growth of any organism or group of organisms is limited by the growth itself through factors such as the prevention of light penetrating to the organism. The logic of this law for phytoplankton is seen once it is realised that phytoplankton grow by photosynthesis, and photosynthesis is directly proportional to irradiance. Therefore, the presence of nutrient in the ecosystem will be limited by the amount of this sustaining irradiance. However, this global limiting via the Michaelis–Menten relationship will only apply to the total nutrient and at the microscopic level there is a more complicated relationship governing the nutrient growth in each layer, if the physical model is layered or at each depth if there is a continuous dependence upon depth. In any particular model, once the amount of nutrient reaches a certain level, the Michaelis–Menten limiting relationship dominates other effects. To summarise, the Michaelis–Menten formula contains two parameters, α is the maximum attainable uptake rate corresponding to v_m for the enzyme, and k the nutrient value when the uptake reaches half the maximum, corresponding to the enzyme rate constant k_m above.

There is a loose analogy between k and the half-life of a radioactive substance that might be helpful to physicists.

The Michaelis–Menten relationship or similar style self limiting formulae figure prominently in many ecosystems models. These are the subject of the next section.

9.4   Ecosystem Modelling

There are several features to ecosystems that make their modelling very distinctive compared to the modelling of physical systems. The essential feature of modelling as described so far in earlier chapters of this text is the writing down of balances. These balances arise from physical laws such as the conservation of momentum (Newton's second law), the conservation of mass, and so on. For modelling ecosystems, there are no such universal laws. However, ecosystems modellers do still write about "energy flows" and flows of other chemically based variables such as nitrates or nutrients in general. Older models based entirely on modelling energy were not successful. In recent times, it has been realised that the understanding of how a particular component of an ecosystem behaves arises from understanding those effects that cause it to grow and those effects that cause it to starve and die. Other effects, such as their direct interaction with another component or the inclusion of a source of the component itself, can also be incorporated. Each component thus gives rise to an equation called a rate equation. On the left-hand side is the growth rate of the particular component, and on the right are the terms which influence its growth or its decay. In general, therefore, we have an equation which has the following structure

growth rate of component $x =$ (a positive constant) $\times x$
$+$ (a negative constant) $\times x$
$+$ source terms $+$ interaction terms.

The first term on the right-hand side represents growth factors (e.g. feeding, and growth due to internal metabolism of nutrients). The second term on the right-hand side represents the decay rate (e.g. defecation, loss due to predators, dissipation by internal metabolism and ultimately death). The third term represents new sources of x (these might be conversions from other components or the mobility of x, causing it to migrate into the domain of the problem). Finally, the fourth term represents the fact that what happens to other components can influence what happens to x. This is the same kind of equation that was taken to govern population growth and decay. Indeed, precisely the same equation could be stated now

$$\frac{dP}{dt} = rP\left(1 - \frac{P}{K}\right),$$

but this time P might stand for phytoplankton, and we have a point or zero dimensional ecosystem model governing the growth of phytoplankton with self shadowing. We shall not pursue this, however. Most ecosystem models do not include migration in space. They concentrate on modelling the growth and decay of biological variables and, as such, they are zero dimensional or point models. In almost all cases, the right-hand side will contain some or all of the other variables in the model. It should be apparent after some thought that the right-hand side will contain many unknowns, not just the variable but also parameters. Parameters that need values, and values that need to be obtained, usually from observations and perhaps from experiments. These parameters can be as elusive to pin down as eddy viscosity, so ecosystem modelling does not appear to be in good shape. This is actually a bit harsh, and many parameters are able to be estimated with reasonable accuracy. In the next couple of subsections of this chapter, it will be seen that, although ecosystems modelling can exhibit a complexity that at first glance seems daunting, they can be extremely useful to the understanding of the important interactions, and the sensitivity of these interactions to particular parameters and variables. The next subsection is not too complex though.

9.4.1   *Predator–prey ecosystems*

As we have seen, the simplest type of model is one that involves a single variable. This variable is usually "population", and the name for this kind of model is a population model. A population of a species can evolve in three ways, provided it is not entirely unpredictable. It can grow without limit, settle down to a steady value (perhaps a zero value) or it can exhibit periodic behaviour. The logistic behaviour mentioned before is an example of a population that settles down to a value (its capacity). The right-hand side of the single equation that governs the change of population can be so structured that any one of these outcomes are possible. Mention has already been made of stochastic modelling, and their application through boundary conditions to population studies. In population studies, several algorithms based on random walk techniques of the type already met in Chapter 7 are possible. In these, there is a random element to the precise value of the population at any instant, and although averaging procedures can regain the previously mentioned underlying trends, only estimates of actual population are possible. In more sophisticated stochastic population models, the nature of the randomness itself can influence long term growth or decay.

The simplest biological population model involving two species is called a predator–prey model. The equation for each species takes a form that permits a closed graphical solution, as shown in Figure 9.4. It is instructive to look at this graphical solution alongside the equations themselves, which are written in the form

$$\text{rate of change of } x = \frac{dx}{dt} = ax - cxy,$$

$$\text{rate of change of } y = \frac{dy}{dt} = -by + dxy.$$

The mathematical solution of these two equations is possible in closed form and is the origin $(0, 0)$ which *is* a solution, albeit a trivial one,

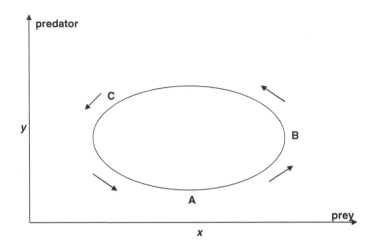

Fig. 9.4   The predator–prey model, no damping.

plus curves of the form

$$a \ln y + b \ln x - cy - dx = \text{constant}.$$

It is a representative of one of these curves that is drawn in Figure 9.4. It will be noticed that the values $x = b/d$ and $y = a/c$ correspond to zero growth of both predator and prey and the point

$$\left(\frac{b}{d}, \frac{a}{c} \right)$$

is at the "centre" of all closed curves of the type displayed in Figure 9.4. It is useful to describe qualitatively what happens to both species as a typical solution curve is traversed. The prey (strictly, either the biomass of the prey or the number of individuals), labelled x, naturally grows (probably by eating and absorbing nutrient, neither of which is included explicitly in the equations). The rate of this growth is represented by the constant a. On the other hand, the predator, labelled y, eats x at a rate governed by the magnitude of the constant c. The total rate of change of x is governed by these two competing effects. The predator naturally dies at a rate governed by the constant b (the death rate). On the other hand, it grows from eating x at a rate governed by the constant d. Note that, although c and d represent the same process (y eating x), they are different because c denotes the effect on the prey, whereas d denotes the effect on the predator. For this simple model, arguments can be proposed that can justify putting $c = d$.

The cycling depicted in Figure 9.4 can be explained in words as follows: as the number of predators is low, to the left of point A on the curve, the prey can increase by grazing without fear of being pounced upon. As the number of prey reaches a maximum (point B), the population of predators also thrives due to the plentiful food supply. The inevitable consequence of this is a decrease in prey until (point C) they become scarce enough to diminish the predator population. Once the predator population has reached a low enough value, the prey thrives and the whole cycle begins again. This is the simplest model. If extra terms are added to the right-hand sides of these equations so that more sophisticated eating habits and more complex relationships between the number of predators, the number of prey and growth rates are represented, then it is possible for the curve to spiral inward towards the point $(\frac{b}{d}, \frac{a}{c})$. Such a point is called an equilibrium point as both rates of change are zero there. It represents a stable point at which fixed numbers of predators and prey can live in perpetual harmony. When more variables are involved, a great variety of different stable, unstable and oscillatory states might well be possible. For a catalogue of possible states in two dimensions, see one of the many elementary textbooks on nonlinear dynamics which have been published in the last few years; the text most relevant to this chapter is Gurney and Nisbet (1998), which has much more mathematical detail than can be fitted in here. Let us now use this simple two dimensional model in order to build a bridge to practical ecosystem models.

9.4.2 *Simple ecosystems models*

The next obvious advance is to include more variables. In the field, the simplest useful models are three variable ecosystem models (PZN models, e.g. Klein and Steele, 1985) and four variable $NPZD$ models. N is nutrient, P phytoplankton, Z zooplankton and D detritus, but let us move more slowly through the steps. Suppose phytoplankton obeys a logistic type equation which is consistent with self limiting or shadowing, then, equation (9.1) gives

$$\frac{dP}{dt} = rP\left(1 - \frac{P}{K}\right).$$

If a second variable Z is added, we might have the two equation system

$$\frac{dP}{dt} = rP\left(1 - \frac{P}{K}\right) - \alpha_h PZ,$$

$$\frac{dZ}{dt} = \alpha_h PZ - \delta_h Z.$$

This is a predator–prey model with a prey with logistic variation and a predator rate that is linear. This is a viable model with two steady states. One the obvious one, zero zooplankton Z and $P = K$ — that is phytoplankton at maximum. The other is less obvious and corresponds to the "centre"

$$\left(\frac{\delta_h}{\alpha_h}, \frac{r}{\alpha_h}\left(1 - \frac{\delta_h}{\alpha_h K}\right)\right).$$

This model has some interesting properties, but let us not dwell on them as in the context of moving to practical ecosystems models we would learn little. Let us advance to a three level system

$$\frac{dP}{dt} = rP\left(1 - \frac{P}{K}\right) - \alpha_h PZ,$$

$$\frac{dZ}{dt} = \alpha_h PZ - \delta_h Z - \alpha_c ZC,$$

$$\frac{dC}{dt} = \alpha_c CZ - \delta_c C.$$

Here, we have added a carnivore C that only eats zooplankton. This time there are three stationary states, the same obvious one as before ($P = K$, $Z = C = 0$) then two others that are essentially equivalent to primary production. The first is as before

$$C = 0, \quad P = \frac{\delta_h}{\alpha_h}, \quad Z = \frac{r}{\alpha_h}\left(1 - \frac{\delta_h}{\alpha_h K}\right),$$

but with zero carnivore. The second is

$$H = \frac{\delta_h}{\alpha_c}, \quad P = P^* = K \left(1 - \frac{\alpha_h \delta_h}{r \alpha_c} \right),$$

$$C = \frac{1}{\delta_c} \left[r P^* \left(1 - \frac{P^*}{K} \right) - \frac{\delta_h \delta_c}{\alpha_c} \right].$$

An alternative modelling strategy to assuming a linear functional response in Z and C is to assume that these are also self limiting through overcrowding, not enough food or some other criteria. In this case, we make use of the Michaelis–Menten relationships. The two variable model would be

$$\frac{dP}{dt} = r P \left(1 - \frac{P}{K} \right) - \frac{\alpha_h P Z}{P + P_0},$$

$$\frac{dZ}{dt} = \frac{\alpha_h P Z}{P + P_0} - \delta_h Z.$$

Again, this is a system with two stationary states of similar characteristics as before; the details are omitted. The three variable model with self limiting zooplankton and carnivore is more complex but holds no surprises.

Let us now move on to consider a four variable system, the $NPZD$ model alluded to at the beginning of this section. The food web diagram of Figure 9.5 shows the simple flows. In this figure, there are

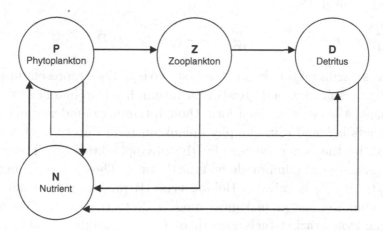

Fig. 9.5   The nutrient flow in an NPZD model: the arrows are explained in the text.

arrows that represent the processes expressed as terms on the right-hand side of the four rate equations, one for each of the variables. It is useful to describe these rather than give the equations baldly without comment. Starting with zooplankton, these gain mass by grazing phytoplankton and grazing detritus (you may have to stop eating while reading this.) Mass is lost from zooplankton through egestion (defecation, excretion) and dying (mortality) and this mass can go to either nutrient or detritus. Turning to the plants — phytoplankton — these grow by absorbing nutrient, usually via photosynthesis, but lose mass through respiration to nutrient, through dying to detritus and through being eaten by zooplankton. The only other arrow left to explain is the remineralisation of detritus to nutrient; this can be a complex biochemical process and has to be caricatured in the model equations. The equations that shall be used are taken from the paper by Edwards (2001). Obviously that paper has many more details than we can give here, but the rate equations are as follows

$$\frac{dN}{dt} = -\frac{N}{e+N}\frac{aP}{b+cP} + \frac{\beta\lambda P^2 Z}{\mu^2 + P^2} + \gamma dZ^2 + \phi D + k(N_0 - N),$$

$$\frac{dP}{dt} = \frac{N}{e+N}\frac{aP}{b+cP} - rP - \frac{\lambda P^2 Z}{\mu^2 + P^2} + (s+k)P,$$

$$\frac{dZ}{dt} = \frac{\alpha\lambda P^2 Z}{\mu^2 + P^2} - dZ^2,$$

$$\frac{dD}{dt} = rP + \frac{(1-\alpha-\beta)\lambda P^2 Z}{\mu^2 + P^2} - (\phi + \psi + k)D.$$

The first terms on the right of the first two equations represent uptake through respiration and growth. Each term has the (familiar by now perhaps) Michaelis–Menten form though there are moderating extra constants a, b and c in the phytoplankton term due to light attenuation by the water (b) and by the phytoplankton (c). The ratio a/b gives the maximum daily growth rate. The grazing function $\lambda P^2/(\mu^2 + P^2)$ is called a Holling type III predation function and is in common use in biological models. Note that a fraction α of zooplankton grazing fuels growth of the zooplankton, and a fraction β is excreted by zooplankton and is regenerated immediately into the nutrient compartment. The remaining fraction $(1 - \alpha - \beta)$ represents zooplankton faecal pellets, which enter the detritus compartment. The rP term is a general phytoplankton loss term, incorporating phytoplankton respiration and natural mortality. Models such as this are usually representative of one layer in the sea, and detritus in the form of pellets can sink and are lost. Remineralisation is modelled as a flow of D converting detritus into nutrient, where the rate has units of day^{-1}. Detritus can sink out of the mixed layer, given by the term ψD, where the sinking rate ψ also has units of day^{-1}. Other mixing terms (the linear ones in the above equations) are also included, and amongst these will be further sinking terms and other migration terms into the layers above. Finally, zooplankton grow by consuming phytoplankton and are themselves consumed by higher predators. This mortality is modelled using the quadratic function dZ^2. A proportion γ of the mortality term is regenerated as nutrient via excretion of the higher predators, with the remaining fraction $1 - \gamma$ fuelling the growth of the higher predators.

The eating of detritus by zooplankton does not feature in this model as yet. This is catered for by introducing a second quadratic factor into the zooplankton rate equation of the form ωD^2. This in turn feeds (excuse the pun) into the nutrient and detritus rate equations giving a new set as follows

$$\frac{dN}{dt} = -\frac{N}{e+N}\frac{aP}{b+cP} + \frac{\beta\lambda(P^2 + \omega D^2)Z}{\mu^2 + P^2 + \omega D^2}$$
$$+ \gamma dZ^2 + \phi D + k(N_0 - N),$$

$$\frac{dP}{dt} = \frac{N}{e+N}\frac{aP}{b+cP} - rP - \frac{\lambda P^2 Z}{\mu^2 + P^2 + \omega D^2} + (s+k)P,$$

$$\frac{dZ}{dt} = \frac{\alpha\lambda(P^2 + \omega D^2)Z}{\mu^2 + P^2 + \omega D^2)} - dZ^2,$$

$$\frac{dD}{dt} = rP + \frac{[(1 - \alpha - \beta)P^2 - (\alpha + \beta)\omega D^2]\lambda Z}{\mu^2 + P^2 + \omega D^2} - (\phi + \psi + k)D.$$

This new model gives a more realistic representation of the interactions between detritus, zooplankton and nutrient and is quite typical of the way simple models are built up. The parameter ω represents the relative palatability of detritus as compared to phytoplankton. To this, one can add grazing on bacteria, for example, or various preferences dictated by field observations of various animals, but in this text we shall just do one more complex model.

9.4.3   *Many variable ecosystem models*

Given that this is a book about coastal and offshore processes, let us follow a relevant, more complex example then relate it back to the more generic modelling issues outlined in the last subsection. This example is adapted from the book by Gurney and Nisbet (1998) cited earlier and concerns modelling a fjord ecosystem. For those not familiar with fjords, they are steep sided estuaries or inlets found on mountainous coasts, usually in Norway (fjord is a Norwegian word). In Scotland in the UK, they are called sea lochs, though these are usually a lot less grand. They are interesting from a modelling point of view because they are a mixed freshwater–seawater environment and mixing is less than one might expect because of sheltering. The prime interest in fjords is probably due to man's interaction with them through tourism (great scenery and wildlife) but there is also salmon fishing and there are mussel beds etc. Fjords are also very interesting from a physical oceanographic viewpoint, but we leave that aspect aside here. Suffice it to point out that the hydrodynamics of fjords are complex because they are stratified most of the time and there are significant seasonal variations due to winter freshwater input, which is a consequence of heavy snow (and some rain) and subsequent run-off from the mountains. It is important for any ecosystem model to reflect this. The ecosystem model has seven variables; the food web is expressed in diagrammatic form in Figure 9.6. The arrows need some explanation, so let's run through the interactions that they represent. First of all, the interactions with the sea; these are two way for nitrogen and phytoplankton. Zooplankton ($\mathbf{Z}$) and carnivores ($\mathbf{C}$) get nutrient from the sea but do not contribute directly to it. They defecate and die but this is taken up by the detritus, which we capture as variable $\mathbf{D}$. Phytoplankton ($\mathbf{P}$) sink and form detritus, and detritus is remineralised as bottom

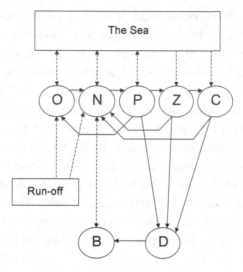

Fig. 9.6 The fjord ecosystem web model; **O** is DON, **N** is dissolved inorganic nitrogen, **P** is phytoplankton, **Z** is zooplankton, **C** is carnivores, **B** is bottom water nitrogen and **D** is detritus. The arrows show interaction and these are explained in the text.

water nitrogen (**B**). We have already modelled carnivores eating zooplankton, and zooplankton grazing phytoplankton: this explains these arrows. Phytoplankton excrete organic nitrogen (**O**), and zooplankton and carnivores excrete inorganic nitrogen (**N**). The organic nitrogen can be remineralised into inorganic nitrogen. The interaction (both ways) between dissolved inorganic nitrogen and bottom water nitrogen happens when it can, principally when conditions physically permit. This is represented by a dotted line in Figure 9.6. Run-off is similarly haphazard as are the interactions with the sea, including the all important nutrient input to the zooplankton and carnivores. Finally, run-off contributes to both types of nitrogen. These are all the interactions in the model. Arrows on a food web model are all very well, but as we saw in the last section, we need to be more precise about these interactions, which means that we write down rate equations for each of the seven variables. It is good to pause and think about the differences between this model and those of the last section. The principal difference is dictated by the environment. The models of the last section were generic, but developed for the ocean. They were point models so that there was no immediate need to parameterise horizontal migration. However, here we have a sea interacting biologically with a sea loch (fjord). This interaction is the prime reason for seven rather than fewer variables. In order not to be over complex, where possible we use linear biological interactions. Only when we consider the various uptakes do we have to go back to Michaelis–Menten type equations to represent the limitations due to self shadowing etc. So, let us now begin to develop the seven rate equations. These will contain many symbols, far too many to remember at once, so we try to smooth the way by explaining them first. There is no change to the left-hand side of each equation; it remains the rate of change of the variable with respect to time. The right-hand side comprises those effects that either increase or decrease this rate of change. The first equation looks at what influences $\mathbf{O}$, the dissolved organic nitrogen (DON). From Figure 9.6 there are four influences: the sea, run-off, and excretion from phytoplankton (all inputs) together with an output into the sea and remineralisation into dissolved inorganic nitrogen N. Taking these in order of simplicity, first deal with the sea. Call T_o the transfer of O due to the sea, S_o the input of O due to run-off, call V the volume of the surface layer and F the volume exchanged with the sea. Then

$$T_o = F \left[S_o - \frac{O}{V} \right].$$

Next, examine the terrestrial wash out W_o. This is quantified in a similar way: denote by f the volume of wash out that exits the system each day, then

$$W_o = f \left[\frac{O}{V} - \rho_o \right],$$

where ρ_o is the concentration of DON. The other two terms are a little trickier, but once mastered, the other six balances do become easier to quantify. The remineralisation process is a complex one, as any biochemist will tell you. There is no way such complexity can be entirely mimicked in a model such as this, but its overall effect can be assumed to be a first order rate process, so if we denote the remineralisation of DON into dissolved organic nitrogen by R_o, then

$$R_o = k_o O,$$

where k_o is a rate constant. This is a gross simplification, but it is acceptable for this model. The remaining term is the excretion from the phytoplankton, which we label E_p as it is coming from the phytoplankton pool P. Let

$$E_p = \epsilon_p U_p + e_p P,$$

where ϵ_p is the fraction of the phytoplankton uptake U_p excreted, and e_p is the fraction of the body mass of the phytoplankton P itself that is excreted. So, we have that

$$\frac{dO}{dt} = E_p - R_o + T_o - W_o,$$

with the right-hand side given by the above expressions. All terms of the right-hand side are either other variables (e.g. N, P etc.) or parameters (e.g. F, f, ϵ_p etc.) apart from U_p, which needs to be considered. We shall do this after looking at the other six equations. The rate equation for dissolved inorganic nitrogen N now is easier to derive. (Only a little, it must be admitted.) The interaction with the sea takes the same form as before in that we write it as T_n and compute it from

$$T_n = F \left[S_n - \frac{O}{V} \right],$$

where this time S_n denotes the input of N due to run-off. Similarly, the wash out W_n is quantified as before through

$$W_n = f \left[\frac{N}{V} - \rho_n \right],$$

where ρ_n is the concentration of dissolved inorganic nitrogen. f, F and V retain their earlier meanings. Now, the remineralisation out of O has to be the same as the remineralisation into N, so a "$+R_o$" term takes care of this. There are two excretion terms, E_c and E_z, denoting the excretion from carnivores and zooplankton, respectively. Using the same notation as before, these are denoted by

$$E_c = \epsilon_c U_c + e_c C$$

for the carnivore and

$$E_z = \epsilon_z U_z + e_z Z$$

for the zooplankton. Once again, ϵ_c and ϵ_z is the proportion of ingestate excreted and e_c and e_z the fraction of nitrogen biomass excreted. The parameterisation of the uptake rates U_c and U_z will be tackled at the same time and place as U_p. In fact, U_p features directly in the rate equation for dissolved inorganic nitrogen N as it is the rate at which nitrogen is fixed directly by the phytoplankton. The only remaining term is the dotted line to bottom water nitrogen, and this is assumed to be

$$M = \phi \left[\frac{B}{V_b} - \frac{N}{V} \right].$$

Here, V_b is the volume of the bottom water and ϕ the rate at which water is effectively exchanged between the surface and the bottom (in $m^3 \, s^{-1}$). So, this is the difference in the concentrations, multiplied by the rate of exchange of water (which could be zero under some circumstances, of course). The rate equation for N is thus

$$\frac{dN}{dt} = E_c + E_z - U_p + R_o + T_n - W_n + M.$$

Now we come to the rate equation for phytoplankton. The uptake of inorganic nitrogen (U_p) has already been mentioned, as has U_z the grazing of phytoplankton by zooplankton. Here is also where E_p features as the exhalation of organic dissolved nitrogen; this is now an output corresponding to the input to the rate of change of O. The sinking to the detritus pool happens through a term L_p. This is quantified via a (measured) mortality rate δ_p, whence we have

$$L_p = \delta_p P.$$

The only remaining terms that demand our attention are T_p, the uptake from seawater, and W_p, the wash out into the same source.

These are a bit different to before because we are now dealing with a living thing, phytoplankton. The model tries to recognise that phytoplankton tend to congregate near the pycnocline by modifying the "expected" rate P/V by a retention factor Ω. So, we have

$$T_{\dot{p}} = F \left[S_p - \frac{\Omega P}{V} \right]$$

and

$$W_p = f \left[\frac{\Omega P}{V} \right],$$

where F and f have already been defined and S_p is the concentration of phytoplankton in the inflowing water. The phytoplankton rate equation is thus

$$\frac{dP}{dt} = U_p - L_p - U_z + T_p - W_p - E_p.$$

Three down and four to go. The zooplankton rate equation is next. All but one term has been considered already. The outstanding effect is labelled J_z and this denotes the rate of migration of zooplankton from neighbouring water — after all, they are little animals and they swim. How to quantify this is another matter entirely. The other factors are listed; U_z is the consummation of phytoplankton (grazing), U_c is predation by carnivores, L_z is loss through defecation and death, quantified as

$$L_z = \delta_z Z + d_z U_z,$$

where the first term contains the *per capita* mortality rate δ_z and the second term is the proportion rejected as faeces d_z multiplied by the uptake rate U_z. E_z is the excretion of DON. So, we are now in a position to state the rate equation as

$$\frac{dZ}{dt} = U_z + J_z - U_c - L_z - E_z.$$

The rate equation for carnivores is more straightforward. The loss term via defecation and death L_c has the same form as the equivalent term in the zooplankton rate equation, that is

$$L_c = \delta_c C + d_c U_c,$$

and further explanation is unnecessary. The only other distinctive term is J_c, which represents the migration to and from neighbouring waters — fish etc. swim too. So, the equation is:

$$\frac{dC}{dt} = U_c + J_c - L_c - E_c.$$

The last two equations are the rate of change of bottom water nitrogen, which is

$$\frac{dB}{dt} = R_d - M$$

and the rate of change of detritus, which is

$$\frac{dD}{dt} = L_p + L_z + L_c - R_d.$$

The only term here that needs discussion is R_d, which represents remineralisation. As has been said, this is a complex process to any biochemist, but can be caricatured here as linear

$$R_d = k_d D,$$

where the coefficient k_d carries all the complex processes. Thus, we have a system of seven rate equations, restated below

$$\frac{dO}{dt} = E_p - R_o + T_o - W_o, \tag{DON}$$

$$\frac{dN}{dt} = E_c + E_z - U_p + R_o + T_n - W_n + M, \tag{DIN}$$

$$\frac{dP}{dt} = U_p - L_p - U_z + T_p - W_p - E_p, \tag{Phytoplankton}$$

$$\frac{dZ}{dt} = U_z + J_z - U_c - L_z - E_z, \tag{Zooplankton}$$

$$\frac{dC}{dt} = U_c + J_c - L_c - E_c, \tag{Carnivores}$$

$$\frac{dB}{dt} = R_d - M, \tag{Bottom Water Nitrogen}$$

$$\frac{dD}{dt} = L_p + L_z + L_c - R_d. \tag{Detritus}$$

Perhaps there has been enough detail for most, but we do need to quantify the uptakes U_p, U_z and U_c before we move on. The zooplankton and carnivore have uptakes that are easier to quantify than that for phytoplankton, and we discuss these first. As expected, the aforementioned Michaelis–Menten relationship comes to our aid here, and it is justifiable to write

$$U_z = \frac{I_z P Z}{P + V H_p} \quad \text{and} \quad U_c = \frac{I_c Z C}{Z + V H_z}.$$

H_p and H_z are half saturation concentrations for phytoplankton and zooplankton, respectively, and I_z and I_c are maximum uptake rates per unit biomass for zooplankton and carnivore, respectively. So the I_z and I_c terms limit the uptake (even sharks can only eat so much.) These maximum uptakes and half saturations are able to be measured, so we have a good model for U_z and U_c; now let us turn our attention to U_p. The difference with plants is that local irradiance must be taken into account together with the internal biology of how nitrogen is fixed by each cell. The simplest route is to assume both are Michaelis–Menten (self limiting) in form so that, if L is the light intensity, n is the local concentration of dissolved inorganic nitrogen and I_p the maximum uptake rate per unit of biomass, then we can set

$$u(n, L) = I_p \left(\frac{n}{n + H_n} \right) \left(\frac{L}{L + H_L} \right),$$

where we have defined the half saturations H_n and H_L as before. $u(n, L)$ is an "uptake density function", varying from place to place due to different light intensity and different dissolved inorganic nitrogen concentration. The surface layer is well mixed, so the dissolved inorganic nitrogen has a concentration $n = N/V$ everywhere there. However, if the surface irradiance is L_s then the local irradiance obeys an exponential profile with depth, the Lambert–Beer law

$$L(h) = L_s e^{-\kappa h},$$

where h is depth (we cannot use the normal z here; there is a conflict of notation). In order to calculate the total uptake, it is necessary to multiply the above by the local phytoplankton concentration P/V and integrate over the whole surface layer volume, remembering that

L is now an exponential function from the Lambert–Beer law. The result is a rather complicated looking expression that contains logarithms (due to the integration). Here it is

$$U_p = \frac{PI_p}{\kappa_d} \left(\frac{N}{N + VH_n} \right) \ln \left(\frac{L_s + H_L}{L_s e^{-\kappa_d} + H_L} \right).$$

We have nearly finished, but the phenomena of self shading affects phytoplankton as they grow and shut out available light. This is catered for in the light equation by altering the coefficient κ by

$$\kappa = \kappa_b + \frac{\kappa_p P}{V}.$$

Seasonality is also a factor that cannot be ignored — fjords in the summer and winter are very different places. Therefore, a "seasonality" function $S(\theta)$ is defined,

$$S(\theta) = 1 - e^{-\theta/\theta_0},$$

that can be used to multiply appropriate variables such as I_p, e_p and I_z, but that is enough detail on this example. For those who have followed all this through, it should be apparent just how complex even a seven variable model is, and importantly how many parameters need to be given a value. Nowadays, ecosystem models can have tens and even hundreds of variables.

We now need to use the model. In order to do this we need a lot of information to fix the parameters. These data were taken from Loch Creran, and some of them are displayed graphically in Figure 9.7. Table 9.2 gives the fixed biological and physical parameters.

One can sensibly ask which of these parameters are generic and which are site specific. In general, it is reasonably safe to assume that the biological parameters are generic because they are averaged over several species, but the physical parameters are site specific and perhaps time specific too. For example, the depth of the mixed layer will be highly variable, varying from place to place and from time to time on several scales. However, this variability, although not captured in the stark value of d given (8, see Table 9.2), is not important to the outputs of the ecosystem model. In general, it is hoped that any uncaptured variability in the physical parameters

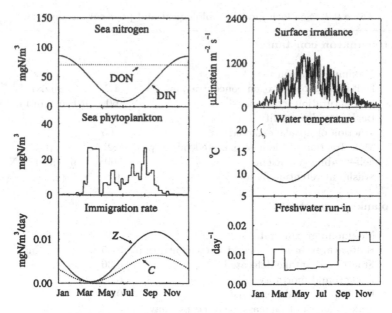

Fig. 9.7   The driving functions for the model; taken from data at Loch Creran, a typical Scottish sea loch.

of the model does not impinge on the workings of the ecosystem model. It is quite likely, if an ecosystem model is to be applied in a different location, the physical parameters will have to be reset to reflect this before the model is run. This is not so true of biological parameters, but the more complex the representation of the biology in an ecosystem model, the more site dependent it is simply because less averaging over species is taking place. So, make a really complex ecosystem model, and although you win on an improved model of one location, you lose on its transportability. This particular model is not too complex, but has been set up using data from Loch Creran in Scotland. The variability has been well captured in Figure 9.7, which gives the driving functions for this model. From this, the seasonal variability of the biological inputs (apart from DON) is obvious and not at all surprising. Physical variables such as temperature and river run off also display seasonal variation, but surface irradiance shows variation on a much shorter scale. A little thought will reveal that this is due to sunny and cloudy periods and these can be averaged

Table 9.2   Sea loch model — biological and physical parameters.

Phytoplankton constants

I_p^m	Maximum fixation rate	1.6	day^{-1}
H_n	Half-saturation nitrogen concentration	4.2	mgNm^{-3}
H_L	Half-saturation irradiance	60	μ Einstein m^{-2} s^{-1}
δ_p	Background mortality	0.1	day^{-1}
ϵ_p	Fraction of uptake excreted	0.05	—
e_p^m	Max. fraction of biomass excreted per day	0.25	day^{-1}
κ_p	Self-shading coefficient	0.008	m^2(mg)$^{-1}$ N^{-1}
Ω	Wash out retention factor	0.5	—

Zooplankton constants

I_c^m	Maximum grazing rate	2	day^{-1}
H_p	Half-saturation phytoplankton concentration	37.5	mgNm^{-3}
d_z	Fraction of uptake defecated	0.36	—
δ_z^b	Background mortality	0.05	day^{-1}
ϵ_z	Fraction of uptake excreted	0.15	—
e_z^m	Max. fraction of biomass excreted per day	0.05	day^{-1}

Carnivore constants

I_c^m	Maximum carnivory rate	15	day^{-1}
H_z	Half-saturation zooplankton concentration	60	mgNm^{-3}
d_z	Fraction of uptake defecated	0.5	—
δ_c	Background mortality	0.05	day^{-1}
ϵ_c	Fraction of uptake excreted	0.2	—
e_c^m	Max. fraction of biomass excreted per day	0.75	day^{-1}

General constants

k_d	Detritus remineralisation rate	0.01	day^{-1}
k_o	DON remineralisation rate	0.02	day^{-1}
θ_0	Seasonal characteristic temperature	10	$^\circ$C

Physical constants

d	Surface layer depth	8	m
F/V	Tidal exchange	0.37	day^{-1}
V_b/V	Bottom layer volume	1.0	—
ϕ/V	Vertical mixing	0.05	day^{-1}
κ_b	Light attenuation	0.22	m^{-1}

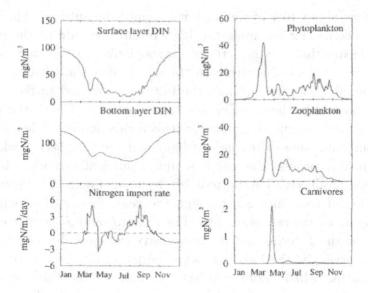

Fig. 9.8   The annual cycle in Loch Creran, a typical Scottish sea loch. All the state variables are shown, with the bottom left a combination of N and O, where DIN is dissolved inorganic nitrogen.

out by taking a moving average over a few days. This way, irradiance data will show only the annual variation that is important to the ecosystem (on this scale).

The response of the biology is shown in Figure 9.8. Despite the seeming complexity of the model, the major features of the biology are quite simply expressed. The primary production over the winter months is very low leading to a decline in all biological populations and an increase in inorganic nitrogen. In the early spring, the irradiance is high enough to cause a bloom in phytoplankton (the spring bloom). The onset of this bloom is critically dependent on the depth of the surface mixed layer. Comparing Figures 9.7 and 9.8 shows that the bloom in the loch occurs about two or three weeks before the corresponding bloom in the sea, shown through the sharp increase in phytoplankton biomass. Once the phytoplankton are available, there are increases in zooplankton numbers and increases in carnivore numbers. The grazing impact of the zooplankton and the predation of the carnivores keeps levels of phytoplankton roughly constant at a level about 10% of the peak value until the autumn. There are some niceties; for example, the high respiration rate of the carnivores means that they only thrive for zooplankton levels that exceed 15 mgNm^{-3}. Thus the carnivore "bloom" is very short lived.

The nitrogen levels in both the bottom water and surface water are depressed in the summer due to the activity of both the plant and animal communities; however, this depression is felt less in the bottom water due to the lack of light and the relatively weak vertical mixing process that stops so much nutrient ascending to the primary producers. The bottom left frame of Figure 9.8 shows the variation of total nitrogen through the year; this can be found by adding up all the rate equations. The terms J_c and J_z are extremely small, so the T terms and W terms carry the most. So, clearly and not surprisingly, the loch is a net exporter of nitrogen in the winter months. A little more surprising, perhaps, is that this is also the case in the summer period of low summer productivity. This exportation is through remineralisation. Nitrogen is imported during the spring and autumn blooms to nourish the flourishing primary producers. More can also be deduced about deep water renewal, about the relationship between surface layer nutrients and primary production, irradiance and grazing efficiency. However, those who want a more detailed discussion can consult the book by Gurney and Nisbet. Enough has been said here to give a flavour of this particular ecosystem model. One question in the minds of many might be "is it possible to simplify the model, given the overall cyclical nature and its insensitivity to a lot of the detailed processes"? The answer is a cautious "yes" and it is this that leads to four, and even three, variable ecosystem models.

Let us first of all turn to a model published more than 15 years ago, Fasham, Ducklow and McKelvie (1990). As can be seen immediately from the title, this is a model that, in contrast to the one just studied, is related to the deep ocean and also restricted in its application to the surface oceanic mixed layer. Nevertheless, it has generic features and is still today cited as the basis for many currently used models. Like the sea loch model, this one too has seven compartments but they are not the same seven. This time the variables

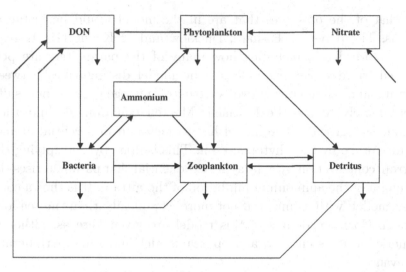

Fig. 9.9 Modelled annual nitrogen fluxes taken from Fasham, Ducklow and McKelvie (1990).

are: phytoplankton, zooplankton, bacteria, nitrate, ammonium, DON and detritus. Thus, the concentration is on smaller animals (bacteria are there but carnivores are not), and we have nitrate, ammonium, DON and detritus with no migration through the water column, which means that this is a "point" model. A skeletal version of the foodweb diagram is shown in Figure 9.9.

The interactions are quantified using several different sets of data in the original paper, but this detail is omitted here as only the general idea of the interactions is required. The arrows "going nowhere" indicate interaction with the deeper ocean (sinking in this case). The whole emphasis of the model is to capture the essence of the nitrogen cycle, hence for example, the presence of bacteria as a variable. Bacteria have a significant role in the breakdown of, and recycling of, dissolved organic matter, and nitrogen is an important part of the remineralisation process. The paper discusses the possibility of urea being a separate variable, but this was ultimately rejected. It is interesting to read that, due to the absence of carnivores, not all dead zooplankton becomes detritus. Instead, that fraction that would be eaten by predators is channelled through ammonium; the justification of this is through consideration of the time scale, which

on that of the processes that are in the model would be instantaneous. The paper by Fasham, Ducklow and McKelvie (1990) gives
a lot more detail, including how many of the parameters are permitted to vary. For example, as the model distinguishes between
nitrate and ammonium, two sources of nitrogen, it is not sufficient merely to use the Michaelis–Menten equation for limitation
of nutrient growth. The model has to factor in preferential ammonium uptake by the phytoplankton. This is done by multiplying the
nitrate contribution by a decaying exponential that parameterises the
strength of the ammonium inhibition of the nitrate. It is thus a complex model, with as much, if not more, complexity than the sea loch
model. No more details of this model are given here as, although
generic, it is essentially a deep sea model and only peripherally
relevant.

What we have here then are two models with the same number of
variables (seven), some similarities but also some differences. These
differences are due to the type of application — deep sea for one, sea
loch for the other. The reason for briefly introducing the Fasham,
Ducklow and McKelvie model now follows. This model, being for the
open sea does not have to distinguish between inshore and offshore
environments. On the other hand, it is a good basis for simplification as it does not contain this awkward migration. This model
is successful in modelling the phytoplankton bloom in the North
Atlantic. Although coastal and offshore engineers may not particularly be interested in the North Atlantic phytoplankton bloom, the
purely biological changes in inshore and coastal waters should be of
interest, if only because an understanding of this is essential to be
able to assess the impact of development on the environment. Let
us see, therefore, how a seven parameter ecosystem can be simplified
without losing too much relevant information. There are several biologically based principles that can guide us towards simplifying an
ecosystem model. One is to examine via the data where the greatest
annual throughput of nitrogen occurs. These pathways are termed
major flows by biologists; nothing to do with fluid flow, of course.
One must be sensible when reducing the number of variables — for
example, reducing a whole ecosystem to zooplankton and detritus is

not sensible, but reduction to phytoplankton P and zooplankton Z might work as this is a major flow and effects such as phytoplankton growth, zooplankton growth, grazing of phytoplankton by zooplankton, limitation of phytoplankton and limitation of zooplankton can all feature explicitly. Several models have been built on P and Z only. There are also some models based on nutrient N and phytoplankton P only, but in order to include the important nutrient limitation of phytoplankton, we would normally turn to PZN models. These so-called three compartment models are still very popular and capture the essential features of many aquatic and terrestrial ecosystems. Let us look at the PZN model and see how it compares with the seven compartment models displayed in Figure 9.9. For a start, ammonium and nitrate are both nutrients as far as phytoplankton are concerned. Secondly, DON and Detritus are simply sinks for phytoplankton and on a PZN model would be an arrow into space. Finally, the ammonium, the detritus and the bacteria input to the zooplankton would simply be represented as "growth". Thus, we arrive at the three compartment model that represents some of the ecosystem dynamics. If the D for detritus compartment is added, then the essential biological lag between the formation of organic nitrogen and its regeneration into inorganic nitrogen can be modelled. There are other possible scenarios for four compartment models, but almost all of them would be $NPZD$. The point is that it is better to build a complex model based on as realistic a biology as possible, then to simplify by combining processes or ignoring them as secondary. The resulting three or four variable model will be more useful as a result. One final comment about these three and four compartment models. With so few variables, recent papers on the development and application of such models have been quite mathematical. With so few connections available, each one has to represent a number of very different processes, this can be a challenge to modellers. Moreover, with correspondingly small number of equations, mathematicians are able to apply analysis on the stability and other properties of the solution to such equations. However, in this text we shall not explore ecosystem modelling in this direction. Instead, let us address recent practical advances, ending with software.

In recent times, it has become more possible to validate ecosystem models than ever before. This is due mainly to advances in satellite technology. For example, large geographic regions can now be monitored for their colour over long time scales and with high spatial resolution. One thing that this has brought home is just how variable the parameters of ecosystem models really are. However, rather than despair, the answer must be to use the newly available data to ascertain the relative importance of the various parameters to important biological phenomenon such as primary productivity, the abundance or lack of various species etc. This way, it may be possible to assess the impact of either local or global change — local change caused by pollution, engineering activity global change caused by factors such as climate change, El Niño or even straightforward seasonality. A paper by Hemmings *et al.* (2004) outlines some of these recent issues. In particular, the simplification of ecosystem models that can be made in the light of these developments. Considering the ocean seven variable model for example, it turns out that the ammonium and nitrogen aspects can be usefully combined for some purposes, and that the microbial loop components can safely be ignored; this then justifies simplification along the lines of the last paragraph. The model is then an example of an $NPZD$ model we met in the last subsection. Of course, it will probably not be precisely the same as the one outlined there as there are options to modify the various rate functions. Nevertheless, it remains useful to simplify complex models so that particular behaviours can be emphasised and less important characteristics suppressed. If, further, the detritus is also simply an exported loss, then the model is reduced to three components: phytoplankton, zooplankton and nutrient — the combined nitrogen and ammonium inputs, ignoring the nitrogen–ammonium–bacteria cycle. Thus, the seven parameter model has been transformed into a PZN model. As in all modelling, judgement is required to ascertain whether the advances in understanding of specific biological dynamics gained through working with this simplified model outweigh the disadvantages of ignoring some processes.

Recent advances have enabled large software packages to be built (for example, (the European Regional Seas Ecosystem Model)

ERSEM, see http://www.pml.ac.uk/ecomodels/ersem.htm but put ERSEM into the search engine Google if the URL cannot be reached by the time you read this). There will be more about ERSEM in the next section. ERSEM, like many of these software packages, has undergone fitness trials, validation exercises and so can receive a kind of certificate of sea-worthiness.

What we have thus been at pains to establish so far in this chapter is an equation for each component and within each equation a right-hand side that contains many parameters that govern the component's growth and decay rate, its sources (if any) and its interaction with its fellow components. It is quickly apparent that even the simplest ecosystem can have many equations. This is another distinctive feature of models of ecosystems: lots of equations to accompany the many free parameters already mentioned. In most ecosystem models, the equations themselves are assumed to be linear when appropriate, obey a Michaelis–Menten or other Holling type relationship. These relationships although not able to be proved as in physics certainly have a secure biological pedigree. If the equations are largely linear, then some analytical progress to solution can be made. However, even the simplest quadratic non-linearity can lead to very complex, behaviour. If there is no reason to be more complex linearity will do if it seems to work.

It is only in the last 20 or so years that there has been recognition that even a single nonlinear term, perhaps xy in the above notation, can lead to the kind of behaviour which has been christened chaos. Chaotic behaviour is easy enough to model and to describe in a general sense; there are many excellent books, usually with nonlinear dynamics in the title. As far as we are concerned, it does not seem to model the short term behaviour of biological populations, so such models are generally avoided.

9.5   Ecosystems Software

In the past few years there have been rapid developments in marine ecosystem models. It is only recently that their application was able to be linked to highly resolved physical models of the marine environment. This has led to the inclusion of a wide range of ecosystem components and processes and the application of advanced data assimilation techniques. Altogether then, we now have access to sophisticated ecosystem models that are linked to models of the underlying hydrodynamics. A conceptual diagram showing how ERSEM is linked to the hydrodynamic model POLCOMS is shown in Figure 9.10.

Let us now describe the components of a good software-based ecosystem model. As our example we choose ERSEM. This model was developed under the Marine Science and Technology (MAST) project 1991–1997 by a group of European scientists — including those at my local research institute, the Plymouth Marine Laboratory — so it is perhaps not surprising that it is the one described here. This description is taken from the PML website and the paper by Blackford, Allen and Gilbert (2004). It should be said, nevertheless, that there are others that are probably equally as good based upon the modelling expertise of other institutions, both in the UK and in other countries. Conceived as a generic model, ERSEM, when coupled to a qualitatively correct physical model, is designed to be capable of correctly simulating the spatial pattern of ecological fluxes throughout the seasonal cycle and across eutrophic (nutrient rich) to oligotrophic (nutrient poor) gradients. There are perhaps three reasons that allow ERSEM this flexibility. Firstly, it includes detailed representations of the benthic system, which are vital for the correct treatment of shelf seas. Secondly, it decouples carbon and nutrient dynamics, which gives a far better approximation to how nutrient limitation acts on cells. Thirdly, it can simulate both the "classical" large cell production and grazing dynamics and the small cell microbial loop, thereby representing the continuum of trophic pathways evident in marine systems.

The model was originally developed for the North Sea; what follows are more details of both the physics and the biology of the model itself. In the model setup, the North Sea is divided into 130 spatial boxes; the boxes are mostly regular $1° \times 1°$ squares. During summer many areas of the North Sea are stratified, and during this period, primary production is mostly confined to the upper layer.

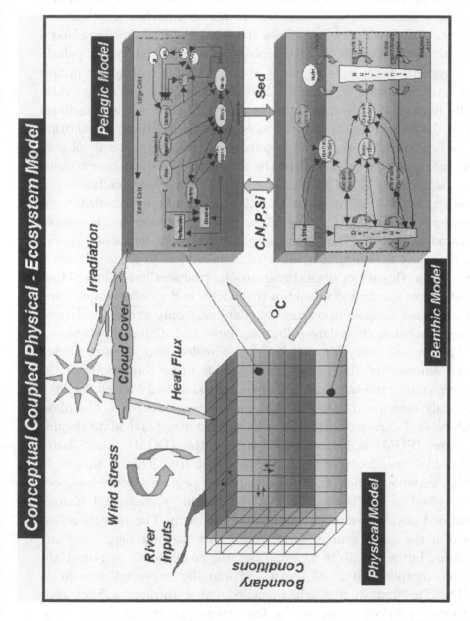

Fig. 9.10   The POLCOMS–ERSEM model.

The boxes were divided in the vertical at the depth of 30 m wherever stratification prevails during summer. This depth was shown to be a reasonable approximation of the thermocline depth in the North Sea. In this way, 45 boxes are separated into an upper and a lower box for the set up. The biology as represented by the pelagic ecosystem model is really a cut above anything met so far in this chapter. For example, there are four phytoplankton species individually represented; picoflagellates (smallest), flagellates, dinoflagellates and diatoms (largest). These, together with irradiance, constitute the primary producers. The flagellates are ordered in terms of size, as it is size that is crucial to the modelled biological interactions. By segregating the phytoplankton in this way, their predation can be represented in greater detail. The zooplankton (labelled "consumers" in ERSEM) are of three types: heterotrophs, microzooplankton and mesozooplankton, again ordered in terms of size. So, although it is tempting to allow all zooplankton to graze all phytoplankton, this is not done. Heterotrophs (the smallest zooplankton) only graze small phytoplankton (flagellates and picoflagellates), and the largest zooplankton (mesozooplankton) only graze the largest phytoplankton (flagellates, dinoflagellates and diatoms). Microzooplankton eat everything. Nutrient is divided into silicate, nitrate and ammonium and is thus one variable more complex than the seven compartment model of Fasham, Ducklow and McKelvie (1990) already mentioned. The ERSEM pelagic ecosystem also has phosphate and carbon dioxide, as well as bacteria, particulate organic matter (POM) and dissolved organic matter (DOM). These latter variables are also partitioned in terms of size. This is thus a 15 plus variable model, and so far only the pelagic system has been described here. To this, we add the benthic system and biological and physical communication between them. The rate equations are of the same general form as we have already met. They are linear, but more likely to have variable co-efficients to reflect the more complex nature of the model and the improved data available. The limitation is still modelled either through a Michaelis–Menten relationship or one of the other appropriate Holling type models.

Considering the physical side of the model, the box set up was chosen as the standard set up during the second phase of the ERSEM project (1993–1996), when ERSEM was applied to the coastal areas. In these areas, the 15-box setup did not satisfy the needs of spatial resolution, thus leading to incorrect river inputs from estuaries along the continental coast and into the central North Sea. The set up is related to the Arakawa B grid (see Chapter 5), i.e. the numerical grid used for the hydrodynamical simulations which are utilized in ERSEM as forcing. The regular box size is three (east–west) by five (north–south) grid points, which corresponds approximately to "rectangles" of 60 km times 100 km side lengths. In addition to the interior boxes 1 to 130, boundary boxes 131 to 155 were defined because they serve for defining boundary conditions for the model. The northern North Sea is closed towards the Atlantic ocean by the boxes 131 to 141 (east to west) in the surface and by the boxes 144 to 154 in the deeper layers. The Skagerrak partly belongs to the interior of the model domain, and the eastern Skagerrak and northern Kattegat serve as boundary box 143 (surface) and box 155 (deep). In the Channel, box 142 was added.

Data were, and are, needed in the ERSEM project for several purposes. For initializing and forcing in the set up of the ecosystem model ERSEM, data are needed in amounts that allow the prescription of the initial situation within the North Sea and the temporally continuous forcing at the open boundaries of the simulation area (i.e. boundary conditions) by statistically meaningful data products. For testing and validating the model, annual cycles of the main state variables are needed to enable a comparison of model outcome with observed system behaviour. Although the model was developed for the North Sea ecosystem, by a consortium of European laboratories, several studies have shown that the model is equally applicable in warm temperate (e.g. Mediterranean) systems and tropical situations (such as the Arabian Sea). The versatility of ERSEM is demonstrated by the range of subjects to which it has been applied. The Blackford, Allen and Gilbert paper uses six very different sites (two contrasting North Sea stations, Catalan Sea, Cretan Sea and two contrasting Arabian Sea stations). Studies of land–ocean interaction have ranged from shallow coastal lagoons to an assessment of riverine influence on the North Sea basin. Basin scale and open ocean applications in one, two and three dimensions have addressed issues varying from the dynamics of viruses to the influence of weather and climate on marine trophodynamics.

To the present state of knowledge, therefore, ERSEM provides a model mesocosm environment that can be expected to react in a qualitatively correct manner to seasonal, regional and inter-annual variations. It has been developed as a commercial product and sold as an ideal test bed with which particular hypotheses can be addressed, from studies of microphytobenthic populations to an exploration of mesozooplankton dynamics in the Irish and Celtic Seas and, most recently, investigations into the interaction of dimethyl sulfide production with the marine ecosystem. (Dimethyl sulfide or DMS is a gas that helps to counteract the greenhouse effect and so could be an important weapon in the fight against global warming.)

One of the more exciting recent developments has been the coupling of ERSEM to the Proudman Oceanographic Laboratory's 12 km resolution hydrodynamic model, covering the entire north–west continental shelf region around the UK and Ireland. This allows, for the first time, the simulation of many mesoscale ecological features that can be tested against data, form the basis of hypotheses and guide future cruise programs. More recently, the 1.5 km eddy resolving scale, which improves the simulation of mixing, has been shown to significantly impact the ecosystem model response. Simulated production is seen to increase by 25%, improving fits with measured data. We conclude, perhaps not surprisingly, that detailed physical and ecological representations are required to produce accurate models, and ERSEM/POLCOMS is beginning to accomplish this. Running this, and other models of a similar nature, gives realistic results, but the distribution of the ecosystem variables are very sensitive to the physics. There is an important message here: the accuracy of an ecosystem model that is embedded in a hydrodynamic model is only as good as the hydrodynamic model. Get the physics wrong and the biology will be horribly wrong. With this warning, another recent development is the coupling of ERSEM to the state-of-the-art GOTM, which allows us to look at ecosystem response to high frequency physics.

In addition to heuristic issues, the ability to predict ecosystem response to climatic change is of great interest to those responsible for environmental policy and management. Any model's ability to predict is directly related to its ability to react correctly to a range of environmental conditions, rather than be constrained to a limited mode of behaviour by excessive simplicity. The ERSEM model's range of processes gives us confidence in its predictive capabilities. For example, recent work has demonstrated that the ERSEM model can reproduce long term inter-annual variations in mesozooplankton biomass seen in the CPR (Continuous Plankton Recorder — developed by Plymouth Marine Laboratory and responsible over many years operation for the gathering of uniquely full data sets on many species of plankton over the North Atlantic waters, as well as those around the United Kingdom) data set. Ongoing work is investigating data assimilation as a technique for producing robust forecasts of ecosystem response to short term climatic influences. In the ERSEM project, three different set ups of the ecosystem model ERSEM for the whole North Sea were defined.

This is thus the current exciting state of ecosystem modelling, but remember that the ecosystem is only modelled correctly once the physics is determined, and some of the most interesting biology occurs precisely where the physics is hardest to model. The final part of this chapter is devoted to modelling single animals. This is done by using a simple biological model of some of its internal biology and using it as a proxy for the entire animal.

9.6   Modelling Animals

The first question to ask is why do we want to model an entire animal? The previous sections of this chapter are about modelling ecosystems of increasing complexity, culminating in the ERSEM software and its interaction with hydrodynamic models. However, all zooplankton are individual animals, and they are alive or dead as individuals. When an estuary or coastal environment gets polluted, animals die. This is usually the first indication that something disastrous has happened. This is an unacceptable and very crude indicator of environmental pollution. A much better solution would be to know how environmental pollution affects particular species on an individual level, and to detect pollution through monitoring rather than clean-up. Prevention is always better than cure. It is important to know when they survive and under what circumstances the level of pollution of their environment becomes toxic for them. If we can build a model of the animal and, through modelling and validation of the modelling, determine these levels then progress has been made. In the future, the kind of ecosystem models already outlined will be informed by such models as the bulk parameters in, for example, the linear and Michaelis–Menten relationships that make up the model will be more accurately determined. In this section, we shall out-line a model of the marine mussel, but before getting into the model proper, there are several biological terms that need introducing.

First of all, we shall be modelling at the cellular level. In par-ticular, lysosomes are important. The lysosomes are more or less the wrecking crew of a cell. They digest macromolecules and break down damaged or old cell parts as well as bacteria. The cytoplasm is the body of the cell, excluding the nucleus, and endocytosis (endo (within) cytosis (cell)) is a process in which a substance gains entry into a cell without passing through the cell membrane. Exocytosis is the reverse of this, i.e. the ejection of matter from the cell without passing through the cell membrane. There are many other parts of a cell, for example the Golgi apparatus for processing proteins could be important to include in a model. Figure 9.11 shows a typical cell with some of its parts. We shall be principally concerned with modelling the lysosome.

Let us now turn to our chosen animal, the marine mussel. Why choose this Marine ecotoxicology has long recognised the utility of the marine mussel as an indicative organism of entire ecological health. The mussel is widely used as an environmental sentinel due to a num-ber of characteristics. This is justified in papers such as Widdows and Donkin (1992). They have a wide geographic distribution and are usually abundant where they do colonise. They are quite robust,

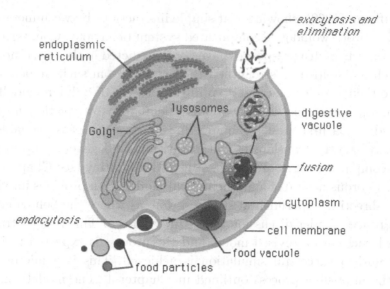

Fig. 9.11 A cell and the processes within. Downloaded from *Encyclopaedia Brittanica* and used with permission.

having relatively high tolerances to contaminants and environmental conditions. As sessile filter feeders, a large proportion of their aqueous environment passes through them, thus exposing them to a high proportion of any associated contaminant. The digestive cell has been chosen as the initial development of the mussel for a number of reasons. It forms part of the continuous interface between the animal and its environment. It is also the destination for food particles, previously sorted and roughly degraded along their passage through the gills and stomach, and a number of toxins are introduced to the organic mass through speciation with colloidal matter ingested as a perceived source of nutrition. In addition, it acts as one of the buffers against toxic stress, having well developed defence mechanisms against such attacks. Cellular modelling has been around for at least 40 years; previous successes range in scope from supercomputer simulations of human hearts and lungs, Hunter, Kohl and Noble (2001), to models of cellular signalling pathways, Bunk (2003), all the way down to models of genetic regulatory networks,

Kauffman (1969), showing that simple interactions between members in a network would give complicated system behaviour. Counterintuitive effects of drug responses have been predicted in the heart model, which has helped to strengthen the confidence in such models and raise their profile. The first step in the process of cellular modelling is a conceptual model; the premise behind an observation is proposed and a first model is built. Once such a model is available it is necessary to translate it mathematically and generate appropriate computer code in order to test it under scrutiny (see Chapter 1). The rigorous nature of a mathematical model often provides impetus into directing experimental study into areas where the behaviour of the process under discussion is not fully known. This is then translated back into a revised model and the process is expected to iterate until a successful simulation is achieved. This is a microcosm of the modelling process outlined in Chapter 1. The model can be tested against another data set in order to determine its capabilities. Certain physical characteristics of the cell can be used as early indicators of reductions in animal fitness. These include lysosomal stability, which correlates well to the proportion of lysosomal volume to cell volume. Thus, if we can accurately model both of these variables under stressful conditions, the result will be an indication of the conditions to which the mussel is susceptible. Lysosomal stability correlates well with other biomarker responses both at the same cellular level and upwards to tissue and animal responses; hence, it can act as an integrated biological stress indicator. If the mussel cell model can accurately predict lysosomal and cellular volume then, by correlation, the health of entire animals can be predicted. It will be necessary to ensure that these correlations exist for all stressful conditions tested.

It is clear that the lysosomal system occupies a central and crucial role in cellular food degradation (intracellular digestion), toxic responses and internal turnover (autophagy) of the hepatopancreatic digestive cell of the marine mussel *Mytilus edulis*. The hepatopancreas is an organ found in fish and other marine life that combines the functions of the liver and the pancreas. Understanding the dynamic response of this system requires factors affecting

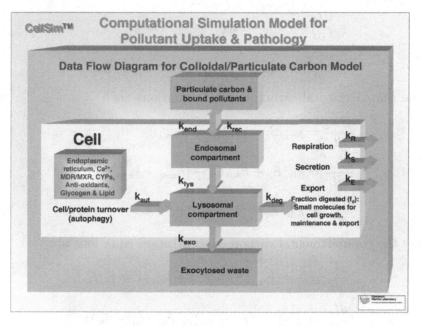

Fig. 9.12   Initial conceptual aggregated discrete compartment model of mussel digestive cell with relevant pathways.

performance (conceived as a function of throughput, the efficiency of degradation and membrane stability) to be defined and quantified. It seems likely that a linear model would be a good first step. We are now ready to set up the first simple model. The flow of carbon through the mussel is indicated in the flowchart displayed in Figure 9.12. In setting up the model, the system of ordinary differential or rate equations usually relies upon the assumption of advective flow between the cellular compartments. A more detailed inspection reveals that this will not, in general, be the case due to the finite volumes of the compartments. A stricter approach gives the set of equations detailed below. The proportion of absorbed material which is digested and passed on to the cytosol (the internal fluid of the cell) can be accounted for by the ratio between the digestion and exocytosis rates. The compartmental volume equations are thus

$$\frac{dE_V}{dt} = k_{\mathrm{end}} - k_{\mathrm{rec}} - k_{\mathrm{lys}},$$

$$\frac{dL_V}{dt} = k_{lys} + k_{aut} - k_{deg} - k_{exo},$$

$$\frac{dC_V}{dt} = k_{deg} - k_{aut} - k_{exp} - k_{sec} - k_{res}.$$

The compartmental carbon content equations are

$$\frac{dE_{xc}}{dt} = k_{end}S_c - (f_r k_{rec} + k_{lys})E_c,$$

$$\frac{dL_{xc}}{dt} = k_{lys}E_c + k_{aut}C_c - (k_{deg} + k_{exo})L_c,$$

$$\frac{dC_{xc}}{dt} = k_{deg}L_c - (k_{aut} + k_{exp} + k_{sec} + k_{res})C_c,$$

and the compartmental carbon concentration equations are

$$\frac{dE_c}{dt} = \frac{1}{E_V}(k_{end}(S_c - E_c) + k_{rec}(1 - f_r)E_c),$$

$$\frac{dL_c}{dt} = \frac{1}{L_V}(k_{lys}(E_c - L_c) + k_{aut}(C_c - L_c)),$$

$$\frac{dC_c}{dt} = \frac{1}{C_V}k_{deg}(L_c - C_c).$$

The rate constants have the following definitions

k_{end} — rate of total endocytosis,
k_{rec} — rate of total recycling,
k_{lys} — rate of total traffic to lysosome,
k_{aut} — rate of total autophagy,
k_{deg} — rate of total digestion,
k_{exo} — rate of total exocytosis,
k_{sec} — rate of secretion,
k_{res} — rate of cellular respiration.

There are various modelling aspects that need clearing up, such as the rate at which solvent enters and leaves the cytosol. These and other details can be found in Dyke and McVeigh (2009), but we give one result of the model due to tidal feeding. Here, the mussel as an intertidal species is subject to square wave input. It feeds when covered by water and starves when not. The square wave indicates that

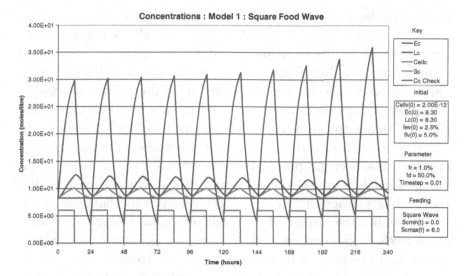

Fig. 9.13 Ten day compartmental concentrations for a median intertidal cell with maximum food concentration when available. The original is in colour so fine distinctions are lost in this monochrome version. However, the periodicity and other general behaviour is clear, in particular the magnitudes.

it always feed at maximum efficiency, an obvious caricature of the truth. This is, after all, a simple first model. The output is shown as Figure 9.13. The model equations (not all given here, but see Dyke and McVeigh, 2009) are solved by the simplest possible numerical scheme — the Euler method. This is so because the results using a more complex scheme were found to be much the same. The response of the mussel as modelled here was found to be reasonable and in accord with observations. We thus have a simple model that is able to be built upon and improved. The first change would be to use some nonlinear effects. However, nonlinear dynamical theory demonstrates that a system of exclusively stable components need not necessarily lead to a stable system; instability may arise from the interactions between the stable components. Previously, nonlinearity in ecosystem models has been rejected as there is not much evidence of it in the field. This is not the case here when modelling individual animals such as the marine mussel. Counterintuitive results can be found often enough to show that identifying individual components and mechanisms should not be the endpoint of scientific investigation. The reductionist history of biology has achieved a wonderful cornucopia of results, insights and data by identifying individual components and their responses, right down to the level of individual proteins. The current challenge facing the biological sciences is to take this myriad of concepts and models and integrate them across all levels to produce a functional model of entire systems. In order to do so, a new integrative multidisciplinary systems biology approach has been proposed. The current work on modelling individual animals is conceived to be part of such an approach within the field of marine environmental toxicology. Environmental toxicology studies the impact of contaminants upon biological systems of all levels in the environment. So complex and subtle are system responses to combinations of environmental stressors that, in the past, the mantra has been that it is impossible for the human mind to calculate or predict them. With increasing computational power available, this is no longer the case as we can copy nature and compute the various known interactions. Past efforts to control negative effects of ecological alteration have only been partially successful; as the direct and indirect consequences of urbanisation, industrial processes and pest control were either unable to be predicted to an effective degree or else were simply not understood or known to the requisite level. What is required is a system which picks up the early warning signals of possible deleterious effects on human and ecological health by means of bioindicators. Under the auspices of "environmental prognostics", it is proposed that this be implemented by a two prong strategy. Initially, identify suitable biomarkers which are sensitive enough to indicate appropriate levels of ecological damage in response to environmental stressors. Secondly, collectively relate these biomarker responses to ecological consequences. In order to achieve the second of these aims, various computational techniques are required to integrate, both horizontally and vertically, these responses. These techniques include multivariate statistics and simulation models of sentinel species, which is the raison d'être of this particular research topic, explicitly the development of a numerical model of a marine mussel digestive cell.

9.7   Summary

This chapter has, it is hoped, given an outline and introduction to ecosystem modelling, culminating in the ERSEM/POLCOMS model. This is an advanced multivariable model with a great deal of detail, but ecosystems modelling is still developing. The last section has been an introduction to the relatively new science of modelling an animal using a simplified biology as a proxy for the entire animal. Attempts have been made previously to describe a mathematical model of pollutant uptake, one such precursor to the marine mussel model which is detailed in Moore and Willows (1998), where both multi-drug resistance as a defence mechanism and the importance of the chemical speciation of the pollutant are stressed. The model outlined in Moore and Allen (2002) illustrates the possibilities that a fully developed model could have. The assumptions driving the mussel model are that the cell can be modelled as three discrete amalgamated compartments; the energy balance of the cell can be simulated by the flow of carbon between these compartments, thus only pathways that carry either appreciable amounts of volume or carbon need be considered; and the control condition is that the concentration of cytosol carbon remains constant throughout — thus providing regulation on cell volume. There is no doubt at all that there is a bright future for the modelling of individual animals and the link between more complex models and projects involving genome research and the gathering of data on all biological species.

9.8   Exercises

(1) State the main three properties that the Michaelis–Menten relationship has that make it suitable for nutrient modelling.
(2) Rework Example 9.1 but use the equation

$$\frac{dP}{dt} = aP - bP^3.$$

Use finite differences and compare the results with those of Table 9.1. In order to facilitate comparison, determine a and b by assuming that $P(1910) = 89$ and $P(1920) = 103.64$.

(3) Consider the three variable ecosystem model

$$\frac{dZ}{dt} = b_2 PZ - dZ,$$

$$\frac{dP}{dt} = aNP - bPZ,$$

$$\frac{dN}{dt} = -aNP + b_1 PZ + dZ,$$

where $b = b_1 + b_2$, a and d are constants, and the units of N, Z and P are milligrams per cubic metre (mg m^{-3}). Time is in days (this model is taken from Klein and Steele, 1985). By adding them, find an equation that links P, N and Z and interpret this. Find two steady state solutions and interpret them in terms of ecosystem dynamics.

(4) A predator–prey system is governed by the equations

$$\frac{dx}{dt} = x\left(a\left(1 - \frac{x}{x_0}\right) - b_1 y\right),$$

$$\frac{dy}{dt} = (b_2 x - c)y,$$

where x is the prey, y is the predator and a, b_1, b_2, c and x_0 are constants. What is the state $x = x_0, y = 0$? Determine a state of peaceful coexistence. Numerical solution of these equations shows that there is an initial oscillatory behaviour in both x and y but that eventually a steady state is reached. Interpret this behaviour.

(5) Comment on the transportability of the sea loch model outlined in section 9.4.3 to estuaries and other places where there is an exchange of fresh and salt water.

(6) Outline the next step after ERSEM in the use of software to model ecosystems.

(7) What are the principal concerns in modelling animals in the way described in section 9.6?

Chapter 10

Climate Change and Large Scale Ocean Models

10.1 Introduction

In this chapter, we shall examine some models of marine areas that, although of interest, fall a little outside what normally is classified as coastal modelling. They are included because of the insight provided that is distinctive from those covered so far. Most of them are in the realm of physical oceanography, but no apology is needed for looking at climate change, which is most assuredly global but the effects of which will be apparent on all scales and more likely sooner rather than later.

The study of climate change is rather different to any of the modelling we have done so far. It involves the analysis of recent climates and estimates of possible future climates of the next decades and centuries. In general, it also addresses the dynamic interaction among various components of the climate system, such as the atmosphere, the hydrosphere (mainly oceans and rivers), the cryosphere (mainly inland ice, permafrost and snow), the terrestrial and marine biosphere, and the pedosphere (mainly soils). Hence, an integrated assessment of the climate system, or natural Earth system, is required. This needs a whole book to do the subject justice therefore we can only give an outline here. The Earth's climate system is an elaborate type of energy flow system (Figure 10.1 displays a cartoon simplified version) in which solar energy enters the system, is absorbed, reflected, stored, transformed, put to work and released back into outer space. The balance between the incoming energy and

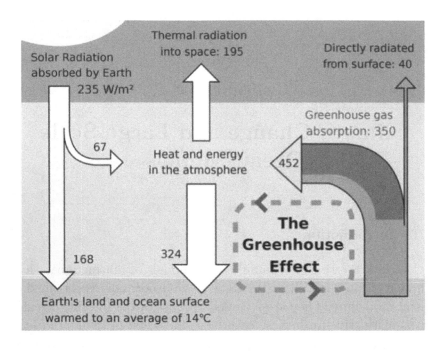

Fig. 10.1   A greatly simplified energy flow of the Earth.

the outgoing energy determines whether the planet becomes cooler, warmer, or stays the same. There has been overwhelming evidence of global warming in the last 50 years of the 20th century, with heat gain from 1955 to 1998 of 1.54×10^{23} joules. About 84% of this increase is stored in the world's ocean, 0.5% in the atmosphere and the rest (15.5%) going to the melting of ice and the warming of the solid earth. In the last 15 years, this has slowed down almost to the point of halting; this has increased the standing of the climate change sceptics amongst the general public. Data supporting the temperature rise emerge from collecting millions of temperature profiles and raiding and interpreting data archives and are quoted in Rhines (2006). The Earth reflects about 34% of the solar energy received; the remainder is used to operate the climate and maintain the temperature of our planet. The Earth also radiates energy back into space — equivalent to 66% of the energy that is received — this implies that there is no net energy gain. Since the amount of energy received approximately equals the amount given back to space, the Earth is approximately in a steady state in terms of energy. As suggested in Figure 10.1, this kind of a steady state is an expected outcome of a system in which the outflow is dependent on the amount of energy stored in the system. In reality, there are temporal and spatial changes in temperature that are very important; some are natural, while others may be due to anthropogenic (that is man-made) modifications of the climate system. One question to ask is has the Earth always been like this? The answer is no, as in the past the climate was very different. For example, during the last ice age the reflection was much greater (for pure snow it can be as much as 90%). We thus expect that the Earth would settle into this "ice age state" at some future date, although precisely when and precisely how long the transition might take is not known. There is also a debate on the interaction between this and the increase of greenhouse gases into the atmosphere. Some say the warming must delay the onset of the next ice age, others say that the melting of the Greenland ice shelf will disrupt the Gulf Stream, causing it to assume a path more directly towards Spain, and that this will trigger the next ice age (because this more southerly route for the Gulf Stream is precisely what happens in an ice age.) Recently, this possible diversion of the Gulf Stream by the melting polar ice cap is thought to be less possible on the grounds of energy balance. So, although the jury's out, this author thinks that the triggering of the next ice age by global warming remains unlikely, simply because, by definition, global warming is precisely that, warming. The weakening of the Gulf Stream may happen, but this is more likely to cause colder winters and climate change in the vicinity of the UK that is different from either the present climate or the ice age. Through a combination of theory, observation and modelling, it now seems most likely that ice ages are triggered by variations in the Earth's orbit that alter the amount of solar energy reaching Earth in both a regional and seasonal way. Although the amounts of energy are small, these changes can be reinforced in a kind of resonance due to snow cover over the northern continents and Arctic Sea. Modelling using the best available data and better knowledge of the processes has to be the answer.

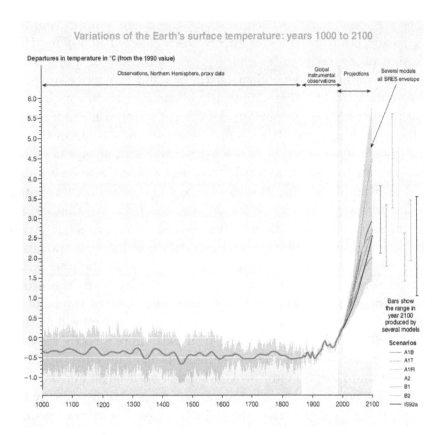

Fig. 10.2   The original "hockey stick" diagram showing the temperature anomaly in the Northern hemisphere — the light grey is the uncertainty.

Take a look at Figure 10.2, the so-called "hockey stick" diagram. There is no modelling here until the projections in the last strip; before the present decade this graph shows data, but how the data have been displayed in this graph is controversial. The main controversy is not about the upturn in temperature since the start of the new millennium, it is more about the flatness of the historical record. To this author, this is rather like arguing about the colour of the walls whilst the house is falling down. Having said this, the revised version shown in Figure 10.3 has a medieval bulge that corresponds to the warm period as well as the so-called little ice age in the period ~1550 to ~1850. So, the stick part of the hockey stick diagram looks more

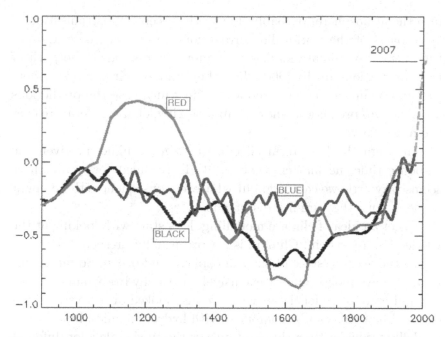

Fig. 10.3   The revised temperature diagram. Red line: rescaled Intergovernmental Panel on Climate Change, 1990, based on work from the Climate Research Unit at the University of East Anglia by Hubert Lamb showing central England temperatures; compared to central England temperatures to 2007, as shown in Jones *et al.* 2009 (green dashed line). Also shown, Mann, Bradley and Hughes (1998) 40 year average used in IPCC TAR 2001 (blue) and Moberg *et al.* 2005 low frequency signal (black). The *y*-axis is temperature anomaly.

sinusoidal. This line is red in the original paper. The other two lines are straighter, the lower one (black) is from Moberg *et al.* (2008), the straighter one (blue) from Mann, Bradley and Hughes (1998).

These bulges in the shaft of the hockey stick could be seen by the climate change sceptics as a reason for debate. Maybe the recent rise is the beginning of another naturally occurring cycle. However, the right-hand rise, corresponding to the blade of the hockey stick, is still there. Both graphs show the temperature anomaly since the year 1000, but the revised version by Jones *et al.* (2009) looks more realistic and is less open to accusations of political manipulation; in particular, the rapid rise in the last 40 or so years that can, in reality, only be due to the injection of man-made greenhouse gases into the atmosphere, probably due to the recent industrialisation of large parts of the world. The projections shown on this graph are the outputs of climate models that more or less follow the path of the observations up to 1900 when they underpredict, unless global warming is included in the models — in which case the predictions are good. *All* predictions show a rapid rise in global temperature over the next 100 years.

Although the long term effects are largely unknown, given the response time, no matter what we do in the next few years there seems very little we can do about whatever consequences there might be; let's hope they are not disastrous.

To get an idea of climate modelling, let's start with looking at the whole climate system. Climate is a word usually preserved for looking at the average state of the atmosphere and ocean. So far in this text, we have looked almost exclusively at the hydrodynamics of the sea and how to model these. We have also looked at ecosystem biology and how this can be combined with hydrodynamics, but climate modelling is wider than this. Not only is the time scale a lot different (often many centuries) but also one needs to take into account the entire biosphere and, in particular, the cryosphere (the name given to all the ice covered parts of the earth, polar caps, frozen sea and rivers, frozen ground including permafrost). So, although the weather and the ocean are governed by very complicated physics, chemistry and biology, in order to model climate, a lot of averaging has to take place.

For the coastal engineer and others involved in flood protection, erosion etc., it is local events that are the most important. The storm surge caused by weather, flooding caused by a local river that is overflowing — that sort of thing. The extreme events that cause such phenomena cannot be predicted with certainty, but using the correct statistics, some prediction is possible once the processes underlying the causes are understood. Since it is these processes that form the basis for all models, including the longer term climate models, climate models should eventually lead to improved predictions of short term extreme events. This leads to a discussion of model hierarchy.

10.2   Model Hierarchy

When modelling is undertaken, a decision has to be made about what to include and what to exclude. If a model needs to address long term climate change, then we really need to integrate our present knowledge of the climate system. In order to do this, we start by discussing process models of differing kinds. For example, we have already looked at shallow sea models that predict tidal and wind-driven flow, so we ask what role might these have on the climate? The answer is a bit. Models of wind driven currents do form a crucial part of understanding the whole system, for it is this when taken as a whole that tells us how the energy and momentum tied up in the winds of the lower atmosphere are imparted into the ocean. These ocean currents include the Gulf Stream and other major ocean currents. Fifty years or more ago, there was a concentration on building models that could predict large ocean currents. These models were geometrically and dynamically simple but they did capture the essential physics of ocean circulation. When large computing power became available in the 1970s, it was these models that drove the codes and produced the first simulations of the major ocean currents. What emerged was that there were really no steady currents; instead the world oceans were crossed by currents that varied in magnitude and direction. Some, such as the Gulf Stream were largely dependable, others such as the Somali Current (off the coast of Somalia in the Indian Ocean), are seasonal, and still others, such as the Equatorial Countercurrent in the southern Pacific Ocean, vary with a crucial uncertainty. When equatorial countercurrent is strong, the phenomenon known as El Niño takes place.

In the last 20 years, the focus has turned to these important variations in ocean currents and the role they play in climate change. Some think that the more frequent and more prolonged El Niño is indeed related to global warming. It is true that the trigger for El Niño does seem to be warmer than usual water south of the Philippines driving the countercurrent eastward and pushing the cool, nutrient rich Humboldt Current and its associated upwelling away from the South American coast. What role El Niño has on the "conveyor belt" that is central in distributing energy about the Earth is questionable, as this conveyor belt takes thousands of years to traverse. What seems the case, however, is that El Niño forms part of a larger system now called ENSO (El Niño Southern Oscillation) that is associated with other worldwide oceanic and atmospheric phenomena.

What is clear is that tidal models only contribute to the large ocean flows at the margins; through nonlinear interaction and through internal waves, which is interaction with stratification. The latest research does seem to indicate that internal waves do play a more important role than hitherto expected. This is largely because of the much longer time scales associated with these waves that make it easier for them to interact with long term processes. The book by Vlasenko, Stashchuk and Hutter (2005) gives a good account of baroclinic tides. Most of the world ocean has a thermocline or halocline, and hence there will be waves that propagate along this density interface. Moreover, as demonstrated in Vlasenko, Stashchuk and Hutter (2005) these waves are capable of propagating energy large distances without very much decay and could play a central role in the distribution of tidal energy into turbulence. This is a very important active area of current research that is only now receiving the attention it deserves.

10.3   Planetary Vorticity

Before looking at the details of ocean modelling, it is worth deriving a general relationship valid on the rotating Earth as long as friction (turbulence) is negligible, which means that what will be derived here will not be valid very close to either coasts or the sea bed. With this in mind, let us turn now to deriving this relationship. Normally a vector treatment is preferred, but for this problem, unusually, it is rather clumsy. This is due to the special place occupied by the vertical in ocean physics. So, here we revert to local Cartesian co-ordinates where the approximations concerning the Coriolis acceleration have already been made. Recall that twice the vertical component of the Earth's rotation is labelled f. Retaining the total derivative notation, the two horizontal equations of motion become

$$\frac{Du}{Dt} - fv = -\frac{1}{\rho}\frac{\partial p}{\partial x},$$

$$\frac{Dv}{Dt} + fu = -\frac{1}{\rho}\frac{\partial p}{\partial y},$$

where once more all frictional effects are ignored. In order to eliminate the pressure, one takes the y derivative of the first equation and the x derivative of the second and subtracts. This is precisely equivalent to taking the curl of the vector equation of motion. More important to realise is that, because the total derivative is not linear, the cross differentiation has to be done carefully, term by term. The details of the algebra are omitted, but the result is

$$\frac{\partial}{\partial x}\frac{Dv}{Dt} - \frac{\partial}{\partial y}\frac{Du}{Dt} = \frac{D}{Dt}\left(\frac{\partial v}{\partial x} - \frac{\partial u}{\partial y}\right) + \left(\frac{\partial v}{\partial x} - \frac{\partial u}{\partial y}\right)\left(\frac{\partial u}{\partial x} + \frac{\partial v}{\partial y}\right).$$

$$(10.1)$$

The quantity

$$\zeta = \left(\frac{\partial v}{\partial x} - \frac{\partial u}{\partial y}\right)$$

is the vertical component of the vorticity of the (two dimensional) flow, and it is the only one that is non-zero given the approximations made (namely $w \ll u, v$ and hydrostatic balance in the vertical). The result of the cross differentiation is therefore

$$\frac{D\zeta}{Dt} + \zeta\left(\frac{\partial u}{\partial x} + \frac{\partial v}{\partial y}\right) + \frac{\partial}{\partial x}(fu) + \frac{\partial}{\partial y}(fv) = 0,$$

which can be written

$$\frac{D}{Dt}(\zeta + f) + (\zeta + f)\left(\frac{\partial u}{\partial x} + \frac{\partial v}{\partial y}\right) = 0.$$

(This uses the fact that

$$\frac{Df}{Dt} = v\frac{\partial f}{\partial y},$$

which of course is actually zero for models with constant f.) The quantity $\zeta + f$ is referred to as the absolute vorticity. In order to eliminate the horizontal divergence term ($\frac{\partial u}{\partial x} + \frac{\partial v}{\partial y}$), it is necessary to do some manipulating with the equation of continuity, namely vertically integrating it. This was done with complete generality in Chapter 3 — see equation (3.1) and the version we need here, equation (3.2). The sea surface is given by the equation $z = \eta(x, y, t)$ and the sea bed by the equation $z = -h(x, y)$, then integrating vertically between these two surfaces, assuming that neither u nor v depend on z, gives

$$(\eta + h) \left(\frac{\partial u}{\partial x} + \frac{\partial v}{\partial y} \right) + [w]^{\eta(x,y,t)}_{-h(x,y)} = 0.$$

Realising that at the surface $w = \frac{D\eta}{Dt}$ due to fluid particles at the surface always remaining there (kinematic condition) and that at the bed $w = -\frac{Dh}{Dt}$ (no flow through the bed itself), and writing $H(x, y, t) = \eta(x, y, t) + h(x, y)$ to represent the total height gives the alternative continuity equation (3.2), repeated here for convenience.

$$\frac{\partial u}{\partial x} + \frac{\partial v}{\partial y} = -\frac{1}{H} \frac{DH}{Dt}.$$

Inserting this into the above vorticity equation gives the simple (looking) equation

$$\frac{D}{Dt} \left(\frac{\zeta + f}{H} \right) = 0. \tag{10.2}$$

This is the standard form of the potential vorticity equation. In those parts of the ocean where there is little fluid shear, the local value of ζ is virtually zero. At these places, the vertically integrated flow follows contours of f/H (called potential vorticity contours). In the early days of oceanography either side of the Second World War, this fact was used extensively to infer deep circulation. In particular, that there must be a deep return flow under the Gulf Stream from Cape Hatteras to the Florida Keys along the continental slope.

In the 1930s, when it was thought that the ocean was completely driven by thermohaline effects, contours of f/H were thought to be the main contours by which the ocean currents flow around the ocean. Since then, ocean modellers have come to realise that it is wind-driven flows that dominate. However, it is now thought to be the case that, although wind-driven currents such as the Gulf Stream in the Atlantic and its equivalent the Kuroshio in the Pacific drive the surface flows, these high density currents sink in high latitudes and there is a "conveyor belt" return flow at lower latitudes that encompasses the entire Earth. The contours of this belt, as the flow is slow and independent of surface effects and friction, will closely follow the geostrophic contours, f/H.

10.4   Ocean Models

Models of the currents on an ocean basin scale date from the Second World War and just before. In the 1930s, most models of the ocean were based on thermal equilibrium, but it was found that the convection currents generated to ensure equilibrium were too small by at least two orders of magnitude. In the 1940s, the idea that ocean currents are primarily wind-driven was proposed and the first successful model of the wind-driven ocean circulation (due to Stommel in 1948) justified this, see Stommel (1965). Stommel's model was beautifully simple; a square ocean with linearity, the simplest quadratic friction and an ideal wind stress (a sinusoidal forcing).

Figure 10.4 shows two sets of streamlines: one with a constant Coriolis parameter and the other with a Coriolis parameter that varies linearly with latitude. This is the only difference, and it is therefore clearly apparent that the western intensification of the ocean currents must be due to this variation in Coriolis parameter with latitude. Despite its age, this remains one of the purest and best examples of modelling; how a simple model uncluttered by complex friction or geometry has value and can tell us in unequivocal terms precisely what must be causing an observed effect — in this case the western intensification of ocean currents.

Stommel actually used what is called the β-plane approximation. Here, the sinusoidal variation of the Coriolis parameter with latitude is replaced locally by a linear variation. The mathematics is as

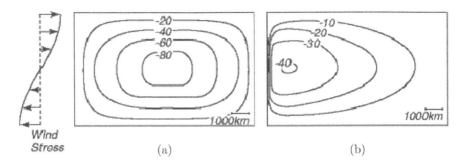

Fig. 10.4   Flow patterns for simplified wind-driven flow with (a) Coriolis acceleration zero or constant and (b) Coriolis acceleration increasing linearly with latitude. Adapted from Stommel (1948).

follows and demands no more than either trigonometric identities or a knowledge of Taylor expansions. Suppose θ_0 is an adjacent latitude — it might be 40° for Stommel's model. Then, at an adjacent latitude, we write

$$f = 2\Omega \sin\theta = 2\Omega \sin(\theta_0 + \delta\theta),$$

where $\delta\theta$ is in some sense small. In reality, it can be as much as 15° but it is a variable. Using either the rule

$$\sin(A + B) = \sin A \cos B + \sin B \cos A$$

with $A = \theta_0$ and $B = \delta\theta$ then putting $\cos B = 1$ and $\sin B = B$ as B is small, or expanding using Taylor series

$$\sin(\theta_0 + \delta\theta) = \sin\theta_0 + \delta\theta \cos\theta_0 + \cdots,$$

then ignoring the terms represented by the dots as negligibly small we get, either way,

$$\sin(\theta_0 + \delta\theta) \approx \sin\theta_0 + \delta\theta \cos\theta_0.$$

From the geometry of the Earth, the small angle $\delta\theta$ is given by y/R, where y is the northwards distance between latitude θ_0 and latitude $\theta_0 + \delta\theta$ on the Earth's surface and R is the radius of the Earth. So, saying the y/R is this angle $\delta\theta$, we can therefore write

$$\sin(\theta_0 + \delta\theta) \approx \sin\theta_0 + \frac{y}{R}\cos\theta_0,$$

whence

$$f = 2\Omega \sin\theta = 2\Omega \sin(\theta_0 + \delta\theta) \approx 2\Omega \sin\theta_0 + \beta y,$$

where

$$\beta = \frac{2\Omega}{R} \cos\theta_0$$

with magnitude about 10^{-11} and dimensions of $L^{-1}T^{-1}$.

This β-plane approximation enables Cartesian co-ordinates to be employed and yet allows us to include in our model dynamics that spring from the spherical nature of the Earth. In analytical models this is still useful and has given many insights into the dynamic processes that dominate in much of the ocean. These include Rossby waves that represent a vorticity balance and underlie important ocean basin wide physics. Through the 1950s and 1960s there were many refinements to theories of ocean circulation, but the crucial advance came with numerical models. These really came into their own in the late-1970s and through the 1980s. First of all, there were large co-operative measurement exercises. One of the first was the GARP Atlantic Tropical Experiment (GATE) in 1974, where GARP stands for the Global Atmospheric Research Program — so we have an acronym within an acronym. The experiment took place in the summer of 1974 in an experimental area that covered the tropical Atlantic Ocean from Africa to South America and incorporated length scales from 100 km to 100 m, covering intermediate length scales. The work was truly international in scope, and involved 40 research ships, 12 research aircraft, numerous buoys from 20 countries, all equipped to obtain the observations specified in the scientific plan. The operations were directed by the International Project Office located in Senegal. The Project Office staff were seconded by the nations involved. The Scientific Director was from the United States and the Deputy Scientific Director was from the then Soviet Union.

An operational plan was developed each day based on the meteorological situation and each ship and aircraft carried out the plan. The data collected were processed by nations participating in accordance with an overall plan and made available without restrictions to all scientists in the world. Research using these data still goes on today, over 40 years later, and it is estimated that over a thousand papers have been published based on the data collected during this short period in 1974.

The experiment involved the world's best scientists, all types of engineers, technicians, pilots, ship captains, logistics specialists, computer specialists, as well as senior policy makers from science agencies and foreign ministries in a large number of countries. A high percentage of the individuals involved are still active and contribute their views.

One suspects much of the funding came from "defence". Out of this and the following observational exercises emerged a very different picture of the earth — atmosphere system. One of the early findings was that 90% of the ocean's energy was tied up in what have now become known as mesoscale eddies. These are the oceanic counterpart of the atmospheric low pressure centre, except that at sea they are 1/30th the size (about 30 km diameter) and last 30 times as long; the recurrence of the number 30 being due to the square root of the ratio of the density of water to that of air. Two experiments followed Mid-Ocean Dynamics Experiment (MODE) and its successor POLYMODE, both of which were securely USA driven (by WHOI — Woods Hole Oceanographic Institute) and looked at the Gulf Stream off Boston/Maine. World Ocean Circulation Experiment (WOCE) was formed and many other experiments — observational, computational and combined — have been instigated since under this umbrella. The emergence of ocean basin and even global scale numerical models became possible at this stage as computing advanced, with better validation coming on the back of the international exercises. One interesting example is Fine Resolution Antarctic Model (FRAM), which is a model of the Antarctic Circumpolar Current, in particular the modelling of the net transport between the Indian, Atlantic and Pacific Oceans and the role played by bottom topography. Today, there are large computer models available either through the web as freeware or else that can be downloaded for a fee. They are based on numerical methods and reflect current knowledge of the dynamic processes.

To step back a little, let us look more closely at some of the developments in our knowledge of the ocean circulation that has informed these numerical models. Models that include the β effect can predict western intensification, but the actual wind forcing together with the geometry of the basin also dictate the form of the ocean currents. For example, in the global ocean, the north Pacific Ocean contains the Kuroshio as the counterpart to the Atlantic's Gulf Stream. It does not travel so far north, hence the west coast of Canada gets less benefit of its warmth in the winter than does the UK from the Gulf Stream. It is, however, a stronger current, as the following figures demonstrate.

Gulf Stream volume flux varies between 5 and 8.5 Sverdrups,

Kuroshio volume flux varies between 19 and 30 Sverdrups (one Sverdrup equals 10^6 m^3 s^{-1}).

The Gulf Stream is also more intense with current speeds exceeding 1.5 m s^{-1}, whereas the Kuroshio Current is more typically between 0.4 m s^{-1} and 1.2 m s^{-1}, even though it transports more mass. This lower momentum together with the closed geography of the northern Pacific Ocean contribute to its more southerly path. A feature of Figure 10.5 is the variability of the ocean currents along the coast of Somalia. In the monsoon season, the Somali

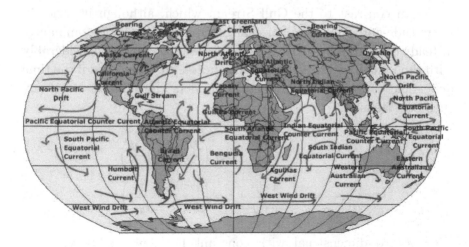

Fig. 10.5   A cartoon of the global ocean circulation in January.

Current is a prominent western boundary current in the northern Indian Ocean. During the northeast monsoon season, the Somali Current flows southward from 5°N–1°N in December, expanding to 10°N–4°S in January–February and contracting again to 4°N–1°S in March. It is then fed from the North Equatorial Current and discharges into the Equatorial Countercurrent. During all these months its speed is 0.7–1.0 m s^{-1}. During the southwest monsoon, the current develops into an intense northward jet with extreme surface speeds; 2 m s^{-1} have been reported for May and 3.5 m s^{-1} for June. The jet is fed from the South Equatorial Current and flows along the eastern coast of the Horn of Africa; part of it continues along the Arabian Peninsula as the East Arabian Current. South of 5°N the jet is shallow; southward flow continues below a depth of 150 m. North of 5°N, the jet deepens and embraces the permanent thermocline. During its northward phase, the Somali Current is associated with strong upwelling between 2°N and 10°N. The upwelled cold water turns offshore near Ras Hafun (11°N), forming a large anticyclonic eddy with a diameter of about 500 km known as the Great Whirl. Eventually, the water from the Somali Current enters the Southwest Monsoon Current. All this is absent from the caricature depicted as Figure 10.5 as this is a January snapshot and cannot depict such a seasonal current.

So, in contrast to the Gulf Stream which, although it varies in magnitude and exhibits minor changes in direction, is predominantly a steady current, the Somali Current is seasonal, varying considerably with time. Such currents can only be successfully modelled if the time dependent terms are included in the model. The simplest set of any such equations is

$$\frac{\partial u}{\partial t} - fv = -\frac{1}{\rho}\frac{\partial p}{\partial x} + \nu_H \left(\frac{\partial^2 u}{\partial x^2} + \frac{\partial^2 u}{\partial y^2} \right),$$

$$\frac{\partial v}{\partial t} + fu = -\frac{1}{\rho}\frac{\partial p}{\partial y} + \nu_H \left(\frac{\partial^2 v}{\partial x^2} + \frac{\partial^2 v}{\partial y^2} \right).$$

This is two dimensional with constant horizontal eddy viscosity and enables us to cross-differentiate to eliminate pressure. In effect obtaining a vorticity equation

$$\frac{\partial}{\partial t}\left(\frac{\partial u}{\partial y} - \frac{\partial v}{\partial x}\right) - v\frac{\partial f}{\partial y} = \nu_H\left(\frac{\partial^2}{\partial x^2} + \frac{\partial^2}{\partial y^2}\right)\left(\frac{\partial u}{\partial y} - \frac{\partial v}{\partial x}\right).$$

In the mid-latitudes, the β-plane approximation is valid, so

$$\frac{\partial f}{\partial y} = \beta, \quad \text{and also} \quad \frac{\partial u}{\partial x} + \frac{\partial v}{\partial y} = 0.$$

This last equation (mass conservation in two dimensions) enables us to define the streamfunction ψ such that

$$v = \frac{\partial \psi}{\partial x} \quad \text{and} \quad u = -\frac{\partial \psi}{\partial y}$$

whence

$$\frac{\partial u}{\partial y} - \frac{\partial v}{\partial x} = -\frac{\partial^2 \psi}{\partial x^2} - \frac{\partial^2 \psi}{\partial y^2} = -\nabla^2 \psi.$$

The vorticity equation then reduces to the neater form

$$\nabla^2 \psi_t + \beta \psi_x = \nu_H \nabla^4 \psi,$$

where the suffix denotes derivative. If friction is ignored entirely, then we have

$$\nabla^2 \psi_t + \beta \psi_x = 0, \tag{10.3}$$

which is the classic Rossby wave equation that describes the evolution of these long period waves. Rossby waves have unusual properties, namely they always propagate energy westward, and they can only exist if the Coriolis acceleration is allowed to vary. Physically, they are restored by vorticity being conserved. Contrast this with water waves that are restored by gravity.

Figure 10.6 is a picture of the jet stream, which is a fine example of a Rossby wave. In the sea, the coasts get in the way and the energy of the Rossby wave feeds the western boundary current, which is why these currents are so strong.

If the time dependent terms are neglected and the friction terms retained, we get a balance that is valid near coasts where the friction is important. In order to analyse the western coast, we retain changes

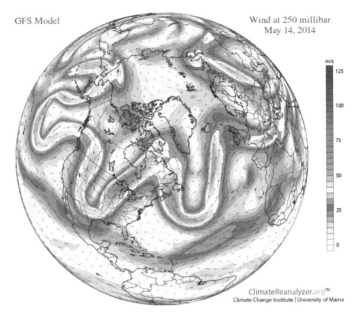

GFS Model

Wind at 250 millibar
May 14, 2014

m/s

125

100

75

50

25

0

ClimateReanalyzer.org™
Climate Change Institute | University of Maine

Fig. 10.6   A Rossby wave in the atmosphere shown as the jet stream. The output of a climate model at the Climate Research Institute, University of Maine, USA.

perpendicular to the coast (x) but ignore those along the coast (y) whence

$$\beta \frac{\partial \psi}{\partial x} = \nu_H \frac{\partial^4 \psi}{\partial x^4}.$$

It is this balance that when solved gives the currents in the western boundary, first done in 1950 by Walter Munk (1917–). In terms of dimensions only, if L is a typical length then this balance implies

$$\frac{\beta}{L} = \frac{\nu_H}{L^4},$$

which gives

$$L = \left(\frac{\nu_H}{\beta} \right)^{1/3}.$$

Munk's values are

$$\beta = 1.9 \times 10^{-11} \mathrm{m}^{-1} \mathrm{s}^{-1}, \quad \nu_H = 5 \times 10^3 \mathrm{m}^2 \mathrm{s}^{-1}$$

which gives

$$L = \left(\frac{5 \times 10^3}{1.9 \times 10^{-11}} \right)^{1/3} \approx 60\,\text{km},$$

which is a reasonable estimate of the width of the Gulf Stream on the Florida coast. It is also possible to estimate the magnitude of the interior flow by assuming an appropriate wind stress acting over the bulk of the ocean (this is called Sverdrup balance, but it is not done here as it is securely large scale physical oceanography, but see Exercise 10.1, at the end of this chapter).

It is sometimes said that the ocean is "tall". This means that if flow near the sea bed is distorted by topography then this distortion tends to be maintained through the depth. This is not precisely true because of friction and the curvature of the Earth, but it is approximately true. It is a consequence of the Taylor Proudman theorem, which is worth deriving as the derivation is short. If, despite all that has gone before, vectors still leave you cold, it is the result that is more important than how it is derived, though seeing the derivation additionally informs you of the assumptions required for its validity. First of all, assume geostrophy, which in vector form is

$$2\mathbf{\Omega} \times \mathbf{u} = -\frac{1}{\rho}\mathbf{\nabla}p.$$

Second, we take the curl of this equation to get

$$2\mathbf{\nabla} \times (\mathbf{\Omega} \times \mathbf{u}) = -\mathbf{\nabla} \times \frac{1}{\rho}\mathbf{\nabla}p = \mathbf{0},$$

where the right-hand side zero results from the vector identity $\mathbf{\nabla} \times \mathbf{\nabla}\phi = \mathbf{0}$ for any scalar function ϕ (assuming that the density ρ is also constant). The left-hand side can be expanded using one of those vector identities

$$\mathbf{\nabla} \times (\mathbf{\Omega} \times \mathbf{u}) = (\mathbf{\Omega} \cdot \mathbf{\nabla})\mathbf{u} - (\mathbf{u} \cdot \mathbf{\nabla})\mathbf{\Omega} + \mathbf{u}(\mathbf{\nabla} \cdot \mathbf{\Omega}) - \mathbf{\Omega}(\mathbf{\nabla} \cdot \mathbf{u}) = \mathbf{0}.$$

Now, if it is assumed that $\mathbf{\Omega}$ is constant, then the middle two terms must be zero as they involve $\mathbf{\Omega}$ being differentiated. Furthermore, the last term is also zero because $\nabla \cdot \mathbf{u} = 0$ is the continuity equation (conservation of mass). This leaves us with

$$(\mathbf{\Omega} \cdot \nabla)\mathbf{u} = \mathbf{0}.$$

Now, only the vertical component of $\mathbf{\Omega}$ is of significance, so we can write $2\mathbf{\Omega} = (0, 0, f)$, where $f = 2|\mathbf{\Omega}|\sin(\text{latitude})$ is the Coriolis parameter. Thus, the scalar product in this last equation picks out the vertical derivative as follows

$$2(\mathbf{\Omega} \cdot \nabla)\mathbf{u} = (0, 0, f) \cdot \left(\frac{\partial}{\partial x}, \frac{\partial}{\partial y}, \frac{\partial}{\partial z}\right)\mathbf{u} = f\frac{\partial \mathbf{u}}{\partial z} = \mathbf{0}.$$

This last equation implies that all components of the current are independent of the vertical co-ordinate z, and this is the Taylor–Proudman theorem. Of course, although $\mathbf{\Omega}$ is actually a constant, approximating it by its vertical component in local Cartesian co-ordinates renders it varying with latitude. In fact, it now depends on the sine of the latitude. Also, geostrophy has been assumed. Nevertheless, this theorem does influence the motion of the large ocean currents to some extent and so it is well worth stating here. It can be well demonstrated in the laboratory by a rotating dishpan; an obstacle on the bottom of the pan causes the flow there to divert around it, and this obstacle is treated by the fluid as if it extends like a cylinder throughout the depth as a consequence of the Taylor–Proudman theorem not allowing any z dependence. Some think that the great red spot on Jupiter is also a manifestation of this theorem. It is tempting to give details of one of the earlier numerical models now, those by Semtner, Jr. (1974) or Gent *et al.* (1998) perhaps, but given that this is principally a modelling text and not a history of modelling, let us instead go straight for a modern model that utilises adaptive grids. Adaptive grids are mentioned in Chapter 4 but got short shrift in Chapter 5. Time to do them more justice.

10.5   Numerical Models with Adaptive Grids

In Chapter 4, some mention was made of unstructured grids and also the application of moving (adaptive) grids to solving problems that result from partial differential equations. Although this is still an application that is very much in its infancy, there are some recent advances that are worth outlining here. These remarks are based on the paper Piggott *et al.* (2008). The equations upon which the computations are based are no surprise

$$\frac{\partial \mathbf{u}}{\partial t} + ((\mathbf{u} - \hat{\mathbf{u}}) \cdot \nabla)\mathbf{u} + 2\Omega \times \mathbf{u} = -\nabla p - g\nabla\eta - \rho g\mathbf{k} + \nabla \cdot \boldsymbol{\tau} + \mathbf{F},$$

$$\nabla \cdot \mathbf{u} = 0,$$

$$\frac{\partial T}{\partial t} + ((\mathbf{u} - \hat{\mathbf{u}}) \cdot \nabla)T = \nabla \cdot (\kappa_T \nabla T),$$

$$\rho \equiv \rho(T),$$

but the notation does require some explanation. The first equation is reasonably standard, but the vector $\hat{\mathbf{u}}$ is there to account for a moving reference frame; the grid that eventually covers the domain of the problem will have nodes that move. The second equation is the conservation of mass. The third, an energy equation written as the conservation of heat. The fourth and last is an equation of state through which density is related to temperature (and sometimes salinity, but not here as salinity does not feature in large scale models.) Other differences in notation are the $\boldsymbol{\tau}$ and κ terms. These express turbulent viscous and thermal effects, respectively, in a most general way using tensors, but the details are not important here. Finally, $\mathbf{F}$ denotes any external forcing, such as that due to tides. This too is zero for the applications dealt with here.

The numerical method employed is to discretise the domain into tetrahedra (pyramids) rather than prisms. The Crank–Nicolson semi-implicit method is used in time, and the variables are expressed as finite sums of shape functions summed over all the elements, as outlined in Chapter 4. The details are given in Piggott *et al.* (2008); it would serve no purpose to repeat the mathematics here. However, it is pertinent to ask under what grounds are the elements changed, new elements created and nodes moved. Put succinctly, the mesh adapts so as to minimise truncation error. Figure 10.7 shows this. The first figure in this diagram, marked (a), is regular with tetrahedra of equal sizes. The diagram on the right, marked (b), is shaded to represent the error the lighter the shade, the greater the error, so there are

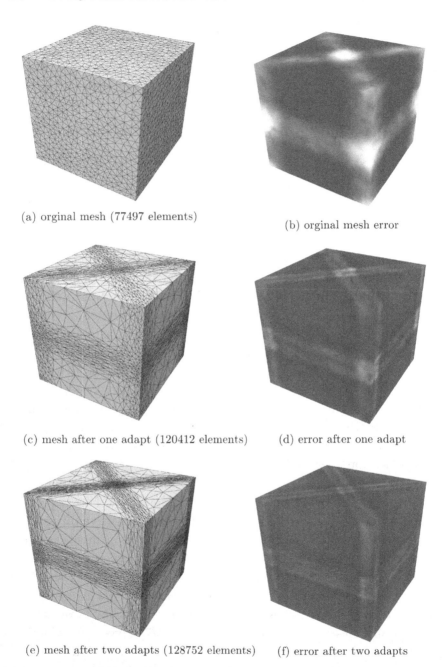

(a) orginal mesh (77497 elements)

(b) orginal mesh error

(c) mesh after one adapt (120412 elements)

(d) error after one adapt

(e) mesh after two adapts (128752 elements)

(f) error after two adapts

Fig. 10.7   Evolution of mesh and solution errors to accurately represent the three ridges.

large errors at the centre of the top face and in the middle of the front edges. As the algorithm proceeds, therefore, it is these areas that gain a concentration of elements at the expense of regions such as the middle of the top of the front left face at which location there is less error and as the elements become large there, this remains the case. The paper by Pain *et al.* (2001) uses a somewhat idealised initial state, but it serves to show how the mesh elements move so as to lessen, and in fact minimise, the error. It is this algorithm that is used to move the grid in the Piggott *et al.* (2008) paper. The paper by Piggott *et al.* written with much generality; for example, all kinds of geometries and scales can be catered for, in particular the effects of the rotation of the Earth through the β-plane effect in simulating Rossby waves. To give a flavour of this particular application, the results are given here in Figure 10.8. There is also a smoothing routine that repositions the centres of mass of the nodes in the same way as demanding Laplace's equation is valid in a region gives a smooth solution (see the end of section 4.3) but this is changed by the presence of diffusion and possible geometric distortion. In the language of section 4.9, this is both an h adaptive grid and an r adaptive grid.

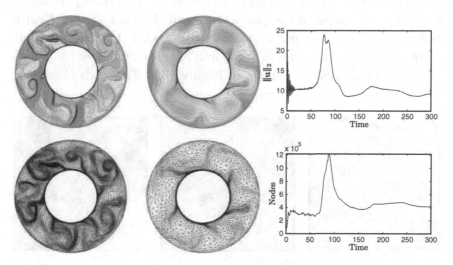

Fig. 10.8   Temperature field and mesh at the upper surface for the rotating annulus problem at times 87 s and 200 s. Also shown are the two norm of the velocity field and total number of nodes.

It is sometimes called an *hr* adaptive grid, but such terms are newer than the publication date (2001). In the application to Rossby waves, Figure 10.8, note how the grid structure follows the temperature pattern. As the temperature field evolves, so the mesh itself bends and refines itself so that it can capture in an optimal way the greatest variations. In particular, note that after 87 s the grid looks more complex and convoluted than at 200 s. At this later time there is a definite sign of a Rossby meander, well captured by this temperature field. The lower graph on the right gives the number of nodes against time, a measure of the complexity. This graph shows a peak at about the time of 87 s. After this time, a more settled number of nodes indicates that the flow is less chaotic and there is (perhaps) a long term or steady state Rossby wave, as illustrated at 200 s. The flow seems to be stabilising at around a wave number of 4 or perhaps 5. The temperatures at the inner and outer ring are set as boundary conditions and it is the difference between these temperatures, as well as the magnitude of the flow (the upper graph on the right), that governs the wave number of the Rossby wave that one might expect. This kind of experiment can be set up physically using a rotating dishpan, but doing it numerically is easier these days. It is also cheaper, allows for greater variation of the parameters and does not involve Health and Safety. Other real world situations are modelled by Piggott *et al.* (2008), including the gyres in the North Atlantic and three snapshots of this evolving grid are shown in Figure 10.9. The grids show how

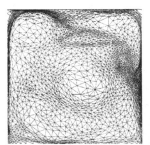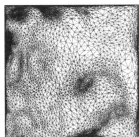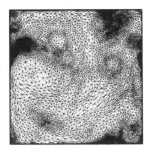

Fig. 10.9   Three North Atlantic simulations showing the evolution of the grid at Reynolds numbers 625, 2500 and 10,000. Meshes are shown after approximately one year of simulation time and contain 5,609, 14,817 and 36,848 nodes.

the forming gyres cause the grid refinement algorithm to concentrate elements where the gyres cause the largest gradients. From recent numerical models, a good idea of the large current systems of the ocean can be gained. What is less certain is how sensitive these are to changes in the driving mechanisms, especially those caused by man-made greenhouse gases in the atmosphere. To current complex models must be added models of the carbon flux which in turn will influence heat balance and ultimately give answers to questions such as how will sea level rise and will the Gulf Stream alter its course?

In Chapter 9, ecosystem modelling was discussed, and as part of this modelling, the flux of carbon was incorporated, although this feature was not particularly dwelt upon. Here is a summary of the role CO_2 (carbon dioxide) plays in the global ocean. It is taken from the UK Meteorological Office website. Carbon dioxide from the atmosphere dissolves in the surface waters. On entering the ocean, carbon dioxide undergoes rapid chemical reactions with the water and only a small fraction remains as carbon dioxide. The carbon dioxide and the associated chemical forms are collectively known as dissolved inorganic carbon or DIC. This chemical partitioning of DIC ("buffering") affects the air–sea transfer of carbon dioxide, as only the unreacted carbon dioxide fraction in the sea–water takes part in ocean atmosphere interaction.

The DIC is transported by ocean currents. Near the poles, cold dense waters sink towards the bottom of the ocean and subsequently spread through the ocean basins. These waters return to the surface hundreds of years later. As more carbon dioxide can dissolve in cold water than in warm, these cold dense waters sinking at high latitudes are rich in carbon and act to move large quantities of carbon from the surface to deep waters. This mechanism is known as the "solubility pump".

As well as being transported around the ocean, DIC is also used by ocean biology. In the surface waters, drifting microscopic oceanic plants, the phytoplankton, grow. As with land-based plants, phytoplankton take in carbon dioxide during growth and convert it to complex organic forms. The phytoplankton are eaten by drifting oceanic animals known as zooplankton, which themselves are preyed upon by other zooplankton, fish or even whales. During these biological processes, some of the carbon taken in during growth of the phytoplankton is broken down from the organic forms of the biology back to inorganic forms (DIC). If, between the carbon uptake by phytoplankton and the subsequent return of the carbon to DIC, the biological material has been transported to depth, for example by the sinking of large biologically formed particles, there is a net transfer of carbon from the surface to depth. This process is termed the "biological pump". The carbon can also sink as skeletal structures of the biology, which is known as the "carbonate pump".

10.6   Exercises

(1) Starting with the geostrophic equations plus continuity in the form

$$-fv = -\frac{1}{\rho}\frac{\partial p}{\partial x},$$

$$fu = -\frac{1}{\rho}\frac{\partial p}{\partial y},$$

$$\frac{\partial u}{\partial x} + \frac{\partial v}{\partial y} + \frac{\partial w}{\partial z} = 0,$$

assuming a constant density but that $f = f_0 + \beta y$, cross differentiate the first two to eliminate pressure p and use the last to deduce that

$$\beta v = f\frac{\partial w}{\partial z}.$$

This is called Sverdrup balance. If a horizontal divergence of magnitude 9×10^{-9} s^{-1} has been induced by the (climatological) wind (a wind stress curl, technically), using $f = 1.1 \times 10^{-4}$ s^{-1} and $\beta = 10^{-11}$ m^{-1} s^{-1} find the magnitude of the southerly flow produced in the ocean interior.

(2) Establish the equation (10.1) by carrying out the differentiation carefully.

(3) Deduce the Rossby wave equation (10.3) directly from the conservation of planetary vorticity equation (10.2).

(4) Substitute the wave $\psi \approx \exp(ikx + ily - i\omega t)$ into the free Rossby wave equation

$$\nabla^2 \psi_t + \beta\psi_x = 0$$

and derive a dispersion relation. Hence, deduce that the wave numbers of all Rossby waves must have a positive westerly component.

(5) Comment on the possibility of using high order difference schemes with adaptive grids for ocean wide models.

(6) Are adaptive grids useful for climate change models?

Chapter 11

Estuarial Flow

11.1 Introduction

Estuaries are very important parts of our coastline both in terms of industry and recreation. In a modelling book such as this the kind of breadth that is contained in specialist texts such as Dyer (1997) cannot hoped to be duplicated. Instead, there is a concentration on the modelling of the currents in estuaries and inlets and nothing about the astonishing variety of geological and biological environments that are found. Physically, an estuary is easy to describe; it is where a river, usually with a substantial flow, meets the sea and so this means that there are features of estuaries that are unique, and it is this that gives a challenge to modellers. In this text so far the Coriolis effect has dominated. When modelling estuaries, it is almost always secondary if not entirely negligible. The modelling here is thus different from most of the rest of the book.

11.2 Estuarial Circulation

River mouths are not only on a smaller scale than the phenomena treated so far in this book, they are places where saltwater and freshwater meet. It is true that the circulation in the more saline mouths of many river outlets resembles that of a coastal sea. The Moray Firth and Firth of Forth in Scotland, and the mouths of the larger Norwegian fjords, are good examples. In these examples, one has to travel further upstream for the presence of freshwater to exert a greater influence on the circulation.

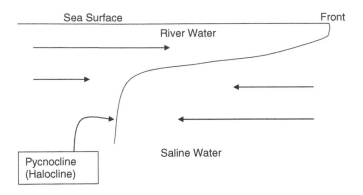

Fig. 11.1   Saline wedge in a stratified estuary.

Saltwater is heavier than freshwater. This gives rise to the "saline wedge" picture often taken as typical of an estuary (see Figure 11.1). Modelling estuarial flows has a long tradition. The first approach, still of some use today, is to use box models (see Section 7.3). In these models, the domain of the estuary is divided into conveniently sized boxes (either two-dimensional vertical or horizontal sections, or three-dimensional boxes), and an estimation is made of the transportation of various water properties across the boundaries of each box. These properties can include salinity, temperature, various chemical and biological constituents as well as flow. Once the boxes have their assigned parameter values, and each box contains water with properties that are consistent with the properties of the water in adjacent boxes and consistent with an overall agreed picture, then the model can be exploited. For example, an inflow at one point of the estuary will influence the water properties in one box or two, but will also probably influence the behaviour of the water in surrounding boxes through having to satisfy the conditions at the boundary of the boxes. These box models are still useful under two circumstances: first, if an approximate picture is required quickly, then a box model can give some insights; and second, if there is no appropriate software available to build a more elaborate model and, more importantly, only a sketchy knowledge in terms of observations in certain areas, then a box model can provide a rough first model. Box models can often be the initial models that trigger additional

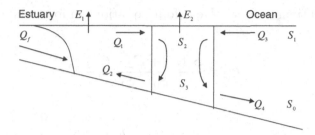

Fig. 11.2 The internal estuarine circulation.

observation programmes or experiments. In an educational context too, box models can be a useful vehicle for introducing the rudiments of modelling.

Here is an example of a box model of an Australian estuary. Figure 11.2 shows the circulation of an idealised estuary as a number of boxes. The river flux is Q_f and the internal fluxes are labelled Q_1, Q_2, Q_3 and Q_4. If $Q_f = 10$ m^3 s^{-1} and the following salinity and evaporation values have been measured $S_0 = 29.5$ ppt, $S_1 = 29.3$ ppt, $S_2 = 35.6$ ppt, $S_3 = 35.8$ ppt, $E_1 = 2$ m^3 s^{-1} and $E_2 = 250$ m^3 s^{-1}, (ppt signifies parts per thousand). Conservation laws will be used to calculate the various fluxes. [Adapted from Wolanski (1988).]

For this problem, we simply observe that both salt and water must be conserved. It is a traditional box model of the type used before numerical methods were available. For the estuary to the left of the maximum salinity zone, the conservation of water means that we must have

$$Q_f + Q_2 = Q_1 + E_1.$$

Salt is neither created nor destroyed in the maximum salinity zone, therefore

$$Q_1 S_2 = Q_2 S_3.$$

Considering the overall balance in the estuary, the conservation of water gives

$$Q_f + Q_3 = Q_4 + E_1 + E_2$$

and the salt balance in the mouth gives

$$Q_3 S_1 = Q_4 S_0.$$

These equations can be solved (four equations in four unknowns). In particular

$$Q_3 = \frac{S_0(E_1 + E_2 - Q_f)}{S_0 - S_1}$$

and

$$Q_1 = \frac{S_3(Q_f - E_1)}{S_3 - S_2}.$$

Putting in the figures given yields the results

$$Q_1 = 1.43 \times 10^3 \, \text{m}^3 \, \text{s}^{-1}, \quad Q_2 = 1.42 \times 10^3 \, \text{m}^3 \, \text{s}^{-1},$$
$$Q_3 = 3.57 \times 10^4 \, \text{m}^3 \, \text{s}^{-1}, \quad \text{and} \quad Q_4 = 3.54 \times 10^4 \, \text{m}^3 \, \text{s}^{-1}.$$

However, it must be said that if reliable predictions are the order of the day, then more sophisticated models that make use of software would normally be required. Some elementary dimensional analysis is also helpful. As estuaries are places where there are often density differences in the vertical, one might expect the Richardson number to be a guide to which processes to model. This is indeed true, except that there are various types of Richardson number in addition to that defined in Chapter 2. The Richardson number R_i is defined as the ratio of the square of buoyancy frequency to shear

$$R_i = \frac{N^2}{|\partial u / \partial z|^2} = -\frac{1}{\rho_0} \frac{\partial \rho_0}{\partial z} \bigg/ \bigg| \frac{\partial u}{\partial z} \bigg|^2.$$

In an estuarial context, Fischer (1972) has defined the *Estuarine Richardson number*, $R_i^{(e)}$ through the formula

$$R_i^{(e)} = \frac{\Delta \rho}{\rho} \frac{g u_r}{b \bar{u}^3}.$$

Here, u_r is the river flow averaged over the cross-section of the estuary, b is the breadth of the estuary and $\bar{u}$ is the root mean square tidal flow. We have used $\Delta \rho$ to denote the change in density between seawater and freshwater, with ρ being the density of the former. Another dimensionless quantity, used more when one wishes to assess the stability of interfacial waves in the presence of currents, is the densimetric Froude number F_d, defined by

$$F_d = \frac{u_r}{\sqrt{gh(\Delta\rho/\rho)}}.$$

This is clearly the ratio of river flow to the speed (celerity) of the interfacial or baroclinic wave. Where F_d changes from greater to less than one is often the definition of where the river ends and the estuary begins. The densimetric Froude number and the Estuarine Richardson number are the two dimensionless groupings that emerge from dimensional analysis. The direct application of the Buckingham Pi theorem is tricky due to the presence of the already dimensionless ratio of the densities $\Delta\rho/\rho$ and the presence of several lengths and speeds.

The two principal reasons for modelling estuaries are firstly to understand the tidal propagation, its interaction with river flow and density currents in order to forecast sediment flows, coastal changes and other similar matters of interest to the general public. This is usually the province of the civil engineer. Secondly, the quality of the water is extremely important. It therefore falls to the modeller to answer important questions concerning likely concentrations of pollutants, movements of water borne material etc. which might arise due to adjacent industry, recreation or accident. In other words, the modeller plays an important role in the overall management of any given estuary. Normally, a one-dimensional model is built using finite differences or finite elements which is then used to predict cross-stream averages. On the other hand, there are some theoretical models which can be mathematically quite demanding, see the works of Ron Smith. Smith (1980), in particular, takes the above stratified estuarial model a good deal further and derives various estuarial regimes. At some locations longitudinal diffusion is important, in others it is the unsteady horizontal mixing that dominates. The important point is that Smith (1980) derives these with mathematical rigor. In the paper Smith (1979) some reasonably accessible results are derived, and these are now outlined. The problem is to assess the combined effects of buoyancy and diffusion, so reference to Chapter 7 is first in order. There, the advection–diffusion equation was, if not exactly "derived" let us say it was explained physically. It takes the form

$$\frac{\partial \overline{c}}{\partial t} + u \frac{\partial \overline{c}}{\partial x} = \frac{\partial}{\partial x} \left(\kappa_x \frac{\partial \overline{c}}{\partial x} \right) + \frac{\partial}{\partial y} \left(\kappa_y \frac{\partial \overline{c}}{\partial y} \right)$$

in a river or estuary situation. It is largely one dimensional with x along the centre line of the river or estuary and y across the river or estuary. The vertically averaged concentration of the contaminant is denoted by $\overline{c}$ and κ_x and κ_y are diffusion co-efficients. Later in that chapter, successful models that followed particles were described. If axes are chosen that move with the centre of mass of the contaminant, then it is possible to derive a nonlinear equation

$$\frac{\partial \overline{c}}{\partial t} = \frac{\partial}{\partial y} \left\{ \left[k_1 + \left(\frac{g \Delta \rho}{\rho} \frac{\partial \overline{c}}{\partial y} \right)^2 k_2 \right] \frac{\partial \overline{c}}{\partial y} \right\}.$$

In this equation, only spreading in the y (cross-stream) direction is considered and k_1 and k_2 are constants which experimental scientists relate to friction, velocity etc. The expressions $k_1 = h \Delta \rho u^* / \rho$ and $k_2 = h^5 g^2 / 96 u^{*3} k^3$ have been used, where h is the depth, k von Karman's constant and u^* a friction velocity. This equation is due to Erdogan and Chatwin (1967) and is referred to as the Erdogan–Chatwin equation. What has happened here is the advection term (nonlinear) has been swept up into a modified nonlinear diffusion term which leads to a different, and often better, description of the process. Solving this equation is, however, very difficult but in-roads have been made by Smith (1978). The ratio of eddy viscosity to eddy diffusivity (κ_y above) is dimensionless and referred to as the Schmidt number. The Erdogan–Chatwin description of diffusion and buoyancy was really only good for large Schmidt numbers. Various other formulations appropriate to estuaries are available, see the review Chatwin and Allen (1985). For a full appreciation of what balances are valid with various parameter ranges, Figure 5 of Smith (1979) or Figure 5 of Smith (1980) should be consulted. No complicated mathematics will be done here.

One of the problems in using dimensional analysis here is hinted at above. As mass never features, it is only present in the dimensionless ratio $\Delta\rho/\rho$; what can be deduced is limited. Take, for example, the Erdogan–Chatwin equation just mentioned

$$\frac{\partial \overline{c}}{\partial t} = \frac{\partial}{\partial y} \left\{ \left[k_1 + \left(\frac{g\Delta\rho}{\rho} \frac{\partial \overline{c}}{\partial y} \right)^2 k_2 \right] \frac{\partial \overline{c}}{\partial y} \right\}$$

and propose the scalings

$$\frac{g\Delta\rho}{\rho} \sim \frac{U^2}{h} \left(\frac{u^*}{U} \right)^A \text{ and cross-stream length scale} \sim \frac{1}{h} \left(\frac{u^*}{U} \right)^B,$$

where U is a typical estuarial flow speed, u^* is the friction velocity, h the depth and A and B are constants to be determined. This is not strictly a dimensional analysis, but is an estimation of scales derived from it. If the above expressions are substituted into the Erdogan–Chatwin equation, then the two terms on the right are of the same order if $A = -B$. Smith (1979) gives arguments to support putting $A = 1$ but only if the downstream flow term which is absent from the Erdogan–Chatwin equation is also included. Note also that it is assumed that the ratio u^*/U is a small parameter. It turns out that the above scaling returns us to linear diffusion, neglecting the interaction terms. Putting $A = 1$ and $B = -1$ into the above scalings reveals that the buoyancy acceleration is proportional to u^*U/h and cross stream gradients proportional to $u^*/(Uh)$. The modelling assumption that the friction and buoyancy are of the same order is all too apparent here.

11.3 Other Models of Estuaries

The analytical approach outlined above is one way of modelling the detail. It is one dimensional as estuaries are dominated by movement in the direction of the flow and so are very suitable candidates for the one-dimensional approximation. The first numerical models were also one dimensional, but one dimensionality cannot hide the inbuilt complexities of the estuary. In terms of physics, currents in estuaries are due to an awkward combination of tidal flow, wind-driven flow and density-driven flow. Before the onset of any serious numerical modelling, however, it was Pritchard (1955) who classified estuaries into four types. In simplistic terms, a "general estuary" was described in section 11.1, but this is an over simplification. According to Pritchard (1955), type A estuaries are well mixed and dominated by diffusive processes, strong tidal currents and weak or strong river discharge; they are different in detail but both are dominated by diffusive processes. The second kind of estuary exhibits reverses of current with depth — call it type B. They have well developed vertical advective circulations as well as diffusive processes and this models most temperate estuaries, and is closest to the salt wedge estuary of Figure 11.1. Type C estuaries are fjords and Scottish lochs, deep with virtually zero mean flow over the depth, but strong surface and, therefore, also reverse deep flow. Finally, type D estuaries have seaward flow with weak vertical structure. These estuaries also have both advective and diffusive processes to model. This kind of classification helps us to understand the broad structure of the estuary but also places the estuary in a geographical context. Other classifications are also possible, as In Figure 11.3; these are principally geographical and certainly have their place as they form part of the general description of a landscape. However, they are not particularly useful for detailed numerical modelling. When enhanced computing power became available, therefore, the opportunity was taken for models to

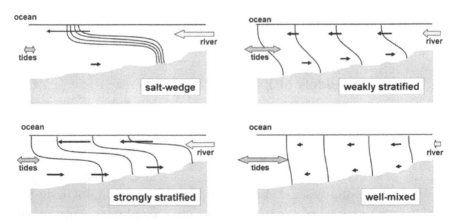

Fig. 11.3   Four different types of estuary, based on vertical structure. Bottom right is the same as type A, and top left is a salt wedge estuary, but otherwise the classification of Pritchard (1955) is subtly different.

include all that was possible and vary parameters so that the correct processes were modelled. Here is a non numerical example of an estuarial model that springs out of Pritchard's classification.

The paper by Oey (1983) models estuaries and takes a lead from the Pritchard (1955) classification. The general salt (S) balance equation is

$$\frac{\partial S}{\partial t} + u\frac{\partial S}{\partial x} + v\frac{\partial S}{\partial y} + w\frac{\partial S}{\partial z}$$
$$= \frac{\partial}{\partial x}\left(K_1\frac{\partial S}{\partial x}\right) + \frac{\partial}{\partial y}\left(K_2\frac{\partial S}{\partial y}\right) + \frac{\partial}{\partial z}\left(K_3\frac{\partial S}{\partial z}\right)$$

but this considerably simplified to

$$u\frac{\partial S}{\partial x} + w\frac{\partial S}{\partial z} = 0$$

once it is assumed that the flow is steady and purely advective, so no time dependence or diffusion (K_1, K_2 and K_3 all zero) and no y dependence. If the discharge is high, then an extra term is added to the right

$$u\frac{\partial S}{\partial x} + w\frac{\partial S}{\partial z} = \frac{\partial}{\partial z}\left(K_3\frac{\partial S}{\partial z}\right).$$

On the other hand, in order to model what Pritchard calls a type D estuary, the model needs some longitudinal diffusion, so the salt balance is now

$$u\frac{\partial S}{\partial x} + w\frac{\partial S}{\partial z} = \frac{\partial}{\partial x}\left(K_1\frac{\partial S}{\partial x}\right).$$

So it is possible to link quantitative modelling to the Pritchard classification. Oey then goes on to calculate parameter ranges that determine which of the estuary types a particular example falls into (he does not consider type C which are well mixed and may need Coriolis effects). This is not a numerical paper, but uses expansion in terms of non-dimensional variables, the Rayleigh number R_a and the densimetric Froude number F_m, defined by the two expressions

$$R_a = \frac{gk\Delta S D_0^3}{A_0 K_0} \quad \text{and} \quad F_m = \frac{U}{\sqrt{(\Delta\rho/\rho)gH}},$$

where k is a thermal expansion co-efficient, D_0 is a representative depth, ΔS is the difference in salinity between salt and fresh(er) water, K_0 is a representative longitudinal salt diffusion co-efficient, A_0 is a similar representative value of turbulent diffusion. The modelling is a quantitative model of the two papers by Hansen and Rattray, Jr. (1965) and (1966) who used dimensionless numbers and ratios of current speed to render the estuary classification of Pritchard (1955) more precise. Oey (1984) goes one stage further and actually computes currents analytically in an attempt to model the Mersey (Liverpool, UK), the Hudsen River estuary (New York, USA) amongst many others. In fact, the 1980s and 1990s were full of good analytical models of estuaries. They usually used idealised geometry and expansions in various combinations and products of carefully defined dimensionless variables. Oey (1984) is but one example. We are now ready to study a modern numerical example.

11.4   Models of Inlets

As a complete contrast, here is an outline of a model of an estuarial inlet (and coastal sea) published by Chen, Liu and Beardsley (2003). This is a modern, numerically based model that uses the full set of equations to start with then utilises the most up to date unstructured grid. Here are the equations; there are seven of them, so hold on to your hats:

$$\frac{\partial u}{\partial t} + u\frac{\partial u}{\partial x} + v\frac{\partial u}{\partial y} + w\frac{\partial u}{\partial z} - fv = -\frac{1}{\rho_0}\frac{\partial p}{\partial x} + \frac{\partial}{\partial z}\left(K_m\frac{\partial u}{\partial z}\right) + F_u,$$

$$(11.1)$$

$$\frac{\partial v}{\partial t} + u\frac{\partial v}{\partial x} + v\frac{\partial v}{\partial y} + w\frac{\partial v}{\partial z} + fu = -\frac{1}{\rho_0}\frac{\partial p}{\partial y} + \frac{\partial}{\partial z}\left(K_m\frac{\partial v}{\partial z}\right) + F_v,$$

$$(11.2)$$

$$\frac{\partial p}{\partial z} = -\rho g, \qquad (11.3)$$

$$\frac{\partial u}{\partial x} + \frac{\partial v}{\partial y} + \frac{\partial w}{\partial z} = 0, \qquad (11.4)$$

$$\frac{\partial \theta}{\partial t} + u\frac{\partial \theta}{\partial x} + v\frac{\partial \theta}{\partial y} + w\frac{\partial \theta}{\partial z} = \frac{\partial}{\partial z}\left(K_h \frac{\partial \theta}{\partial z}\right) + F_\theta, \qquad (11.5)$$

$$\frac{\partial s}{\partial t} + u\frac{\partial s}{\partial x} + v\frac{\partial s}{\partial y} + w\frac{\partial s}{\partial z} = \frac{\partial}{\partial z}\left(K_h \frac{\partial s}{\partial z}\right) + F_s, \qquad (11.6)$$

$$\rho = \rho(\theta, s). \qquad (11.7)$$

The first three equations are the two horizontal equations of motion together with hydrostatic balance. The density in the first two equations is the reference density ρ_0, which means that the Boussinesq approximation has been applied (see section 3.2). Equation (11.4) is the conservation of mass or continuity equation, and the next two equations express conservation laws of potential temperature θ and salinity s. Without giving a lecture on thermodynamics, potential temperature is the temperature referred back to a reference pressure. This means it is a more sensible quantity in a model of the dynamics as the potential temperature is immune to the changes that would occur through changes in pressure. The terms K_m and K_h are eddy viscosities, specifically K_m the vertical eddy viscosity, K_h and the thermal vertical eddy diffusion co-efficient. The terms F_u, F_v, F_θ and F_s are also diffusion co-efficients, but this time horizontal momentum (the first two), thermal (the third) and salt (the last). The eddy viscosities are not just constant; those modelling days have gone. Now, it is reasonably standard to incorporate a more complex closure scheme, and this research uses the Mellor and Yamada (1974), Blumberg and Mellor (1987) schemes whereby

$$\frac{\partial q^2}{\partial t} + u\frac{\partial q^2}{\partial x} + v\frac{\partial q^2}{\partial y} + w\frac{\partial q^2}{\partial z}$$

$$= 2(P_s + P_b - \epsilon) + \frac{\partial}{\partial z}\left(K_q\frac{\partial q^2}{\partial z}\right) + F_q,$$

$$\frac{\partial(q^2 l)}{\partial t} + u\frac{\partial(q^2 l)}{\partial x} + v\frac{\partial(q^2 l)}{\partial y} + w\frac{\partial(q^2 l)}{\partial z}$$

$$= lE_1\left(P_s + P_b - \frac{W}{E_1}\epsilon\right) + \frac{\partial}{\partial z}\left(K_q\frac{\partial(q^2 l)}{\partial z}\right) + F_l.$$

The terms in these equations are defined as follows

$$q^2 = \frac{1}{2}(u^2 + v^2) \quad \text{turbulent kinetic energy,}$$

$$P_b = \frac{gK_h}{\rho_0}\frac{\partial \rho}{\partial z}, \quad P_s = K_m\left\{\left(\frac{\partial u}{\partial z}\right)^2 + \left(\frac{\partial v}{\partial z}\right)^2\right\}.$$

$\epsilon = \frac{q^3}{16.6l}$ turbulent kinetic energy dissipation rate. This arises from a Kolmogorov homogeneous turbulence assumption and l is a turbulent macroscale. W is termed a "wall proximity function" and is given by

$$W = 1 + \frac{E_2 l^2}{(\kappa L)^2}, \quad \text{where} \quad \frac{1}{L} = \frac{1}{\eta - z} + \frac{1}{z + H}.$$

$\kappa = 0.4$ is von Karman's constant, and η and H retain their usual meaning (sea surface elevation and mean depth, respectively). The diffusion co-efficients F_q and F_l are kept as small as possible. Finally, the system is closed by $K_m = lqS_m, K_h = lqS_h$ and $K_q = 0.2lq$ with

$$S_m = \frac{0.4275 - 3.354G_h}{(1 - 34.676G_h)(1 - 6.127G_h)},$$

$$S_h = \frac{0.494}{1 - 34.676G_h}$$

and

$$G_h = \frac{l^2 g}{q^2 \rho_0}\frac{\partial \rho}{\partial z}.$$

There are a bewildering number of parameters here, very typical of all turbulence closure schemes of recent times. The paper by Chen, Liu and Beardsley (2003) gives a little more discussion on the ranges of some of the parameters and the reasons behind these ranges. Much more detailed discussion is to be found in the book by Mohammadi and Pironneau (1994). The boundary conditions are those appropriate for any model that contains friction. At the surface $z = \eta(x, y, t)$ we have

$$K_m\left(\frac{\partial u}{\partial z}, \frac{\partial v}{\partial z}\right) = \frac{1}{\rho_0}(\tau_{sx}, \tau_{sy}),$$

$$w = \frac{\partial \eta}{\partial t} + u\frac{\partial \eta}{\partial x} + v\frac{\partial \eta}{\partial y}$$

and at the sea bed $z = -H(x,y)$ we have

$$K_m \left(\frac{\partial u}{\partial z}, \frac{\partial v}{\partial z} \right) = \frac{1}{\rho_0} (\tau_{bx}, \tau_{by}),$$

$$w = -u \frac{\partial H}{\partial x} - v \frac{\partial H}{\partial y}.$$

The stresses are given by quadratic laws at the bed and surface, and the usual co-efficient C_D — instead of just being a constant is — at the sea bed, related to a logarithmic layer adjacent to the bed. As for heat, there is a no flow (zero gradient) condition at the sea bed. At the surface, there is a more sophisticated flux condition with an exponential profile in z dependent on the wavelength of the radiation. For salinity, again zero flux through the bed but precipitation and evaporation are catered for at the sea surface.

Having set up this elaborate model, the whole is transformed into σ co-ordinates through

$$\sigma = \frac{z - \eta}{\eta + H},$$

so the equations have to be redrafted and look even more complex. Although the discretisations in terms of differences are reasonably standard in that they are explicit but use a time splitting technique, the grid is unstructured: see Figure 11.4. This unstructured grid is applied to model the Bohai Sea, which is an embayment on the east coast of China: see Figure 11.5. The results are discussed and compared to the output from the Princeton Ocean Model which uses a conventional finite difference grid, albeit curvilinear in part so that the boundary of the model has a more realistic and less "steppy" coastline. Here, some examples of the output are shown in a few graphs, see Figures 11.8–11.10.

The actual unstructured grid filling the sea is shown in Figure 11.6. To lessen the number of graphs, the modelling output from POM is not shown here. Looking at the paper by Chen, Liu and Beardsley (2003), the unstructured grid really comes into its own in modelling fine structure, particularly the river system. Under POM these rivers look virtually blank but using the unstructured grid there is a lot of structure. Another graph, not shown here, where there is a

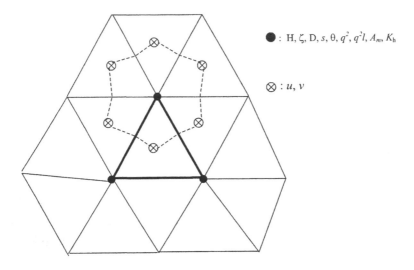

Fig. 11.4   The layout of the unstructured grid.

lot of difference is in the residual flow. The POM output shows very little variation or detailed structure. The unstructured grid model showed more variation and more structure, this is particularly true around the all important islands where storm surges and flooding could be important. The paper also applied the model to the Satilla River, Georgia in the USA and here the contrasts are, if anything, greater. The tides are reasonably modelled by either technique apart from the amphidrome near the top of the bay, which is apparent in the output of the unstructured grid shown here in Figure 11.7 but absent in the POM output. Which is right is open to controversy, but the model pictured here is probably the more accurate. Temperature looks to be well modelled. In the vertical, although both outputs look similar, the unstructured grid consistently predicted lower temperature. There looks to be less contrast in the surface outputs, but there is some. As in the tidal maps, there is more structure in the unstructured grid.

It is a shame that a model that carefully models the salinity using a parameterised diffusion of salinity really has no salt variation. It is a fact that at the mouth of an estuary there typically is a front, one side of which is freshwater and the other side of which is salty

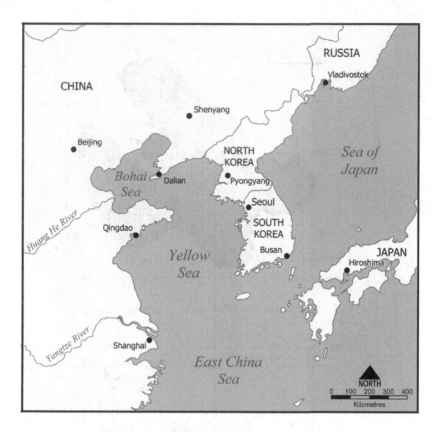

Fig. 11.5   The Bohai Sea and environs.

water. These fronts are poorly resolved by even sophisticated models like the one presented in this paper. Usually, in order to capture the behaviour of a front in a tidal regime, there has to be a correction to the output every time step otherwise the sharpness of the front would be lost. This is the role of the kind of modelling outlined in Chapter 7, section 7.5. It also clear to see the value of theoretical modelling of the type done by Ron Smith. Such models help us to understand the physics of the interaction between these two bodies of water and consequent flow patterns. A better understanding then results in more accurate modelling since modelling without understanding means an inability to adjust for unwanted spreading, which is still inevitable even using the most sophisticated of modern numerical

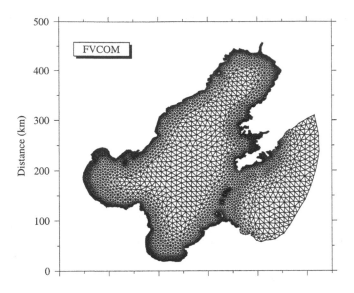

Fig. 11.6   The Bohai Sea filled with an unstructured grid.

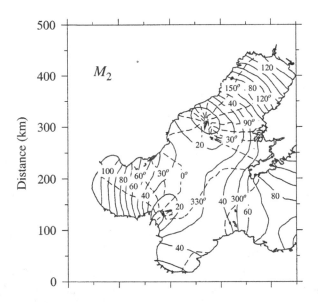

Fig. 11.7   The Bohai Sea M_2 tide modelled using the unstructured grid.

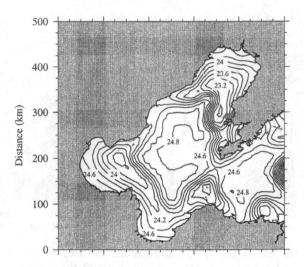

Fig. 11.8   The Bohai Sea modelled horizontal temperature distribution.

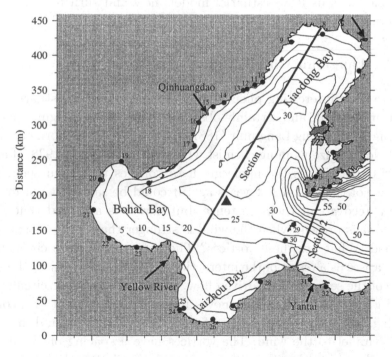

Fig. 11.9   The Bohai Sea showing depth contour and two transects for modelling
the temperature variation in the vertical.

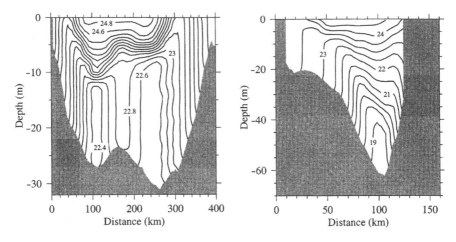

Fig. 11.10   The Bohai Sea temperature along the two transects indicated in Figure 11.9; the left is the first transect and the right the second.

schemes. Let us leave estuarial models now and turn to something different.

11.5   Langmuir Circulation

There is an interesting phenomenon, not restricted to estuaries but of a similar scale and certainly observed there, called Langmuir circulation (after Irving Langmuir (1881–1957) a New York born engineer and Nobel prize winning chemist who observed them in 1938). Langmuir circulation is the name given to vortices that occur near the surface of the sea. A few years ago, it could be said that not everyone agreed about their existence, but now it is accepted that they are always present when the wind is stronger than seven metres per second. What are these vortices? They are near-surface circulation that arises as a result of the interaction between the waves on the surface of the sea and wind-driven flow. In typical Langmuir circulation (see Figure 11.11) the fluid moves in a path reminiscent of a coiled spring, spiraling with the axis of the spring horizontal and more or less parallel to the wind. The vortices have a spacing of about 200 m in the sea, although analogous circulation is found in lakes and here the spacing is much less, in fact only of the order of tens of

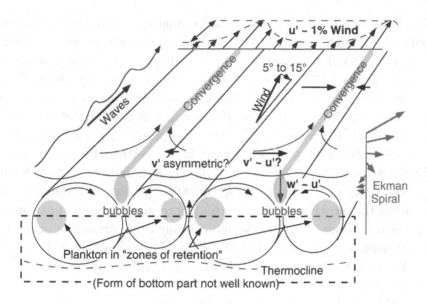

Fig. 11.11   Langmuir circulation.

metres. Detailed modelling of Langmuir circulation is not very easy
since very different mechanisms need to be incorporated to include
both wind-driven currents and currents due to waves, however some
progress using LES has been made recently (see Chapter 5, section
5.6). In these models, direct simulation of the vortices is possible and
coherent turbulence like structures are able to be simulated that are
distinct from embedded turbulence modelling artifacts. These simulations have been verified by careful measurements using Doppler
sonar devices. So, progress is being made and knowledge gained, but
we are still far from understanding all about Langmuir circulations.
When modelling waves, the first assumption is that the profile of the
waves themselves can be approximated by a sinusoidal shape under
which the water particles follow closed circular (or nearly circular)
paths; see Figure 11.11. Unfortunately, such a model of waves cannot
predict any net movement of water or indeed waterborne material.
This was covered in section 3.6.2. The assumption of a purely sinusoidal surface wave in fact eradicates any wave drift — there is no
wave drift under a sinusoidal sea. It is thus essential to include the

nonlinear or advective terms in any model that is to predict transport of material. It is this combination of a nonlinear wave theory, which can predict a drift (the wave drift or Stokes drift), see section 8.4.2 and 12.3.1, with a direct wind drift that can successfully model Langmuir circulation. The dimensionless number associated with wave theories is the wave slope, which is the ratio of the wave amplitude to the wave length. An appropriate dimensionless number associated with pure wind drift is the vertical Ekman number for flows whose vertical scale is typically hundreds of metres. Such a scale is, however, usually inappropriate. The only alternative is to examine the pure wind-driven current in terms of the direct action of the air on the sea. This can be done through the stress-rate of strain relationships of the type used to define eddy viscosity (see Chapter 3). We then look at the ratio of this wind drift to the pure wave or Stokes drift.

If this number is around unity, then we can expect both of these effects to be equally important. A model that includes all the relevant nonlinear wave and wind drift terms is then built, and Langmuir circulation is predicted. Another approach is to use the Froude number, which denotes the relative importance of current (in this case wind-driven flow) and wave speed. This then leads to a stability criterion for Langmuir circulations and takes us outside the scope of this text. Perhaps the only good news is that the scale is such that the Coriolis terms can be safely ignored since the typical length scale, even of the largest Langmuir cells, is still deemed small when compared with the length scale associated with the Coriolis acceleration (this is Rossby radius of deformation, U/f, where U is a typical speed and f is the Coriolis parameter, which is of the order of hundreds of kilometres; see section 3.6).

11.5.1   *Models of Langmuir circulation*

Given the very complexity of Langmuir circulation, it should come as no surprise that there are no simple models. The passage at the end of the last paragraph is an attempt at describing what is going on physically. We give here a basic model due to Craik and Leibovich

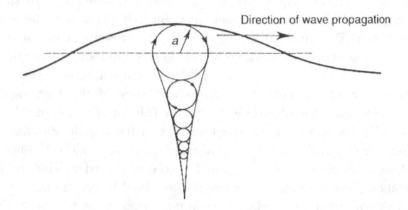

Fig. 11.12 Particle motion under a deep water wave: a is the wave amplitude.

(1976) and include comments from more recent numerical experiments that have included turbulence closure. The dominant motion is assumed to be surface (gravity waves) which have an amplitude which is small compared with depth, see Figure 11.12. Friction is in the form of a constant eddy viscosity ν_v, and it is necessary to expand variables in terms of the (small) wave slope parameter ε, so

$$\mathbf{u} = \varepsilon \mathbf{u}_0 + \varepsilon^2 \mathbf{u}_1,$$

where $\mathbf{u}_0$ is the velocity due to linear water waves over infinite depth and $\mathbf{u}_1$ is the second order correction. The linear water wave equations thus give rise to a vorticity balance

$$\frac{\partial \boldsymbol{\omega}}{\partial t} = \boldsymbol{\nabla} \times (\mathbf{u} \times \boldsymbol{\omega}) + \nu_v \boldsymbol{\nabla}^2 \boldsymbol{\omega},$$

where $\boldsymbol{\omega} = \boldsymbol{\nabla} \times \mathbf{u}$ is the vorticity vector. Since $\mathbf{u}_0 = \boldsymbol{\nabla}\phi$ with $\nabla^2\phi = 0$, which encapsulates linear water wave theory, systematic expansion in powers of ε is possible. Doing this, we have

$$\boldsymbol{\omega} = \varepsilon^2 \boldsymbol{\nabla} \times \mathbf{u}_1,$$

$$\mathbf{u}_1 = \mathbf{u}_{10} + \varepsilon \mathbf{u}_{11},$$

$$\boldsymbol{\nabla} \times \mathbf{u}_1 = \boldsymbol{\omega}_0 + \varepsilon \boldsymbol{\omega}_1 + \cdots .$$

This enables $\mathbf{u}_{10}$, $\mathbf{u}_{11}$, ω_0 and ω_1 to be separated into mean and fluctuating parts. The equations are then followed through with time averaging taking place. The essential point being that it is the non-linear term $\nabla \times (\mathbf{u} \times \omega)$ which fuels all the subsequent interaction and is the source of the corkscrew motion characteristic of Langmuir circulation. Later papers provide evidence of a connection between this theory and a more comprehensive theory of the Lagrangian mean. The inclusion of turbulence closure (Mellor–Yamada and $k - \epsilon$ methods) has been done in recent years. Our basic understanding of Langmuir circulation has been confirmed; in these models the effects of Langmuir circulation are limited to the increased production of turbulent kinetic energy throughout the mixed layer, the net effect resulting in greater boundary layer mixing from wind forcing in the mixed layer. One important conclusion is that the inclusion of Langmuir circulation in models gives a non-local mixing mechanism, an enhanced shear production and more rigorous entrainment. The non-local effects are difficult to include in even the latest models and this remains a subject for research.

11.6   The Firth of Forth and Forth Estuary, a Case Study

Estuaries are indeed special places and deserve special models. As will now be obvious to anyone who has read what has gone before in this chapter, there are texts entirely devoted to the physical and dynamical description of estuaries e.g. Dyer (1997). A brief case study is now presented that cannot compete with such specialist texts or papers such as Hansen and Rattray, Jr. (1965); instead we point out some of the key points of modelling this and similar estuaries, so it is a learning model. The Firth of Forth is a wide estuary located in the south east of Scotland. It is at the mouth of the river Forth, a major Scottish river, but from the road and rail bridges at Queensferry to the North Sea the salinity is virtually the same as the North Sea (see Dyke, 1987). Oceanographically, it more closely resembles an inlet than an estuary. In fact, the name "Forth Estuary" is usually preserved for that part of the estuary upstream of the bridges to

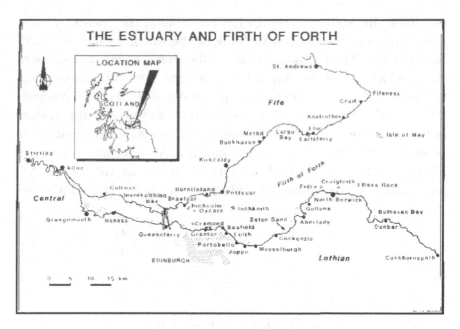

Fig. 11.13   The Forth Estuary and the Firth of Forth. Reprinted by permission of the Royal Society of Edinburgh.

Alloa (see the location map, Figure 11.13, and Webb and Metcalfe, 1987).

Because of this geographical split, models of the estuary tend to be built in two parts, one covering the estuary to the east of the bridges and the second to the west of them. There are no complete models of either. Engineering companies have built very localised models for specific purposes, for example managing the environment in terms the siting of marinas and sewage outfalls. These models have been especially commissioned by commercial outfits and therefore are not available in the public domain.

Models of the estuary upstream of the bridges follows the norm and we have a freshwater outflow over the deeper saline water which is pulled up as a saline wedge. These days, most of these models are numerical and include the ability to model possible spillage from outfalls and the like that are situated along the bank. Again, these models are built for companies that have an interest in monitoring and modelling spillages and are not available in the open literature. Downstream of the bridges, the bulk of the water is saline at North Sea salinities. The freshwater river outflow tends to hug the southern shore, which is generally taken to be as a result of the Coriolis effect (see Dyer, 1997). The two layer simple model of Dyke (1980) confirms this general picture. Two layer models are, however, only appropriate in the outer Forth some of the time. There is a seasonal thermocline present only in the late spring and summer. At other times the outer Firth of Forth is well mixed. It is useful to look at the general two layer model. The equations valid in a rectangular two layered inlet are

$$\frac{\partial u'}{\partial t} - fv' = -g\frac{\partial \eta'}{\partial x},$$

$$\frac{\partial v'}{\partial t} + fu' = -g\frac{\partial \eta'}{\partial y},$$

$$\frac{\partial}{\partial t}(\eta' - \eta'') + h'\left(\frac{\partial u'}{\partial x} + \frac{\partial v'}{\partial y}\right) = 0$$

for the upper layer and a slightly more complicated set

$$\frac{\partial u''}{\partial t} - fv'' = -g\frac{\partial}{\partial x}\left(\frac{\rho'}{\rho''}\eta' + \frac{\rho'' - \rho'}{\rho''}\eta''\right),$$

$$\frac{\partial v''}{\partial t} + fu'' = -g\frac{\partial}{\partial y}\left(\frac{\rho'}{\rho''}\eta' + \frac{\rho'' - \rho'}{\rho''}\eta''\right),$$

$$\frac{\partial \eta''}{\partial t} + h''\left(\frac{\partial u''}{\partial x} + \frac{\partial v''}{\partial y}\right) = 0$$

for the lower layer. The notation is straightforward: a single dash for the upper layer, a double one for the lower layer. (u, v) is the velocity, h the undisturbed layer depth, η the displacement of surface or interface from equilibrium and ρ the density. The deduction from them follows the more general arguments of Gill and Clarke (1974). Namely, that the flow can be put in terms of barotropic and baroclinic modes $(u^{(1)}, v^{(1)})$ and $(u^{(2)}, v^{(2)})$. These are related to the variables above via

$$(u^{(1)}, v^{(1)}) = (u', v') + \frac{h''}{h'}(u'', v''),$$

$$(u^{(2)}, v^{(2)}) = (u', v') - (u'', v''),$$

$$h^{(1)} = \frac{h'h''}{h' + h''}.$$

The second baroclinic mode is associated with the baroclinic radius of deformation c_1/f, where c_1 is the internal wave speed $\sqrt{(\rho'' - \rho')gh^{(1)}/\rho''}$ and f is the Coriolis parameter. As $\rho'' - \rho'$ is a difference in density and the ratio $(\rho'' - \rho')/\rho''$ about 1/1000, c_1 is typically one thirtieth of the barotropic value. One important consequence of this is that the speed of travel of the first baroclinic wave mode is smaller by a factor of 30, which means in turn that the characteristic length scale associated with this first baroclinic mode is 1/30th of that associated with the barotropic scale. Hence, the Coriolis effect through baroclinicity can be felt on estuarial length scales (~10 km) which is the first internal (first baroclinic) radius of deformation. The boundary conditions appropriate to a rectangular estuary are that there is no flow through the closed end and that there is no flow through the sides. Theoretical models are very simple and serve only to establish general principles (but see, for example, Smith (1980) for a not so simple theoretical model). In order to build a workable numerical model, we would need one that copes with modes. One option is a time splitting technique with a marching method working with both barotropic and first baroclinic modes simultaneously, but this will only work if there is no interaction. As more modes are taken into account, this interaction will unfortunately increase and the time splitting technique will not work. Another option is to use a continuously stratified model; neither of these have been done in the eastern part of the Forth.

11.7 Exercises

(1) Describe the features of estuaries that make them difficult to model.

(2) Compare and contrast four methods for modelling estuaries: (a) box models; (b) the use of dimensionless numbers; (c) analytical models, e.g. Smith (1980); (d) numerical models.

(3) The Grashof number is defined by

$$Gr = \frac{gk\Delta T L^3}{\nu^2}.$$

Starting with an equation that governs motion in an environment where changes in temperature ΔT are important

$$u\frac{\partial u}{\partial x} + v\frac{\partial u}{\partial y} = \frac{\partial}{\partial x}\left(\nu\frac{\partial u}{\partial x}\right) + kg\Delta T,$$

where all the symbols have their usual meaning with k a (dimensionless) thermal expansion co-efficient. Non-dimensionalise introducing the Reynolds number Re, where

$$Re = \frac{UL}{\nu},$$

and show that the magnitude of the Grashof number governs the importance of the thermal term.

(4) When is the Grashof number (see the last exercise) important in estuarial modelling?

(5) Give advantages and disadvantages of models such as that outlined in section 11.4.

(6) What are the effects of stratification on Langmuir circulation?

(7) What is the next step to improve the Firth of Forth model of section 11.6?

Chapter 12

Trapped Waves and Currents

12.1 Introduction

In the history of research into physical oceanography, the modelling of waves or currents that are trapped against a coast or at the equator feature prominently. In large scale oceanography, one immediately thinks of the equatorial currents so important to modelling the ENSO and much else besides. Today, these trapped waves are seen to play an important part in the understanding of the effects of climate change too. More locally there are the longshore currents that play such an important role in the dynamics of upwelling, itself a vital ingredient in biological production and the maintenance of vital feeding grounds in places like the seas off Peru and Nigeria. The modelling of waves and currents trapped to a coastline is important. These waves and currents feature along many of the world's continental shelf edges and they need to be understood by engineers interested in exploiting the sea bed thereabouts for hydrocarbons and other minerals. They also play a role in the distribution of fish. Islands are another place where waves can be trapped. They are intriguing to model and can also be used to detect signals of a seismic or meteorological nature. In this chapter, trapped waves of different kinds will be modelled. It will be seen that the agents that force a wave to be trapped can be many and various, but the Coriolis acceleration always plays an important part in the trapping of waves in regimes where length scales exceed tens of kilometres.

501

12.2   Continental Shelf Waves

There is an intriguing type of current where the change of depth at the continental slope and the fluid vorticity add, so that the movement of the current up and down the slope as it travels like a sidewinder along the slope itself ensures potential vorticity is conserved. The wave in this case is a current meander and there is little or no surface elevation associated with it. This is the continental shelf wave, which should really be called a continental slope wave. These waves were first predicted by Robinson (1964), and since the mid-1970s their importance to the computation of extreme currents in slope regions has meant continued interest and more sophisticated models. In this section, the basic balances will be described in as simple terms as possible and complexities gradually introduced. The simplest equation set for motion on the continental slope is

$$\frac{\partial u}{\partial t} - fv = -\frac{1}{\rho}\frac{\partial p}{\partial x},$$

$$\frac{\partial v}{\partial t} + fu = -\frac{1}{\rho}\frac{\partial p}{\partial y},$$

$$\frac{\partial}{\partial x}(Hu) + \frac{\partial}{\partial y}(Hv) = 0.$$

If the pressure is eliminated from the first two by cross differentiation, a potential vorticity balance is obtained of the type examined in Chapter 10 when the general circulation of the ocean was discussed. That is

$$\frac{\partial}{\partial t}\left(\frac{\partial u}{\partial y} - \frac{\partial v}{\partial x}\right) + f\left(\frac{\partial u}{\partial x} + \frac{\partial v}{\partial y}\right) = 0$$

and the continuity equation ensures that it is possible to define a streamfunction ψ such that

$$Hu = -\frac{\partial \psi}{\partial y}, \quad \text{and} \quad Hv = \frac{\partial \psi}{\partial x}.$$

Writing

$$\frac{\partial u}{\partial x} + \frac{\partial v}{\partial y} = -\frac{1}{H}\frac{\partial H}{\partial x}u,$$

where it has been assumed that $H = H(x)$, the vorticity balance becomes

$$\frac{\partial}{\partial t}\left(\frac{\partial u}{\partial y} - \frac{\partial v}{\partial x}\right) = \frac{fu}{H}\frac{\partial H}{\partial x}.$$

Here, it is assumed that the coast runs north–south and is west facing. Axes are y north and x east as usual, and the equations are then in the same frame as those used in Chapter 3. With this simplification, the single equation in terms of the streamfunction ψ is

$$\frac{\partial}{\partial t}\left[\frac{\partial}{\partial y}\left(\frac{1}{H}\frac{\partial \psi}{\partial y}\right) + \frac{\partial}{\partial x}\left(\frac{1}{H}\frac{\partial \psi}{\partial x}\right)\right] - \frac{f}{H^2}\frac{\partial H}{\partial x}\frac{\partial \psi}{\partial y} = 0.$$

The way forward now is to look for wave-like solutions that propagate in the y direction. If we let

$$\psi = H^{1/2}\phi(x)\exp(ily - i\omega t)$$

then it turns out that ϕ satisfies the ordinary differential equation

$$\frac{d^2\phi}{dx^2} + \left\{\frac{d}{dx}\left(\frac{1}{2H}\frac{dH}{dx}\right) - \left(\frac{1}{2H}\frac{dH}{dx}\right)^2 - l^2 - \frac{fl}{\omega}\frac{1}{H}\frac{dH}{dx}\right\}\phi = 0.$$

It is at this stage that actual profiles for $H(x)$ could be inserted into this equation, however the most instructive thing to do is to use the exponential depth profile

$$H = H_0\exp(-2\lambda x),$$

in which case the equation for ϕ simplifies considerably. The solution $\phi(x) = \phi_0 \sin kx$ is permitted and there is a dispersion relation

$$\omega = \frac{2fl\lambda}{k^2 + l^2 + \lambda^2}.$$

Classical solutions in the paper by Buchwald and Adams (1968) are for shelf profiles that are exponential then flat

$$\psi = \begin{cases} e^{\lambda(-x-B)}\sin kx\, e^{i(ly-\omega t)} & \text{for } 0 \le -x \le B, \\ e^{\lambda(x+B)}\sin kB\, e^{i(ly-\omega t)} & \text{for } -x > B. \end{cases}$$

Note that the sea is in the domain $0 \ge x \ge -\infty$ so x is negative. It is emphasised here that there is no surface elevation associated with

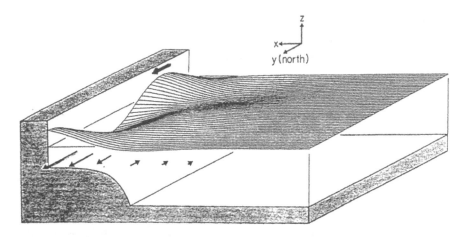

Fig. 12.1   Trapped waves over idealised topography (not the exponential profile shown in the text); these are hybrid Kelvin/continental shelf waves.

this wave; the continuity equation

$$\frac{\partial}{\partial x}(Hu) + \frac{\partial}{\partial y}(Hv) = 0$$

shows this explicitly. When expanded in the form

$$\frac{\partial u}{\partial x} + \frac{\partial v}{\partial y} = -\frac{1}{H}\frac{\partial H}{\partial x}u$$

it also shows that the offshore depth change gives rise to an offshore current u that, because the alongshore flow is quasi geostrophic, meanders up and down the continental slope due to vorticity conservation. Figure 12.1 shows a trapped wave where there is both topography and a surface elevation. The surface elevation would be there if the sea bed was flat; it is trapped due to Coriolis acceleration. The presence of topography enhances the currents, shown schematically through the arrows. So, what we are showing is a hybrid Kelvin/continental shelf wave.

It may not be obvious, but these waves are quite large scale. The idealised geometry above contains the parameter λ, the reciprocal of which can indicate the shelf width, which is typically 30 km. The frequency ω is always less than f, typically $\omega = 0.7f$ (for the case $k = \lambda$, Gill, 1982). In this case also, $f/\lambda = 3$ m s^{-1}. So far, nothing has been said about how these waves are forced. The above solution has been found simply by solving the basic unforced equations. It will come as no surprise to learn that it is surface wind stress that provides the forcing. The balance is expressed by the equations

$$-fv = -g\frac{\partial \eta}{\partial x},$$

$$\frac{\partial v}{\partial t} + fu = -g\frac{\partial \eta}{\partial y} + \frac{1}{\rho H}\tau^y(y, t),$$

$$\frac{\partial}{\partial x}(Hu) + \frac{\partial}{\partial y}(Hv) = 0.$$

In the above set of equations, advantage has been taken of the previous unforced analysis to approximate longshore flow (cross-shelf momentum of course) by geostrophic balance. The term $\frac{1}{\rho H}\tau^y(y, t)$ is due to the action of the wind stress, and only the component along the shore is effective in generating these waves and associated currents. This is reasonable given the geometry; the shelf is simply not wide enough for the cross shore component to grip the sea in any significant way. It is also supported by observational evidence. With these equations, progress via a streamfunction is possible as before; this time, however, a single forcing frequency is not appropriate as wind is seldom that helpful. The single equation for ψ is now

$$\frac{\partial^2}{\partial x \partial t}\left(\frac{1}{H}\frac{\partial \psi}{\partial x}\right) - \frac{f}{H^2}\frac{\partial H}{\partial x}\frac{\partial \psi}{\partial y} = -\frac{\tau^y}{\rho H^2}\frac{\partial H}{\partial x}.$$

The forcing at all sorts of frequencies means that a series of the form

$$\psi = \sum_{n=1}^{\infty} A_n(y, t)H^{1/2}\phi_n(x)$$

is sought, which leads to what is termed an eigenvalue problem, namely

$$\frac{d^2\phi_n}{dx^2} + \left\{\frac{d}{dx}\left(\frac{1}{2H}\frac{dH}{dx}\right) - \left(\frac{1}{2H}\frac{dH}{dx}\right)^2 - \frac{f}{c_nH}\frac{dH}{dx}\right\}\phi_n = 0.$$

Eigenvalue problems were met in Chapters 1 and 2 in the context of PCA and EOFs but the principle here is the same. In this equation,

c_n is the wave speed and the amplitude functions A_n satisfy the equation

$$\frac{1}{c_n}\frac{\partial A_n}{\partial t} + \frac{\partial A_n}{\partial y} = -\frac{b_n \tau^y(y,t)}{\rho f},$$

where the constant b_n emerges from expanding the term $-\frac{\tau^y}{\rho H^2}\frac{\partial H}{\partial x}$ (which is on the right-hand side of the governing equation for ψ) in a series of the eigenfunctions $\phi_n(x)$. This is a general method, first formalised in an oceanographic context by Gill and Clarke (1974).

Stepping back from all this mathematical detail, what we have is a wind stress forcing a train of continental shelf waves along the slope. The modes that are forced depend on the frequencies present in the wind stress forcing and are obtained by superposition, but once the waves have been forced the train propagates and must disperse. The original source for this model is Gill and Schumann (1974) and a few further details are reproduced in Gill (1982) (beware of the different definitions of x and y). In fact, the books by Gill (1982) and Leblond and Mysak (1978) are recommended for further general reading about continental shelf waves. In this text, it has been decided to devote extra space later in this chapter to trapped waves around islands because these have some interesting features. The trapped nature of continental shelf waves should perhaps be emphasised here as it may have been missed. All the wave-like characteristics are along the coast and away from the coast — both the current and any surface elevation that might be present simply decay. As mentioned earlier, the surface elevation is zero only in the ideal case. Where there is a forcing wind, there is an elevation — it is simply dynamically unimportant for continental shelf waves. Kelvin waves have already been mentioned here and they are analysed in Chapter 6. In Chapter 8 there is more detail about smaller scale coastal trapped waves called edge waves and it is shown that they also have the property of being entirely trapped, despite being independent of Coriolis effects. Kelvin waves exist over a flat sea bed and are quasi geostrophic; it is the sea surface elevation that is essential this time as it provides the essential balance between sea surface slope and Coriolis acceleration. This is what is meant by quasi geostrophy, that is geostrophy but including time dependence (see the detailed Kelvin wave model in Chapter 6).

In reality, there is friction that dampens continental shelf waves, and the water is usually stratified at eastern edges of the ocean. Let us briefly describe models that take this into account. The kind of general modal decomposition model of the type outlined in Gill (1982) and given in more detail in Gill and Clarke (1974) cannot be used here as horizontal boundaries are difficult to incorporate. One alternative is to use continuous stratification, as in the models of Brink (1991). The other alternative is to use layered models. Layered models consist of dividing the sea into a number of layers vertically. In practice, most models have concentrated on just two layers as this closely matches reality, with a mixed layer at the surface, beneath which there is a thin thermocline (more generally called a pycnocline) that divides the surface layer from the deep or abyssal layer. The dynamics of a two layer model are outlined briefly in the case study on the Firth of Forth at the end of the last chapter. In general, two layer models have many distinctive features, the most obvious arising from the interface that can house interfacial waves that can transmit energy. These waves travel at much slower speeds and have associated with them much shorter time scales, which is why most books and articles that describe the dynamics of these waves ignore the Earth's rotation.

Amongst the distinctions between a layered model and a continuously stratified model is the lack of a buoyancy frequency, which is in some ways a disadvantage. Another difference is in a continuously stratified sea there is no potential problem of the layers interacting with the topography — this is certainly an advantage. Given these differences, it is worth examining a continuously stratified model in a little detail. First of all, some discussion of dimensionless numbers gives an indication of underlying dynamics. In a density stratified fluid, one can define the buoyancy frequency, N as

$$N^2 = -\frac{g}{\rho}\frac{\partial \rho}{\partial z},$$

as in Chapter 3. This is the frequency at which a neutrally buoyant float would oscillate vertically if displaced. In a density stratified sea with topography, there are a number of dimensionless numbers that can be important. The most used sometimes rather loosely by civil engineers is the Froude number; this is defined as the ratio of current to long wave speed

$$F_r = \frac{U}{\sqrt{gD}},$$

where U is a typical speed and D a typical depth. It is this form of the Froude number that is commonly used by civil engineers in the analysis of the hydraulic behaviour of rivers and estuaries, particularly the ability of waves to propagate upstream and the existence of tidal bores. However, in reality the Froude number is not well-defined as D can be the whole depth, as it usually is for a river, but also might be a smaller depth associated with the way density changes with depth or with the height of topography, or indeed, it might be a horizontal length. If the Froude number is defined using as a typical depth one associated with the vertical density structure then the Froude number is a "densimetric" Froude number or one of a number of possible internal Froude numbers. These can be used in determining the stability or otherwise of internal waves. There are other useful dimensionless numbers, in particular, when there is flow over topography in the presence of stratification, the parameter

$$\frac{Nh}{U}$$

certainly affects the nature of the flow. The larger this parameter, the shorter the wavelength of the lee waves until, when this parameter reaches a value of 1.5, they seem to break. There is also evidence of a stagnant patch within the wave field (see the book by Baines, 1995). A wider discussion of appropriate dimensionless numbers is in order here as there are different physical effects that need characterising. Back in Chapter 3, the Rossby number was defined as the ratio of the advection acceleration to the Coriolis acceleration, and it took the form

$$R_o = \frac{U}{fL}$$

for typical length scale L, current speed U and Coriolis parameter f. Now this can be thought of as the ratio of a typical speed U to $f\lambda$, which is a wave speed associated with rotation, say an inertial wave speed (see Chapter 3). This looks like a rotational Froude number though there is nothing about stratification, instead the wave part is due to rotation. If the Rossby number is zero, then geostrophic balance is most likely and any attempt to force the flow up a slope fails due to the Taylor–Proudman theorem (see Chapter 10), therefore a small non-zero Rossby number implies a non-geostrophic balance and, therefore, continuity will also imply a small vertical motion. Now suppose both rotation and stratification are present, and there are waves. The first question to ask is whether Taylor columns can still exist. The continuity equation is

$$\frac{\partial u}{\partial x} + \frac{\partial v}{\partial y} + \frac{\partial w}{\partial z} = 0$$

and the last term, though zero for zero Rossby number, can be characterised by the ratio W/H, where W is a typical vertical speed and H a vertical length scale for non-zero Rossby number. It is therefore reasonable to identify the ratio

$$\frac{W/H}{U/L} = R_o,$$

provided it is recognised that this difference is due to rotational (Coriolis) effects.

Now let us look at the Froude number again, but first some basics. If the change in density with depth is significant, then write

$$\Delta\rho = \left|\frac{\partial\rho}{\partial z}\right| \Delta z$$

and note that using the definition of buoyancy frequency N together with dimensionality

$$\left|\frac{\partial\rho}{\partial z}\right| = \frac{\rho_0 N^2}{g} \quad \text{and} \quad \Delta z = \frac{WL}{U},$$

so that

$$\Delta\rho = \frac{\rho_0 N^2 WL}{gU}.$$

Hydrostatic balance thus implies that

$$P = gH\Delta\rho$$
$$= \frac{\rho_0 N^2 HLW}{U}.$$

However, also dimensionally, and in terms of energy (Bernoulli's equation)

$$P = \rho_0 U^2,$$

so equating these two expressions for P gives

$$U^2 = \frac{N^2 HLW}{U},$$

which on rearranging gives

$$\frac{W/H}{U/L} = \frac{U^2}{N^2 H^2}.$$

This expression deserves some attention. If $U \ll NH$ then immediately $W/H \ll U/L$, which implies that the horizontal divergence U/L cannot be satisfied by the vertical convergence W/H, so physically the flow must dominantly be horizontal. The stronger the stratification, the smaller U is compared to NH and the weaker the vertical velocity. The Froude number associated with this is

$$F_r = \frac{U}{NH}$$

and the smaller F_r the more important stratification effects. However, from above, the ratio

$$\frac{W/H}{U/L}$$

can also be associated with the Rossby number under different circumstances. Now suppose both stratification and rotation are present, so that geostrophy can be assumed, which in dimensional terms is

$$fU = \frac{P}{\rho_0 L}.$$

What the combined effects ensure are an increase in vertical velocity scale, but it must still be the case that

$$\frac{W}{H} \leq \frac{U}{L}$$

as vertical convergence cannot exist without horizontal divergence. Combining the stratification and rotation effects gives

$$\frac{W/H}{U/L} = \frac{U^2}{N^2 H^2} \times \frac{fL}{U} = \frac{F_r^2}{R_o}$$

and so on physical grounds $F_r^2 \leq R_o$, which is

$$\frac{U}{NH} \leq \frac{NH}{fL}.$$

So, if the rotation is fixed as it is on the Earth, the above inequality gives an upper limit on the horizontal velocity scale. In particular,

$$R_o = \frac{U}{fL} \leq \frac{(NH)^2}{(fL)^2}.$$

The ratio NH/fL is a dimensionless number that expresses the relative importance of these two effects that both contribute to vertical velocity. This number has been called the Burger number after Alewyn P. Burger (1927–2003) a rather unsung Dutch mathematical modeller of the atmosphere. Dependent on the relative magnitude of F_r, R_o or B_u tells us whether stratification effects (F_r), rotational effects (R_o) or both (B_u) govern vertical motion. High Burger number and the flow over obstacles tends to suppress Taylor columns and allow some vertical motion. Low Burger number and Taylor columns are formed, suppressing vertical motion. It remains true that dimensional analysis only gives a broad brush picture of the physics and detailed numerical models are required to give fine structure. With this in mind, let us move on to do this.

As we have seen, the dynamics of stratified trapped waves can be very complex. The approach chosen here follows Brink (1991), which has the merits of being relatively straightforward; well, as straightforward as it is possible to be and still be correct. The sea is assumed to be governed by the following linear system of equations

$$\frac{\partial u}{\partial t} - fv = -\frac{1}{\rho_0}\frac{\partial p}{\partial x} + \frac{1}{\rho_0}\frac{\partial \tau^x}{\partial z},$$

$$\frac{\partial v}{\partial t} + fu = -\frac{1}{\rho_0}\frac{\partial p}{\partial y} + \frac{1}{\rho_0}\frac{\partial \tau^y}{\partial z},$$

$$0 = -\frac{\partial p}{\partial z} - g\rho',$$

$$\frac{\partial u}{\partial x} + \frac{\partial v}{\partial y} + \frac{\partial w}{\partial z} = 0,$$

$$\frac{\partial \rho'}{\partial t} + w\frac{\partial \bar{p}}{\partial z} = 0.$$

In this formulation, which is Boussinesq (which means that changes in density have no dynamic interaction with currents, so that ρ is constant and equals ρ_0 in the momentum equations), the density assumes the form

$$\rho(x, y, z, t) = \rho_0 + \bar{\rho}(z) + \rho'(x, y, z, t)$$

and

$$|\rho'| \ll \bar{\rho} \ll \rho_0.$$

The single equation for p can be derived

$$\frac{\partial}{\partial t}\left(\frac{\partial^2 p}{\partial x^2} + \frac{\partial^2 p}{\partial y^2}\right) + \left(f^2 + \frac{\partial^2}{\partial t^2}\right)\frac{\partial}{\partial z}\left(\frac{1}{N^2}\frac{\partial^2 p}{\partial z\partial t}\right) = 0$$

for the case where there is no wind stress. The buoyancy frequency is N, where

$$N^2 = -\frac{g}{\rho_0}\frac{\partial \bar{\rho}}{\partial z}.$$

The equation for p is, in fact, the equation for vorticity for this stratified system. The complications introduced by stratification are all too obvious. In order to apply this equation to coastal trapped waves, it is assumed that variables (specifically p as all others can be expressed in terms of p) have wave-like behaviour

$$p \sim \exp[i(\omega t + ly)].$$

It remains true that the equation for p, even in only the two variables x and z, has to be solved numerically, but some progress can be made by assuming that the waves along the coast have a long length scale compared to that offshore; this is called the "long wave limit". The problem then simplifies a little, and the response of such a regime to wind driving and frictional drag through the sea bed is expressed in the following eigenvalue problem

$$\frac{\partial^2 p}{\partial x^2} + f^2\frac{\partial}{\partial z}\left(\frac{1}{N^2}\frac{\partial^2 p}{\partial z\partial t}\right) = 0,$$

$$\frac{\partial^2 p}{\partial x \partial t} + f\frac{\partial p}{\partial y} = 0 \quad \text{at } x = 0,$$

$$\frac{\partial p}{\partial z} + \frac{N^2}{g}p = 0 \quad \text{at } z = 0,$$

$$f^2\left(\frac{1}{N^2}\frac{\partial^2 p}{\partial z \partial t}\right) + \frac{\partial h}{\partial x}\left(\frac{\partial^2 p}{\partial x \partial t} + f\frac{\partial p}{\partial y}\right) = 0 \quad \text{at } z = -h(x)$$

$$\text{and} \quad \frac{\partial p}{\partial x} \to 0 \quad \text{as } x \to \infty.$$

The solutions to this problem can be written

$$p = P_n(y,t)F_n(x,z),$$

where

$$\frac{\partial P_n}{\partial t} - c_n\frac{\partial P_n}{\partial y} = 0,$$

where c_n is the celerity for mode n. If the wind stress takes the form

$$\boldsymbol{\tau} = (0, \tau^y(y,t)),$$

then the solution is a superposition of the form

$$p = \sum_n P_n(y,t)F_n(x,z)$$

with

$$\frac{1}{c_n}\frac{\partial P_n}{\partial t} - \frac{\partial P_n}{\partial y} = b_n\tau_0^y,$$

where $fb_n = -F_n(0,0)$ and the b_n represent coupling co-efficients between mode n and the wind stress (the mixed layer is here assumed infinitesimally thin). In dimensional terms, the appropriate scaling is given by the Burger number

$$S = \left(\frac{N_0 H}{fL}\right)^2,$$

where H, N_0 and L are typical values of depth, buoyancy frequency and cross-shelf horizontal length, respectively. The other assumption that is implicit in this formulation is that the frequency of the disturbances is much less than the local value of the Coriolis parameter.

The depth dependence of the vertical velocity arises solely from the stratification and vanishes if the Burger number is allowed to tend to zero. The trapped waves that result from solving the above eigenvalue problem need a depth that increases with x, although modification due to the presence of (say) trenches is not impossible to make. An increase in the stratification results in the frequency of these waves increasing towards the Coriolis parameter. Well before this happens, of course, the equations need to be modified to take this into account.

The numerical methods used to solve the original equation for $p(x, y, z, t)$ are based on a semi-implicit scheme. The solutions represent trapped waves that are neither classical continental shelf waves (because of the stratification) but nor are they Kelvin waves (no free surface elevation). Figure 12.2 shows some of these waves and their trapping, as represented by contours of pressure. In a

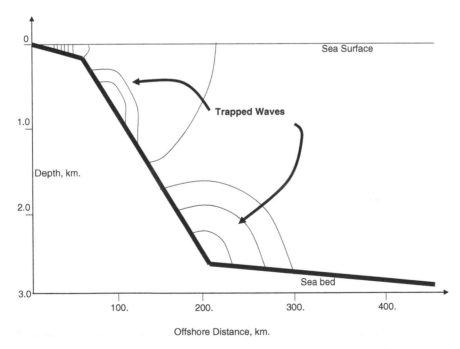

Fig. 12.2   This is a cartoon of the numerical output that uses the equations of Brink (1991) given in the text. The contours are those of pressure; the trapped waves indicated are entirely sub-surface and are propagating out of the page.

stratified regime, the distinction between (baroclinic) Kelvin waves and continental shelf waves becomes artificial and not worth making. For forced waves where the single equation for p cannot be derived, the numerical procedure has to make use of an eigenfunction expansion as in the layered solution above, but this remains a research topic yet to be fully resolved.

12.3 Waves and Flows Around Islands

In this section, we shall look at the modelling of waves and then flows which are trapped around islands. The history of modelling the diffraction of waves around islands really started in the civil engineering community with the modelling of the diffraction of waves by cylindrical piles. The cylindrical geometry means a different coordinate system, but the mathematics is reasonably easy to solve if the depth of the sea is large compared with the diameter of the cylinder. The model is a little trickier if this is not the case, but exact solution is still possible. As we want to discuss both the consequences and some extensions of this model, we shall give some technical details. The factors that are important to consider are the variations in depth, which lead to the refraction of waves and can ultimately, singlehandedly as it were, lead to trapping. Those familiar with Snell's law will understand how refraction can entrap waves. Then there is the Coriolis acceleration. This leads to the distortion of waves and the prevention of standing waves. If the frequency of the wave (σ) is large compared to the Coriolis parameter (f) then perfect trapping is not possible in a sea of constant depth. The depth needs to increase at a greater rate than inversely as the square of the radial distance from the island for trapping to result. In this case, waves are refracted back towards the island and cannot escape. However, in reality, depth cannot fall away this rapidly over long distances; $h \rightarrow$ constant is far more likely. If $\sigma \geq f$, then even if the depth is constant, trapping is possible due to rotation alone. If there are asymmetries, then Poincaré waves may be possible. These will carry energy away from the island and thus prevent perfect trapping. If the sea is stratified, then a spectrum of waves is possible and trapping also becomes less likely. If the dimensions of the island are large enough, then the variation of f with latitude will allow Rossby waves to propagate energy westward and once more trapping is prevented. Finally, in the super-inertial frequency case $\sigma < f$, rotation dominates and perfect trapping is once more possible. This is a brief summary of the mechanisms that can trap waves around islands. Therefore, there are two mechanisms that aid the trapping of waves — depth increase away from the island and Coriolis acceleration — and wave trapping is certainly important as it leads to increased amplitudes as well as the amplification of previously small offshore signals. What follows is a brief look at models that explain the trapping of waves and their associated currents. Those interested in a more extensive treatment are directed towards the texts by Leblond and Mysak (1978) and Dingemans (1997) and the more recent paper by Dyke (2005).

We are concerned with islands, therefore it is natural to use plane polar (r, θ) co-ordinates. The shallow water equations are

$$\frac{\partial u}{\partial t} - fv = -g\frac{\partial \eta}{\partial r},$$

$$\frac{\partial v}{\partial t} + fu = -\frac{g}{r}\frac{\partial \eta}{\partial \theta},$$

$$\frac{\partial \eta}{\partial t} + \frac{1}{r}\left\{\frac{\partial}{\partial r}(hru) + \frac{\partial}{\partial \theta}(hv)\right\} = 0.$$

The nonlinear terms are assumed negligible and there is no friction. If the time dependence takes the form of a wave of single frequency ω, perhaps a tide, then

$$\eta = A(r, \theta)e^{i\omega t}$$

and we have that

$$\frac{\partial \eta}{\partial t} = i\omega \eta.$$

In plane polar co-ordinates (r, θ), we have

$$\nabla^2 \eta = \frac{\partial^2 \eta}{\partial r^2} + \frac{1}{r}\frac{\partial \eta}{\partial r} + \frac{1}{r^2}\frac{\partial^2 \eta}{\partial \theta^2}$$

and the equation for η turns out to be the rather lengthy

$$\nabla^2\eta + \frac{\omega^2 - f^2}{gh}\eta + \frac{1}{h}\frac{\partial h}{\partial r}\frac{\partial \eta}{\partial r} + \frac{1}{r^2 h}\frac{\partial h}{\partial \theta}\frac{\partial \eta}{\partial \theta} + \frac{if}{\omega h r}\left[\frac{\partial h}{\partial \theta}\frac{\partial \eta}{\partial r} - \frac{\partial h}{\partial r}\frac{\partial \eta}{\partial \theta}\right] = 0.$$

The flow that results from this wave is quasi geostrophic, that is basically geostrophic (balance between pressure gradient force and force resulting from Coriolis acceleration) but also slowly varying. Since the time dependence does only consist of oscillation at a single frequency, we can easily derive the following equations for radial current u and transverse current v

$$u = \frac{g}{\omega^2 - f^2}\left[\frac{f}{r}\frac{\partial \eta}{\partial \theta} + i\omega\frac{\partial \eta}{\partial r}\right],$$

$$v = \frac{g}{\omega^2 - f^2}\left[\frac{i\omega}{r}\frac{\partial \eta}{\partial \theta} - f\frac{\partial \eta}{\partial r}\right].$$

Some of you might still be worried about the presence of the imaginary unit $i = \sqrt{-1}$ in these expressions. All this means is that there are phase differences between $\eta(r,\theta)$, $u(r,\theta)$ and $v(r,\theta)$. The presence of rotation has this effect. If we set $f = 0$ in the long equation for η, we can re-derive an equation that can yield solutions of interest to civil engineers which is applicable to waves diffracted around small islands as well as piers and other engineering type constructions of a cylindrical nature. The smallness being defined by the radius being much less than the local Rossby radius of deformation but is also related to the frequency ω being small enough to compare with the local value of the Coriolis parameter f. If the diameter of the island is a, then the Rossby number is U/fa, where U is a typical current speed. The Rossby radius of deformation is $\sqrt{gh}/f$, and in order to ignore f, we require

$$a \ll \sqrt{gh}/f \quad \text{or} \quad af \ll \sqrt{gh}.$$

We have already assumed that U/fa is negligibly small. If the denominator (fa) is replaced by the much larger $\sqrt{gh}$, the Froude number $U/\sqrt{gh}$ must be even smaller. Conversely, even though the Froude number is small (waves more important than currents) it may not be the case that the nonlinear terms can be ignored. The current associated with a wave is of magnitude $Ag/\omega\lambda$, where A is the amplitude of the wave and λ its wavelength. The Froude number is thus $Ag/\omega\lambda\sqrt{gh}$ and its smallness is another way of saying that, after using the dispersion relation for $h \ll \lambda$, that $A \ll h$ or the amplitude is much less than the water depth. This is the classic small amplitude wave assumption.

In order to progress, it is also assumed that the depth profile is axisymmetric of the form $h(r)$, so that all terms involving the derivatives of h with respect to θ are zero. The solution can now be obtained in terms of what is a standard technique of separating the variables

$$\eta(r,\theta) = Z(r)\Theta(\theta).$$

Note that even though there is no variation of depth with tangential co-ordinate θ, it certainly cannot be assumed that the surface elevation η is also independent of θ. The separation technique yields the two equations for $Z(r)$ and $\Theta(\theta)$

$$\frac{d^2Z}{dr^2} + \frac{1}{r}\frac{dZ}{dr} + \frac{1}{h}\frac{dh}{dr}\frac{dZ}{dr} + \left(\frac{\omega}{gh} - \frac{n^2}{r^2}\right)Z = 0$$

and

$$\frac{d^2\Theta}{d\theta^2} = -n^2\Theta,$$

where the separation constant n is an integer. The negative sign in the second of these equations ensures that the solution $\Theta(\theta)$ is sinusoidal and n is an integer because there must be an exact number of waves around the island. The equation for $Z(r)$ is more interesting. Writing it in self adjoint form gives

$$\frac{d}{dr}\left[rh\frac{dZ}{dr}\right] + r\left[\frac{\omega^2}{g} - \frac{n^2}{r^2}h\right]Z = 0.$$

This enables us to deduce that as long as the inequality

$$\frac{\omega^2}{g} > \frac{n^2h}{r^2}$$

holds then the positive nature of the co-efficient of Z indicates that the solution $Z(r)$ will assume an oscillatory character as r increases.

On the other hand, if

$$\frac{\omega^2}{g} < \frac{n^2 h}{r^2},$$

$Z(r)$ will behave exponentially as r increases. The increasing term is rejected on physical grounds, therefore $Z(r)$ will, in this case, decay to zero as r increases. We call this *wave trapping* and trapping is thus assured as long as $h(r)$ is a function that grows faster than r^2. In this case, a value of r will eventually be reached for which the solution decays exponentially. The zero mode ($n = 0$) is special and there will never be complete trapping in this case. As mentioned earlier, from a practical point of view, $h(r)$ is never going to increase as rapidly as r^2, except perhaps very locally. In fact, over a large distance the depth will vary stochastically about the mean depth of the ocean thus ensuring $h(r)$ is, in fact, bounded for large r. Perfect trapping thus does not occur in practice. In order to say something about the flow, we note that, for the case of no Coriolis acceleration,

$$u = \frac{ig}{\omega} \frac{\partial \eta}{\partial r},$$

$$v = -\frac{ng}{r\omega} \eta.$$

If $n = 0$ (zero mode), then $v = 0$ and all flow is in the direction radially away or towards the centre of the island. Note that this is the "ripples due to a stone thrown into a pond" solution. The elevation is 180° out of phase with the (radial) current that is common to all one dimensional progressive waves. The generally varying $h(r)$ case is still not possible to solve analytically, but if it assumed that h is a constant then it is possible to express the elevation η as follows

$$\eta(r, \theta) = (AJ_n(kr) + BY_n(kr))e^{in\theta - i\omega t},$$

where $k = \frac{\omega}{\sqrt{gh}}$ and J_n, Y_n are the Bessel functions of the first and second kind, respectively. (See the tome by Watson (1922) for all there is to know about Bessel functions. All you actually need to know here is that they are mathematical functions that oscillate and decay as $r \to \infty$ and, as this happens, the period gets larger. Much like a vertical cross section, $\theta = $ constant, through the ripple in a pond due to a thrown stone mentioned above in fact.) It is tempting, and possibly advantageous, at this stage to discuss sea mounts. As these do not break the surface, the wave must be finite at $r = 0$ whence the $Y_n(kr)$ term must be zero. The mathematics is thus simpler; simpler too because there are no island boundary conditions to consider. The temptation will, however, be resisted for reasons of space (see Leblond and Mysak (1978)). The advantage of possessing exact mathematical solutions is that precise deductions are possible. For example, the solution for large n (high mode number) is virtually trapped. This is because once outside a particular critical radius, $k = \frac{n\sqrt{gh}}{\omega}$, the solution exhibits asymptotic behaviour which is that of a decaying exponential. The two sets of waves

$$\eta(r, \theta) = J_n(kr)e^{(in\theta - i\omega t)}$$

and

$$\eta(r, \theta) = Y_n(kr)e^{(in\theta - i\omega t)}$$

represent two sets of spiralling waves that are not refracted (no depth change) and that propagate away from the island (see Figure 12.3). The flow due to this type of wave can be readily deduced since

$$u = \frac{ig}{\omega}\frac{\partial \eta}{\partial r} \quad \text{and} \quad v = -\frac{ng}{r\omega}\eta$$

and, for large, r, this becomes predominantly radial as both $\frac{\partial \eta}{\partial r}$ and η have an $r^{-1/2}$ dependence asymptotically. The steady streaming (Stokes drift) due to sinusoidal waves of this type is to be considered later in this chapter.

So far, nothing has been said about the effects of the Earth's rotation. To see how their inclusion might affect things, let us take advantage of the earlier derivation of the general equation for the elevation $\eta(r, \theta)$ and keep a non-zero Coriolis parameter f but still make the assumption $h = h(r)$, that is the depth contours are axisymmetric. The (only slightly) simplified equation for η is

$$\nabla^2 \eta + \frac{\omega^2 - f^2}{gh}\eta + \frac{1}{h}\frac{\partial h}{\partial r}\frac{\partial \eta}{\partial r} - \frac{if}{\omega h r}\frac{\partial h}{\partial r}\frac{\partial \eta}{\partial \theta} = 0.$$

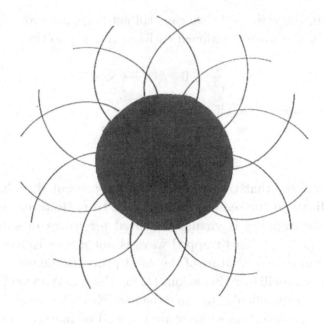

Fig. 12.3   The spiralling waves.

The last term on the left-hand side prevents the use of the method of separation of variables, which is unfortunate. Nevertheless, some progress can be made. As we are still considering the waves around an island, the variation of $\eta(r, \theta)$ must be periodic in θ, hence we can still write

$$\frac{\partial \eta}{\partial \theta} = in\eta.$$

This enables the elimination of the variable θ and the following equation for η can be derived

$$\frac{\partial}{\partial r}\left[rh\frac{\partial \eta}{\partial r}\right] + r\left[\frac{nfh'}{\omega r} + \frac{\omega^2 - f^2}{g} - \frac{n^2}{r^2}h\right]\eta = 0.$$

We have written h' for the derivative of h with respect to r. For trapping, therefore, the criterion is the inequality

$$\frac{nfh'}{r\omega} + \frac{\omega^2 - f^2}{g} < \frac{n^2}{r^2}h.$$

So if $n = 0$, the zero mode, we can still get trapping provided $\omega < f$ (the Kelvin like wave mentioned earlier). As, in practice, we have

$$\frac{h}{r^2} \to 0 \quad \text{as } r \to \infty$$

the above inequality in effect reduces to

$$\frac{nfh'}{r\omega} + \frac{\omega^2}{g} < \frac{f^2}{g}.$$

We can thus say that trapping may still be prevented by large positive gradients in the sea bed with respect to r. However, as $h' \to 0$ as $r \to \infty$, trapping is virtually assured for waves of sub-inertial frequency ($\omega < f$). This trapped wave is not a true Kelvin wave of course because the curvature of the coast prevents the precise decoupling between oscillatory behaviour (along the coast) and exponential behaviour (perpendicular to the coast) which is the usual feature of Kelvin waves. Nevertheless, they are trapped no matter how h varies with r, which is very different from the situation with $f = 0$. When $f < \omega$, it seems there is unlikely to be trapping. The presence of stratification complicates the picture still further. In a continuously stratified sea, it is tempting to try to follow Gill and Clarke (1974) directly (there is a summary of their approach in Gill (1982) section 9.10, page 342) and separate into modes. However, there are two minor difficulties. First of all, the use of polar co-ordinates prevents plane waves. Secondly, expressing the solution in terms of an "elevation" has no obvious meaning in a modal decomposition. Hence, the formulation of Brink (1999) is followed whereby a polar version of his 1991 Cartesian equations are investigated. The equations in polar form are

$$\frac{\partial u}{\partial t} - fv = -\frac{1}{\rho_0}\frac{\partial p}{\partial r}, \tag{12.1}$$

$$\frac{\partial v}{\partial t} + fu = -\frac{1}{r\rho_0}\frac{\partial p}{\partial \theta}, \tag{12.2}$$

$$\frac{\partial}{\partial r}(ru) + \frac{\partial v}{\partial \theta} + \frac{\partial w}{\partial z} = 0, \tag{12.3}$$

$$0 = \frac{\partial p}{\partial z} - g\rho',\tag{12.4}$$

$$0 = \frac{\partial \rho'}{\partial t} + w\frac{\partial \rho_0}{\partial z},\tag{12.5}$$

and using the Boussinesq approximation, as before, we have written the density ρ to take the form

$$\rho(x, y, z, t) = \bar{\rho} + \rho_0(z) + \rho'(x, y, t)\tag{12.6}$$

with

$$\bar{\rho}(x, y, z, t) \gg \rho_0 \gg |\rho'(x, y, t)|.\tag{12.7}$$

The simplifying assumption is that all waves have the single constant frequency σ. This is consistent with other modellers (Wang and Mooers (1976), Brink (1982) etc.) and also follows from measurements of internal tides in stratified seas, see for example Baines (1974). Whether this is also true for meteorologically or seismically generated internal waves is open to question. To progress without this assumption means allowing a spectrum of frequencies which can always be done at a later stage. A note of caution however, the criterion that either $\sigma > f$ or $\sigma < f$ needs to be borne in mind and appropriate solutions used.

Thus let

$$p = P(r, \theta, z)e^{i\sigma t}\tag{12.8}$$

and elimination of other variables in favour of P gives

$$\frac{1}{r}\frac{\partial}{\partial r}\left(r\frac{\partial P}{\partial r}\right) + \frac{1}{r^2}\frac{\partial^2 P}{\partial \theta^2} + (\sigma^2 - f^2)\frac{\partial}{\partial z}\left[\frac{1}{N^2}\frac{\partial P}{\partial z}\right] = 0.\tag{12.9}$$

Here, the buoyancy frequency N has been reintroduced with the usual definition

$$N^2 = -\frac{g}{\rho_0}\frac{\partial \rho}{\partial z}.\tag{12.10}$$

In order to prescribe the vertical structure, it is proposed that

$$\frac{\partial}{\partial z}\left[\frac{1}{N^2}\frac{\partial P}{\partial z}\right] = \frac{P}{gH}\tag{12.11}$$

in which case the differential equation for P becomes

$$\frac{1}{r}\frac{\partial}{\partial r}\left(r\frac{\partial P}{\partial r}\right) + \frac{1}{r^2}\frac{\partial^2 P}{\partial \theta^2} + \lambda^2 P = 0. \qquad (12.12)$$

It is now possible to define modes as follows

$$u = \rho_0 g \sum_{n=0}^{\infty} u_{mn}(\theta, t)\zeta_{mn}(r, z), \quad v = \rho_0 g \sum_{n=0}^{\infty} v_{mn}(\theta, t)\zeta_{mn}(r, z),$$

$$(12.13)$$

where m is the azimuthal wave number so that

$$p(r, \theta, z, t) = P(r, \theta, z)e^{i\sigma t} = \rho_0 g \zeta_{mn}(r, z)e^{i(\sigma t + m\theta)}. \qquad (12.14)$$

Thus the modal decomposition first proposed by Gill and Clarke (1974) has been generalised to polar co-ordinates much in the manner of Brink (1999), but now such that the analytical solution for a homogeneous sea given earlier is directly applicable. Note that the most important consequence of making σ constant is to make the character of the stratified response (whether it be oscillatory or decaying) independent of both vertical mode number n and azimuthal wave number m. This is artificial, but it is a start. Further work is necessary in order to remove this restriction, and to take this stratified model further. For the present, it suffices to say that the results from the homogeneous model can, with care, be applied to a stratified sea that is dominantly hydrostatic and Boussinesq; this is an important conclusion. Thus modal separation indicates that trapping is even more unlikely; a deduction confirmed by the numerical experiments of Brink (1999). Finally, numerical (and some analytical) modelling on the interaction of waves and flows with real, not axisymmetric islands has been done. It is not difficult to deduce (see Leblond and Mysak, 1978) that asymmetry leads to the destruction of the standing nature of the waves. The lowest order modes exhibit a 180° phase difference between points on opposite sides of the island. Analytical models using elliptical islands have been used to represent tides around the Hawaiian Island of Oahu (see Reynolds, 1978) as well as Macquarie Island in Australia (Summerfield, 1969). Measurements on Oahu, on Cook Reef (north west of Australia) and Bermuda in the West Indies all indicate that some wave trapping is possible and that such a wave propagates around the island. Phase differences have also been measured in approximate agreement with models. Of course, validating models is rendered difficult as there is considerable noise so how do you tell whether or not there is a "perfectly" trapped wave? Other reasons for modelling such waves, however, have emerged. Signals from internal waves propagating from Cook Reef (for example) prove important indicators of seismic activity. This could help in the fast detection of underwater earthquakes. Similar indicators could also help us to understand the migration of certain marine organisms and build a better picture of the ecology of the area.

If the island is large enough, it is possible that the change of Coriolis parameter with latitude will be important. In this case, Rossby waves can be generated as waves diffract or as a large current impacts upon an island. In this case, even for wave frequencies that are well below the local value of the Coriolis parameter there is no trapping as the Rossby waves leak their energy westward. Such large scale flows are not really of concern here and interested readers are directed towards Leblond and Mysak (1978). They feature in the behaviour of tsunamis. One deduction from this section is that wave trapping off islands and continental shelves is still an active research area.

12.3.1 *Steady flows*

Flows due to waves will, of course, be primarily oscillatory in character. However, even perfectly sinusoidal waves will give rise to a steady current, albeit a slow one. The reasons for this are not obvious, but the nonlinear nature of fluid dynamics lies behind it. It was the 19th century mathematician and fluid dynamicist G. G. Stokes (1819–1903) who first showed that sinusoidal surface water waves must have a steady drift and the primary (but not the only) drift is named after him. The two ways of modelling the motion of a fluid are called Lagrangian and Eulerian. The Lagrangian model is where the observer is attached to a particular fluid particle and rides around on it (see sections 7.4.2 and 8.3.2). The Eulerian model is the standard one in which there is a fixed origin and axes, and the fluid is modelled relative to these. If $\mathbf{U}_L$ is the Lagrangian flow and $\mathbf{U}_E$ the Eulerian flow then

$$\mathbf{U}_L = \mathbf{U}_E + \mathbf{U}_S,$$

where $\mathbf{U}_S$ is the Stokes drift. This has already been met in the context of surface water waves in Chapter 8. As mentioned there, the account of Longuet-Higgins (1953) is the classic reference for the principal steady flows brought about by water waves. His detailed derivation is highly mathematical and requires looking at marked particles and tracking them as in the modelling of diffusion (see Chapter 7), but the primary steady current due solely to the interaction of first order terms is the Stokes drift, defined by

$$\mathbf{U}_S = \overline{\left(\int \mathbf{u}dt \cdot \boldsymbol{\nabla} \right) \mathbf{u}}.$$

In this expression, $\mathbf{u}$ must be purely oscillatory and the overbar represents the time average over a period of the oscillation. Often, the average over a few periods is taken when estimating the Stokes drift in the field. Before investigating the Stokes drift in detail, it is worth noting that there have been a number of interesting developments regarding the Lagrangian mean flow over the years, most of them stemming from the application of the conservation of potential vorticity and incorporating the change of Coriolis parameter with latitude. If the oscillation is due to a single frequency, then it can be shown (Moore, 1970) that if there are no closed contours, or closed *geostrophic* contours if f varies, then $\mathbf{U}_L$ is zero throughout. This means that $\mathbf{U}_E$ is precisely equal and opposite to the Stokes drift $\mathbf{U}_S$.

Around an island, there *are* closed contours and therefore one would expect $\mathbf{U}_L$ to follow the depth contours. In the calculations performed here, $\mathbf{U}_E$ (and therefore $\mathbf{U}_L$) have not been considered, but the calculation of $\mathbf{U}_S$ is straightforward as it is calculated solely from the primary flow. Of course, if h is a constant, Moore's theory tells us nothing at all about $\mathbf{U}_L$ other than it is arbitrary. For the situation here with non-zero but constant f, $h = h(r)$ and

$$\eta(r, \theta, t) = F(r)e^{i(n\theta - \omega t)},$$

then the calculation of the Stokes drift velocity is possible explicitly. In most papers and textbooks, the exponential form is retained and all functions are permitted to assume complex values, with the real parts being extracted at the end. This is neat but can lead to misunderstandings and make obscure some of the results. Let us proceed here by assuming that η, defined above, is such that $F(r)$ is real. This fixes the phase as zero. The velocity vector (u, v) is given by

$$u(r, \theta, t) = -U(r)e^{i(n\theta - \omega t)},$$
$$v(r, \theta, t) = V(r)e^{i(n\theta - \omega t)},$$

where

$$u = \frac{g}{\omega^2 - f^2}\left[\frac{f}{r}\frac{\partial \eta}{\partial \theta} + i\omega\frac{\partial \eta}{\partial r}\right],$$

$$v = \frac{g}{\omega^2 - f^2}\left[\frac{i\omega}{r}\frac{\partial \eta}{\partial \theta} - f\frac{\partial \eta}{\partial r}\right],$$

and the reason for the minus sign in front of $U(r)$ will be revealed later. So, inserting the above expressions for u and v we have

$$U(r) = -\frac{g}{\omega^2 - f^2}\left[\frac{infF(r)}{r} + i\omega F'(r)\right],$$

$$V(r) = \frac{g}{\omega^2 - f^2}\left[\frac{-n\omega F(r)}{r} - fF'(r)\right].$$

From these equations, $U(r)$ is imaginary ($i \times$ a real quantity) and $V(r)$ is real. So, to have a real velocity it must be the case that

$$(u, v) = (U(r)\sin(n\theta - \omega t), V(r)\cos(n\theta - \omega t))$$

and the reason for the minus sign is clear; it avoids a minus sign in front of $U(r)$ in this final expression. In some books, the expression "in quadrature" is used for this difference in phase between u and v. So $u(r, \theta, t)$ and $v(r, \theta, t)$ are in quadrature, which really means that they represent waves that are 90° out of phase. The same is true for η and u, but η and v are in phase of course. The Stokes drift can now be calculated without complex arithmetic. Start with the definition

$$\mathbf{U}_S = (U_S, V_S) = \overline{\left(\int \mathbf{u} dt . \nabla \right) \mathbf{u}},$$

where the overbar denotes time average over a few wave cycles. Thus, inserting the above expressions,

$$U_S = \overline{\frac{U(r)}{\omega} \cos(n\theta - \omega t)[-U'(r) \sin(n\theta - \omega t)]}$$

$$+ \overline{\frac{V(r)}{r\omega} \sin(n\theta - \omega t)[nU(r) \cos(n\theta - \omega t)]}$$

and

$$V_S = -\overline{\frac{1}{\omega} U(r) \cos(n\theta - \omega t)[V'(r) \cos(n\theta - \omega t)]}$$

$$+ \overline{\frac{V(r)}{r\omega} \sin(n\theta - \omega t)[nV(r) \sin(n\theta - \omega t)]}.$$

Now these expressions are informative. Notice that the equation for U_S is the product of a sine and a cosine of the same argument. Using $\sin \phi \cos \phi = \frac{1}{2} \sin(2\phi)$ averaging will give zero. Thus $U_S = 0$, which is completely logical for islands as there cannot be steady flow radially away — this would invalidate mass continuity. V_S on the other hand is not zero; it is given by

$$V_S = -\frac{1}{2\omega} \left[U \frac{dV}{dr} + \frac{nV^2}{r} \right]$$

once the averaging over the squares of the trigonometric functions has taken place. (The average of both $\sin^2 \phi$ and $\cos^2 \phi$ is $\frac{1}{2}$.) Hence, the Stokes drift can be calculated explicitly using

$$U(r) = \frac{g}{\omega^2 - f^2} \left[\frac{nf F(r)}{r} + \omega F'(r) \right],$$

$$V(r) = \frac{g}{\omega^2 - f^2} \left[\frac{-n\omega F(r)}{r} - f F'(r) \right].$$

If study is restricted to the lowest mode $n = 0$ then we have the expressions

$$U(r) = \frac{g}{\omega^2 - f^2} \omega F'(r),$$

$$V(r) = \frac{-g}{\omega^2 - f^2} f F'(r),$$

$$V_S = -\frac{1}{2\omega} U \frac{dV}{dr},$$

which implies

$$V_S = \frac{g^2 f F' F''}{2(\omega^2 - f^2)^2}.$$

It is not difficult to see that this speed, which is positive and so anticlockwise in the northern hemisphere, clockwise in the southern hemisphere, can be if not large, not entirely negligible. To do some crude calculations, the magnitude of this Stokes drift can be estimated as follows. Suppose the wave has an offshore (non-wave-like) structure with a characteristic length a, the radius of the island. Then the derived Stokes drift will have magnitude

$$V_S \approx \frac{[U]^2}{af},$$

where $[U]$ denotes the magnitude of the current. This could get quite large. Of course, if a is small then the Coriolis effect is negligible, and straight from the formula, the lowest mode Stokes drift will be zero. However, there are mid range islands of radius a few tens of kilometres where V_S would be measurable.

Finally, to go back a step; for those who want to use complex quantities (see the exercises), this can be done, and when the final expression is obtained, it is only the real parts that have any physical significance. There is a useful expression for calculating the average of products of the real parts of quantities such as we have here, and this is

$$\overline{\Re\{z_1\}\Re\{z_2\}} = \Re\left\{\frac{1}{2} z_1 z_2^*\right\} = \Re\left\{\frac{1}{2} z_1^* z_2\right\},$$

where * represents complex conjugate. For the skeptics, here is a proof. Let $z_1 = Ae^{i(n\theta - \omega t)}$, where $A = A_1 + iA_2$ so that

$$\Re\{z_1\} = A_1 \cos(n\theta - \omega t) - A_2 \sin(n\theta - \omega t).$$

Let $z_2 = Be^{i(n\theta - \omega t)}$, where $B = B_1 + iB_2$ so that

$$\Re\{z_2\} = B_1 \cos(n\theta - \omega t) - B_2 \sin(n\theta - \omega t)$$

then

$$\overline{\Re\{z_1\}\Re\{z_2\}} = \frac{1}{2}(A_1 B_1 + A_2 B_2) = \Re\left\{\frac{1}{2}z_1 z_2^*\right\} = \Re\left\{\frac{1}{2}z_1^* z_2\right\}$$

as required. It is possible to deduce similar expressions valid in a two layer sea, but more general extensions to deducing flows in stratified seas are the subject of present research. The work of Brink (1999) is mainly concerned with wave trapping, but the calculation of currents is certainly possible from his numerical models, and it is to numerical models one must look next for progress. For super-inertial frequencies ($\omega > f$), the work of Dale and Sherwin (1996) needs to be adapted to islands, which has not yet been done. One suspects trapping only at very special frequencies, and what currents are induced have yet to be revealed. This then represents a very fertile area for future research; some of you may be tempted to take it up.

12.4   Exercises

(1) State two principal physical effects that enable waves to be trapped against a straight coastline. Does the presence of stratification hinder or help trapping?
(2) Consider the equation

$$\frac{d^2\phi}{dx^2} + \left\{\frac{d}{dx}\left(\frac{1}{2H}\frac{dH}{dx}\right) - \left(\frac{1}{2H}\frac{dH}{dx}\right)^2 - l^2 - \frac{fl}{\omega}\frac{1}{H}\frac{dH}{dx}\right\}\phi = 0$$

and look for solutions with

$$H(x) = H_0 e^{-2\lambda x}$$

and

$$\phi(x, t) = Ce^{ikx}$$

valid for continental shelf waves of the type

$$\psi = H^{1/2}\phi(x)\exp(ily - i\omega t).$$

Hence, determine a dispersion relation. Calculate the velocity field for the solution

$$\psi(x,y,t) = e^{-\lambda x}\sin(kx)\cos(ly - \omega t)$$

and comment on its value for large x.

(3) Brink's (1991) model for stratified trapped waves against a coast derives the following equation for pressure

$$\frac{\partial}{\partial t}\left(\frac{\partial^2 p}{\partial x^2} + \frac{\partial^2 p}{\partial y^2}\right) + \left(f^2 + \frac{\partial^2}{\partial t^2}\right)\frac{\partial}{\partial z}\left(\frac{1}{N^2}\frac{\partial^2 p}{\partial z\partial t}\right) = 0.$$

By non-dimensionalising this equation, show that calculations depend on the magnitude of the Burger number

$$S = \left(\frac{N_0 H}{fL}\right),$$

where N_0 is a typical value of the buoyancy frequency and H, L and f are also typical values (depth, cross-shelf length and Coriolis parameter, respectively). What does this equation indicate at zero stratification?

(4) The following equation was derived in Chapter 5

$$\frac{(\omega^2 - f^2)}{g}\eta + \left[\frac{\partial}{\partial x}\left(H\frac{\partial\eta}{\partial x}\right)\right] + \left[\frac{\partial}{\partial y}\left(H\frac{\partial\eta}{\partial y}\right)\right] + \frac{if}{\omega}\left|\frac{\partial(H,\eta)}{\partial(x,y)}\right| = 0.$$

From this, derive the equation valid for the surface elevation on the continental shelf

$$\frac{d}{dx}\left(H\frac{dF}{dx}\right) + \left(\frac{lf}{\omega}\frac{dH}{dx} + \frac{\omega^2 - f^2}{g} - l^2 H\right)F = 0, \qquad (12.15)$$

where it has been assumed that the surface elevation takes the form of continental shelf waves

$$\eta(x,y,t) = F(x)e^{ily - i\omega t},$$

H is the depth and the Coriolis parameter f is constant. Discuss the following cases:

(a) $lf > 0$,

(b) $lf < 0$,

(c) zero slope ($\frac{dH}{dx} = 0$),

(d) the significance of the term $\frac{\omega^2 - f^2}{g} - l^2 H$.

(5) When modelling flow and waves around islands, what condition can be imposed that is not available to modellers of waves along coasts? What advantage does this condition bring to models?

(6) Compare equation (12.15) with the radial version that governs the trapping of waves around islands

$$\frac{d}{dr}\left[rH\frac{dF}{dr}\right] + r\left[\frac{nf}{\omega r}\frac{dH}{dr} + \frac{\omega^2 - f^2}{g} - \frac{n^2}{r^2}H\right]F = 0,$$

where

$$\eta(r, \theta, t) = F(r)e^{in\theta - i\omega t}.$$

Are the conditions for trapping significantly different?

(7) Assuming lowest mode $n = 0$, derive the Stokes drift for the trapped island waves by using complex arithmetic.

Chapter 13

Conclusion

13.1 Summary

This book has been an attempt at writing an account of how the modelling of coastal and offshore processes has progressed in the last 50 or so years. It is obvious that the subject is not standing still, and that in the intervening time between when these words are written and when they are read, other advances would have been made. It is thus inevitable that a book such as this will date. On the other hand, the use of specific software, tailor-made techniques has been avoided to keep this problem to a minimum. This book is about modelling principles and applications to processes rather than a collection of routines for solving particular problems. The main purpose of this present, final, chapter is to try and bring together some modelling strands that appear disconnected, mainly because they are in separate chapters here.

Arguably the most important motivator for modelling coastal and offshore seas and estuaries lies in environmental protection. This was stated in the first chapter. The six chapters that give prominence to the appropriate modelling are Chapter 6 (storm surges), Chapter 7 (diffusion modelling), Chapter 8 (specifically section 8.5.3, software for erosion modelling), Chapter 9 (ecosystem modelling), Chapter 11 (estuarial modelling) and that part of Chapter 5, section 5.4.1, that deals with the sea bed boundary condition. The opportunity to combine models has been mentioned in Chapter 9, for example ERSEM, which is ecosystems software that emerged in 1995, Baretta, Ebenöh and Ruardij (1995). What has always been the case, but now needs emphasising more than ever, is that it is the physical processes that need modelling correctly first, otherwise — no matter how good the biology may be parameterised — the model will not be useful. Therefore, it is essential to get to grips with the physics of the sea bed, as it is here that many of the important processes occur. In Chapter 7, the process of diffusion was modelled, but in reality much of the transportation of foreign material through estuarial and coastal waters occurs via the sediments. Therefore the bottom boundary condition and how sediment is picked up and settles is very important to understand and model correctly. Therefore, we turn to models such as MIKE 21 that are the best around for modelling sediment transport and coastal erosion. The sophistication of the bottom boundary condition has yet to reach this software, but it will. There is the interaction of these turbulence closure schemes and the LES (large eddy simulations in which the resolution of the finite difference or finite element model is sufficient to resolve most of the eddies that contribute to the turbulent energy) with diffusive processes and, in turn, with ecosystems that has yet to be done seriously, but surely this is the next step after the marrying of ERSEM and POLCOMS. In 2008, the model NEMO was launched, and it is a modelling tool that includes marine physics at all scales, includes estuaries as well as diffusion and ecosystems. This is a state-of-the-art piece of software worth a closer look and certainly a pointer to the future.

13.2   Finding NEMO

Actually, NEMO (short for Nucleus for European Modelling of the Ocean) is very easy to find. The large 367 page book on NEMO is free to download and peruse, NEMO (2012). It is easy to register as a user and everything is cost free. The software was produced by Gurvan Madec and his team at the Laboratoire d'Océanographie et du Climat: Expérimentation et Approches Numériques (LOCEAN), Paris, France, supported by a European grant, is freeware under the French version of the General Public Licence and consequently can be used by anyone who can register. There are now a vast number of publications, on-screen movies and presentations on all aspects of modelling available under the auspices of NEMO. To focus this brief summary of a very large project, let us concentrate on the NEMO Ocean Engine. This manual is 367 pages in length and the detail in it is quite breathtaking. Not only that, but it interfaces with thermodynamics, a sea ice model, a particle tracer and a biogeochemical model. It also links with an atmospheric model, but NEMO is securely a model of the ocean as the O in NEMO indicates. The basic equations are written in curvilinear co-ordinates, and the user has the choice of using constant steps in the vertical (z co-ordinate) or a terrain following s co-ordinate, which is more or less the σ co-ordinate (see Chapter 5) but is a bit more general. The horizontal pressure gradient is expressed as

$$\nabla p|_z = \nabla p|_s - \frac{\partial p}{\partial s} \nabla z|_s$$

and it is the second term on the right that gives the correction to horizontal pressure due to the tilt of the sea bed. The equations assume that the ocean is on a thin spherical shell, the fluid is hydrostatic and Boussinesq (the variation of density with vertical does not interact with horizontal dynamics; it only matters in the vertical where buoyancy is felt). There is also turbulent closure and the whole is incompressible. Before looking at the literature, a word of warning. The mathematics has largely been written by a Frenchman, and this means the notation uses tensors and is beautifully general. It is not easily understood by those without a mathematical background. It is, however, there for reference and can be taken on trust. The equations themselves are written in flux form (see Chapter 4)' including potential temperature, salinity (both are *active tracers* in the language of NEMO) and an equation of state. These equations are discretised in the horizontal on an Arakawa C grid (see Chapter 5), the positions of the Coriolis terms labelled f and the vertical velocity labelled W. Temperature, salinity and pressure are evaluated in the centre at T and all are shown in Figure 13.1. The points f are also used to evaluate the relative vorticity (the curl of the current velocity). Let q be any of the variables, then looking at just one dimension $x = i\Delta x$, the centred difference operator is defined by

$$\delta_i[q] = q_{(i+\frac{1}{2})} - q_{(i-\frac{1}{2})},$$

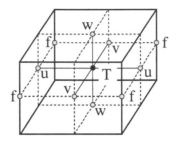

Fig. 13.1   The three-dimensional grid under NEMO.

which is consistent with the definition in Chapter 4, and the mid-point is given by

$$\bar{q}_i = \frac{q_{(i+\frac{1}{2})} + q_{(i-\frac{1}{2})}}{2}.$$

The time stepping follows the general scheme

$$q^{s+1} = q^s + 2\Delta t [\text{RHS}]^{s-1,s,s+1},$$

where the RHS denotes any of the evolution equations in NEMO. In Figure 13.3, there is a flowchart of how the equations in NEMO are solved, however this is taken straight from the manual NEMO (2012) and n not s is used for the time step. The non-diffusive parts of the physics use a leap frog scheme (see Chapter 4), the details are exhibited in Figure 13.2. The leap-frog scheme called the Leap-frog Robert Asselin is as follows

$$q^{s+1} = q^{s-1} + \Delta t (Q^{s+\frac{1}{2}} + Q^{s-\frac{1}{2}}),$$

$$q_F^s = q^s + \gamma [q_F^{s-1} - 2q^s + q^{s+1}] - \gamma \Delta t [Q^{s+\frac{1}{2}} - Q^{s-\frac{1}{2}}].$$

You are reminded that q stands for any variable, the subscript F denotes a filtered value and is named after Robert (1966) and Asselin (1972), and γ is a number called the Asselin co-efficient with a default value of 10^{-3}. This scheme was mentioned in Section 4.5. This is successful for the Coriolis terms, the point acceleration, the adjective acceleration and the pressure gradient terms, but must be avoided when modelling terms that actually represent real physical diffusion as the numerical diffusion so generated by the second-order derivatives cannot be distinguished from the real thing. The corrections

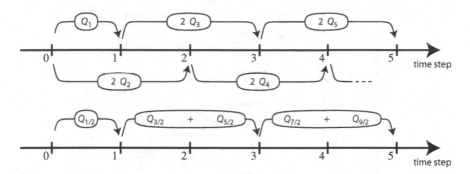

Fig. 13.2 The NEMO leap-frog scheme is shown, together with the more traditional formulation: (top) the forcing is defined at the same time as the variable to which it is applied (integer value of the time step index) and it is applied over a $2\Delta t$ period. Modified formulation (bottom): the forcing is defined in the middle of the time (integer and a half value of the time step index) and the mean of two successive forcing values $(s - \frac{1}{2}, s + \frac{1}{2})$ is also applied over a $2\Delta t$ period. (NEMO uses n and not s; Q_s denotes any variable.)

that sort out possible errors due to using the basic leap-frog algorithm are given in the NEMO manual. The treatment of the pressure gradient term is not shown in the flowchart, Figure 13.3, and is in fact the subject of a special smaller time step. All these carefully researched enhanced numerical methods optimise efficiency and ensure stability according to the latest published knowledge of the applications of these numerical methods. The manual and NEMO software are updated as new knowledge comes to light. Users are informed of updates and the manual is frequently updated to reflect the latest version. Here is an interesting quirk. For the modelling of the passive tracers, the manual NEMO (2012) mentions the TVD scheme

$$\tau_u^{\text{ups}} = \begin{cases} T_{i+1} & \text{if } u_{i+\frac{1}{2}} < 0, \\ T_i & \text{if } u_{i+\frac{1}{2}} \geq 0, \end{cases}$$

$$\tau_u^{\text{tvd}} = \tau_u^{\text{ups}} + c_u(\tau_u^{\text{cen2}} - \tau_u^{\text{ups}}),$$

which is clearly based on upstream differences (the first equation) but the second equation, although labelled TVD, contains the factor c_u, which is usually one but could be less (even zero) in the vicinity of an extremum in the magnitude of the local tracer T. The τ^{cen2}

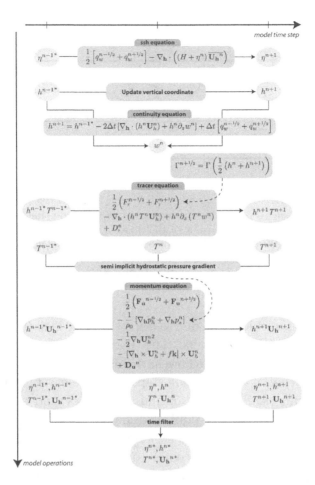

Fig. 13.3   The computational flow in NEMO taken directly from NEMO (2012) and using leap-frogging as given by Leclair and Madec (2009). The top equation labelled "ssh" stands for the sea surface height equation.

denotes the now value of T; the notation stems from a program within NEMO called "cen2". Strangely, however, TVD here denotes Total Variance Dissipation rather than the usual Total Variation Diminishing. Remember, it's not what it's *called* that matters, it's what it *is* (as the great Nobel prize winning physicist and educator Richard Feynman said) and the definition lies in the given equations. The ocean dynamics equations will not be given here as this is not the emphasis of this text, but one interesting point is that they pay heed to what has been termed the "Neptune Effect", which is the name given to how (some) bottom topography can affect the ocean circulation to an unexpected degree due to the input of vorticity. A series of papers by Greg Holloway are worth pursuing by the interested; start with Holloway (1986, 1992). The flux form of the equations, surface boundary conditions, bulk forms, atmospheric pressure and diurnal signals are all treated in minute detail, then there is river run-off and freshwater input as well as other very specific modules that cater for particular geographic areas such as modelling the Mediterranean. At open boundaries, calculations are carefully performed to reduce errors. First of all, the maximum use is made of any available data, then a radiation condition is employed to eliminate reflection. There is also sound advice on how to input the depth data so as to prevent any artificiality, and an option to force the total volume to be constant. The grid itself is restricted to those areas where there is sea; a typical mask is shown in Figure 13.4. Finally, there is a section

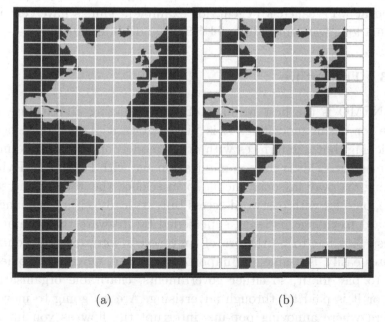

(a)　　　　　　　　　　　(b)

Fig. 13.4 The horizontal boundaries as implemented on the Atlantic Ocean, taken from a project called CLIPPER. The original mesh is cut down so that the white squares shown on the right are not used.

on the inclusion of unstructured open boundary conditions, which minimises the staggered effects present even with the finest square grids. This looks the way forward.

Mixing in the horizontal is treated carefully, but that in the vertical more so. Interestingly, NEMO seems to have gone for the $k - l$ closure rather than the more popular $k - \epsilon$ formulation (see Chapter 5). At the surface, a kind of $k - \epsilon$ method is used based on Mellor and Blumberg (2004). Physically simpler bottom friction (e.g. linear) can be implemented, but the numerical algorithms are complex to ensure efficiency, stability and accuracy.

There is a great deal of care devoted to the output of the model, both in terms of reading it correctly and forms of presentation. Finally, there are sections on comparing the output to observations and how to do this correctly and how to assimilate data into your model (data assimilation).

This ground breaking code is simply ahead of all the rivals seen by this author, though there are others along the same lines (the one based at MIT called MITgcm, for example). Things have certainly come a long way since Stommel (1948) and Munk (1950).

13.3   The Future

So, NEMO is the best we have of the all-singing all-dancing software for ocean modelling. It is biased towards larger scale ocean modelling, but it interfaces with other models and that is the modern message. This is the communication age with Facebook, Twitter leading the way in social media so in science the stand alone software, paid for with password access but, critically, either unchanging or hard to change, is being superseded by freeware that communicates. This has to be the era of easy access software that communicates easily, is usable and importantly is free. Obviously, someone has to pay for it, so either governments, charitable organisations pay or it is paid for through advertising. Are we going to move to an era where annoying pop-ups interrupt the flow as you interrogate an ocean model? Hopefully not, but there might be links to commercial sites embedded in modelling software. A small price to pay in this author's opinion. Perhaps this is all too specialist for commercial companies to be interested. However, climate change is fuelling interest in environmental issues, in particular the quality of our water and how to maintain it where it is good enough and how to improve it where it is not. As the public get more tuned to the science of modelling, commercialisation seems less far fetched. It is vital to continue to model the dissolved substances in estuarial and coastal waters and to validate such models with better data sets. Software like NEMO but focussed on water quality is therefore probably coming soon. Nor has the parameterisation of turbulence been anything like solved. The best commercial software is full of free parameters and there are big gaps in our knowledge on the interaction between stratification and turbulence. In recent years, there have been many interesting developments that were missed in the first edition of this book and must now get a mention. At first thought, stratification and turbulence seem strange bedfellows and not a little contradictory. A strongly stratified flow would seem to inhibit, not to say prevent, the development of turbulence. Not so apparently. Recent numerical and experimental research — see Brethowuer *et al.* (2007) and Waite (2013) — have made in roads. There is an ongoing research project based in the UK at Cambridge and Bristol under the guidance of Paul Linden. It seems there can be defined a stratified Reynolds number R_s (ignore the rotational effects here), see also Bartello and Tobias (2013),

$$R_s = \frac{U^3}{\nu N^2 L}$$

with U typical horizontal speed, ν eddy viscosity, N buoyancy frequency and L typical horizontal length scale. The ranges of R_s define interesting features. Existing experiments have explored values of R_s around or less than one, which means they are restricted to rather atypical regimes as far as geophysics is concerned. Waite (2013) seems to consider values around 10^2 in his general discussion and Paul Linden, in his introduction in von Larcher and Williams (2015), states that oceanic versions are around 10^3. Of course, dimensions can only be crude indicators, but experiments are hinting at bands of turbulence and quieter areas where the stratification inhibits the turbulence but in a structured way. The rather different spectral approach of Augier, Chomaz and Billant (2012) relates their work to known Kolmogorov scales as well as other known ranges. There is also evidence from their numerical simulations of how Kelvin–Helmholtz billows break down locally under gravitational instability and result in local bands of stratified turbulence. They also show how shear breaks down and the resulting instability is non-local and, therefore, probably nonlinear. This is exciting new research and will shortly, I am sure, impact the commercial software like NEMO. As our confidence in the physics increases, the software will be ever more sophisticated, and it is not hard to see an increased use in techniques such as PCA (the use of EOFs) to help us make sense of the data, then to use data assimilation techniques such as Kalman filters to improve the model and enhance the accuracy of its output. Finally, the recent models of internal waves and the understanding of Langmuir circulation and how these interact with the turbulence (see the review of James, 2002) points to possible improvements in the model physics. The importance of internal waves, (see the book by Vlasenko, Stashchuk and Hutter, 2005) has come to most modellers and ocean scientists as a surprise; no doubt there will be other surprises, and keeping abreast with the literature, particularly good reviews like James (2002) and new work as given in research texts such as von Larcher and Williams (2015), is the only way to keep abreast of them.

13.4   Exercises

This last set of exercises is focussed on asking your views on possible futures and their consequences.

(1) Speculate on areas where there are no models at the moment but might house future mathematical models. Once you have an area of future modelling, speculate also on how this interfaces with existing models.

(2) From your knowledge of NEMO, what areas could this branch into?

(3) How will an improved knowledge of stratified turbulence change established models of the sea bed boundary condition, e.g. Blumberg and Mellor (1987), and the modification due to COHERENS (see Chapter 5)

Chapter 14

Answers to and Comments
on the Exercises

14.1 Chapter 1

(1) For the Gulf Stream, for a first model the wind can be ideal, for example all zonal with a sinusoidal variation, we can use a simple square ocean, the Coriolis effect and its variation with latitude have to be present. Estuarial pollution will need tide, river characteristics, a representation of the geometry of the estuary, the diffusion properties of the pollutant. The surface oil slick will need tide and wind, properties of the oil (whether it sinks or evaporates) diffusion.

(2) Essay question. Knowledge of river flow rate and the destiny of any inputs as predicted by a model will help to know where to start the clean-up process.

(3) Most processes are continuous, so a knowledge of calculus is essential for a correct mathematical description.

(4) Pressure, salinity and temperature are scalars; wind, current and force are vectors; stress is also a vector, (although technically it is a second order tensor, having nine rather than three components).

(5) The circulation is defined by

$$\oint_C \mathbf{u} \cdot d\mathbf{s} = \oint_C \boldsymbol{\nabla}\phi \cdot d\mathbf{s} = [\phi]_C = 0$$

because the potential ϕ is a continuous function and so does not change its value when evaluated around a closed curve.

(6) If the χ^2 test is used on the ranked data:

Location	A	B	C	D	E	F
Expt. 1	6	2	1	3	4	4
Expt. 2	3	2	1	4	6	5

with either H_0 "Experiment 1 is true" or "Experiment 2 is true" then χ^2 is small enough (4.3 or 2.55) for either hypothesis to be considered true at the 99% level (using the formula for χ^2 this has to be less than 15.1, which is the case. There are five degrees of freedom here). Now it might be argued that each reading should be taken more at face value as the larger the flow rate the more important it is. Also, one suspects the first reading (0) for location A, the Pentland Firth; more likely an instrument was lost or a reading failed to materialise for some reason. This question needs to be answered with intelligence and not by automatically putting numbers into a χ^2 test.

(7) Maximum probability $p = 1/2$.

(8) The function is $\ln(\ln(p))$

14.2   Chapter 2

(1)

$$Q = k\frac{pa^4}{\mu l},$$

where k is a constant unable to be determined by dimensional analysis (it is $\pi/8$ actually).

(2) Using standard dimensional analysis,

$$MLT^{-2} = L^a(LT^{-1})^b(ML^{-3})^c(ML^{-1}T^{-1})^d.$$

Equating the powers of M, L and T yields

$$c = 1 - d \quad b = 2 - d \quad \text{and} \quad a = 2 - d$$

so

$$F = D^2U^2\left(\frac{\mu}{\rho U D}\right)^d.$$

Since the ratio $\frac{\mu}{\rho U D}$ is the Reynolds number this is just a constant, so

$$F = KD^2 U^2.$$

(3)

$$\nu \frac{\partial^2 u}{\partial z^2} \approx \nu \frac{U}{D^2} = LT^{-2} \text{ and also } fu \approx fU = LT^{-2}.$$

So the ratio is

$$E_V = \frac{\nu}{fD^2}.$$

(4) Trick question; two white noise signals have the same frequency properties, therefore the cross correlation is the same as the auto-correlation which is zero, except at the origin.

(5) Two signals in phase will have a cross correlation that is a positive constant; 90 degrees out of phase will be zero, and 180 degrees out of phase a negative constant. The mathematical answers are $1/2, 0$ and $-1/2$.

(6) They both achieve optimality by learning. A GA operates by searching a database, a NN does not; a NN operates through ever more sophisticated connectivity, a GA does not.

14.3 Chapter 3

(1)

$$\frac{\partial \mathbf{u}}{\partial t} = \mathbf{0} \text{ and } \frac{D\mathbf{u}}{Dt} = -x\mathbf{i} - y\mathbf{j}.$$

$$\nabla p = -\rho(x\mathbf{i} + y\mathbf{j}).$$

Hence,

$$\eta = z - \frac{1}{2g}(x^2 + y^2).$$

(2) Directly from

$$\frac{p}{\rho} + \frac{1}{2}u^2 - gz = \frac{P_A}{\rho},$$

where P_A is atmospheric pressure.

(3) The angular momentum of a rotating fluid element is the tangential component of fluid velocity averaged over the circumference of the circle. If the radius of this circle is a, then this is

$$\frac{1}{2\pi a} \cdot \frac{1}{a} \int_C \mathbf{u} \cdot d\mathbf{s}.$$

Now, using Stokes theorem, this is

$$\frac{1}{2\pi a^2} \int_S \nabla \times \mathbf{u} \cdot d\mathbf{S},$$

where S is the disc enclosed by the circle C. The quantity

$$\frac{1}{\pi a^2} \int_S \nabla \times \mathbf{u} \cdot d\mathbf{S}$$

is simply the vorticity, $\boldsymbol{\zeta} = \nabla \times \mathbf{u}$, hence it has been shown that vorticity is twice the local angular momentum.

(4)

$$x = \frac{1}{\omega} \sin \omega t + 4 \cos \omega t.$$

(5) This is done in Section 10.4, Chapter 10.

(6) Established in Chapter 8 as

$$\frac{1}{2} c \left(1 + \frac{2kh}{\sinh 2kh} \right).$$

(7) The wave equation is derived as

$$\frac{1}{gh} \frac{\partial^2 \eta}{\partial t^2} = \frac{\partial^2 \eta}{\partial x^2}.$$

(8) The three equations reduce to:

$$-i\omega u - fv = -gik\eta,$$
$$ilv + fu = -gil\eta,$$
$$i\omega \eta + ihku + ihlv = 0$$

and the determinant is

$$\begin{vmatrix} igk & i\omega & -f \\ igl & f & i\omega \\ i\omega & ikh & ilh \end{vmatrix} = 0,$$

which when multiplied out gives the result. When $\omega = f$, we have $k^2 + l^2 = 0$, so both wave numbers are zero and there are no plane waves.

(9) Dimensionless time would be $(\beta L)^{-1}$. This gives

$$R_o = \frac{U}{\beta L^2} \quad \text{and} \quad E_H = \frac{\nu_H}{\beta L^3}.$$

The higher power of L in both of these means greater sensitivity to it. Demanding that $E_H \approx 1$ immediately gives $L = (\nu/\beta)^{1/3}$, which is the Munk scale for the Gulf Stream (see Chapter 10). Demanding that $R_o \approx 1$ gives $L = (U/\beta)^{1/2}$, consistent with the width of a free inertial boundary layer, see Pedlosky (1986), page 298.

(10) From the equations in section 3.7, c_V must have the dimensions of energy $ML^{-2}T^{-2}$, the constant κ is in

$$c_V \frac{dT'}{dt} = \frac{\kappa}{\rho} \nabla^2 T',$$

writing T' for temperature here, so in terms of dimensions this is

$$ML^{-2}T^{-2}T^{-1} = [\kappa]M^{-1}L^3L^{-2},$$

so the dimensions of κ must be $M^2L^{-3}T^{-3}$. Q has the dimensions of work, which are energy divided by time or ML^2T^{-3}.

14.4 Chapter 4

(1) Substituting the formula for $u^s_{m,n}$ into the finite difference form of the given equation gives, after some algebra,

$$\lambda^{\Delta t} = 1 - \frac{4\kappa \Delta t}{\Delta x^2} \left(\sin^2 \frac{1}{2}m\Delta x + \sin^2 \frac{1}{2}n\Delta y \right)$$

from which

$$\frac{\kappa \Delta t}{\Delta x^2} < \frac{1}{4(\sin^2 \frac{1}{2}m\Delta x + \sin^2 \frac{1}{2}n\Delta y)} < \frac{1}{4},$$

for stability.

(2) By using the stability analysis of section 4.4, the error term

$$\epsilon_i^s = e^{\alpha s \Delta t} e^{i \pi m \Delta x}$$

satisfies the difference equation

$$\frac{\epsilon_m^{s+1} - \epsilon_m^{s-1}}{2\Delta t} + U \frac{\epsilon_{m+1}^s - \epsilon_{m-1}^s}{2\Delta x} \approx \kappa \frac{\epsilon_{m+1}^s - 2\epsilon_m^s + \epsilon_{m-1}^s}{(\Delta x)^2}.$$

As in the text, the term $\epsilon^{\alpha s \Delta t} \epsilon^{i m \pi \Delta x}$ can be cancelled and the remaining combinations of exponentials turned into trigonometric expressions. Writing

$$p = e^{s \Delta t} \quad \text{and} \quad N = iU \frac{\Delta t}{\Delta x} \sin(\pi \Delta x) + 8K \sin^2 \left(\frac{1}{2} \pi \Delta x \right)$$

and we eventually arrive at the same quadratic as in the text

$$p^2 + Np - 1 = 0$$

that has roots

$$p = -\frac{N}{2} \pm \frac{1}{2} \sqrt{N^2 + 4}$$

which, since $|N|$ is small, gives after taking the modulus (remember, only real numbers obey inequalities, and this time there is an $i = \sqrt{-1}$ in the mix)

$$|p| \approx \left| \frac{N}{2} + 1 \right|$$

to lowest order, as in the text. This time we need an extra step. As N is complex, call it $N = a + ib$ for neatness (the values of a and b are given above). Then

$$|p| \approx \left| \frac{1}{2}(a + ib) + 1 \right| = \frac{1}{2} \sqrt{(2 + a)^2 + b^2} > 1$$

so, as before, instability for any time step is assured. Putting $U = 0$ restores the problem done in Section 4.4. Allowing for the advection term makes the problem more unstable.

(3) Substituting for the finite difference form of ϕ into the telegraph equation gives

$$\frac{\phi_{i+1,j}^s - 2\phi_{i,j}^s + \phi_{i-1,j}^s}{(\Delta x)^2}$$

$$= a\frac{\phi_{i,j}^{s+2} - 2\phi_{i,j}^{s+1} + \phi_{i,j}^s}{(\Delta t)^2} + \frac{b}{\Delta t}(\phi_{i,j}^{s+1} - \phi_{i'j}^s) + c\phi_{i,j}^s.$$

So, the two parameters are

$$\frac{b\Delta x^2}{\Delta t} \quad \text{and} \quad \frac{a\Delta x^2}{\Delta t^2}.$$

As forward differences in time are elected and it is second order, *two* start values are required to commence computation. This can be seen from the presence of s, $s+1$ and $s+2$ in the finite difference formula; both $s = 0$ and $s = 1$ conditions are required.

(4) The Crank–Nicolson scheme leads to the finite difference approximations

$$\frac{\partial^2\phi}{\partial x^2} \approx \frac{\phi_{i+1,j}^s - 2\phi_{i,j}^s + \phi_{i-1,j}^s}{(\Delta x)^2},$$

$$\frac{\partial^2\phi}{\partial t^2} \approx \frac{\phi_{i,j}^{s+1} - 2\phi_{i,j}^s + \phi_{i,j}^{s-1}}{(\Delta t)^2},$$

$$\frac{\partial\phi}{\partial t} \approx \frac{\phi_{i,j}^{s+1} - \phi_{i'j}^{s-1}}{2\Delta t},$$

so the partial differential equation is defined over the six points shown in Figure 4.2. Stability is determined by substituting $\phi \approx \exp(\alpha t + ikx)$. The double meaning of the symbol i is dealt with by using m, n for the integers in the finite difference formula; but this is not a problem here as actual computation (which is lengthy) is not asked for.

(5) First of all, the u^* term is eliminated from the equations

$$u^{s+1} = u^s + \frac{\Delta t}{2}[F(u^s) + F(u^{s-1} + 2\Delta t F(u^s))].$$

Then, it is noted that this is, to the lowest approximation,

$$u^{s+1} = u^s + \frac{\Delta t}{2}[F(u^s) + F(u^{s-1})]$$

for all reasonable functions F. It is then a question of inserting the error term

$$e^{\alpha s \Delta t} e^{i\pi m \Delta x p / L}$$

into this approximation to the finite difference scheme and following the example in section 4.4.

(6) In theory, Zalesak's flux corrected transport algorithm can be applied to any system, and it does work if implemented carefully. There is an element of "how long is a piece of string" to this question. There are specialist FCT schemes that are slightly different from the scheme outlined here where fronts are better preserved, but Zalesak's general scheme works well.

(7) Finite difference schemes use only the values at grid points. Finite volumes use average values throughout a volume. Sometimes these are equated to values at the centre of mass (volume) of each element. Sometimes the methods can be equivalent, but in general the finite volume method will give a smoother solution with the same advantages in incorporating boundary conditions and local refinement as the FEM. FVM seems to have all the advantages of the FEM. For rectangular volumes, FVM and Finite Difference Methods (FDM) can be very similar.

(8) The strong form is the differential equation, say $u_t - \mathcal{L}u = 0$ together with the boundary conditions, that has a unique solution if $\mathcal{L}u$ is linear, and also solutions that can be found approximately otherwise. The weak form of the solution to a finite element problem is demanding that

$$\int_V (u_t - \mathcal{L}u)\alpha_i dV = 0$$

for all i, where $u_t - \mathcal{L}u = 0$ is the same governing set of equations. The weak form is a variational statement; it is called weak because, although the solution to the strong form will always satisfy the weak form, the reverse is not true. The strong form satisfies conservation laws everywhere, the weak form satisfies them on average over a domain. In practice, this can be the best we can do.

(9) Adaptive grids are finite approximations using nodes that can move so that the overall approximate solution is optimally good. It can concentrate nodes at places of maximum variation whilst using few nodes where there is little variation. Nested grids also concentrate nodes where there is greatest variation, but there is no reduction over places where variation is small. It is therefore a less efficient scheme and has been superseded by adaptive grids.

14.5 Chapter 5

(1) This question involves lengthy algebra. Substitution of the given expressions into the three equations, gives three algebraic equations for A, U_0 and V_0. Putting $\Delta x = \Delta y$ and denoting $e^{-\alpha \Delta t} - 1$ by $\gamma(t)$ gives the following consistency condition for the three equations

$$\begin{vmatrix} \gamma(t) & -f\Delta t & \dfrac{g}{c}\sin k\Delta x \\[2mm] f\Delta t & \gamma(t) & \dfrac{g}{c}\sin l\Delta x \\[2mm] \dfrac{h}{c} & \dfrac{h}{c} & \gamma(t) \end{vmatrix} = 0.$$

Multiplying this out and approximating for small Δt by using $\sinh \frac{1}{2}\alpha\Delta t \approx \frac{1}{2}\alpha\Delta t$, ignoring squares, and for Δx by using $\sin k\Delta x \approx k\Delta x$ gives

$$\alpha \approx \frac{k+l}{k-l}.$$

For stability, α has to be negative otherwise variables grow exponentially, and this is only true if $l > k$, which can never be assured. Hence, the scheme is always unstable.

(2) For the D grid, although all variables are evaluated at different places, they are all centred about the middle of each cell.

Therefore, the stability is as for the explicit time, centred space, i.e.

$$\frac{\kappa \Delta t}{d^2} \leq \frac{1}{2}.$$

For the E grid, u is evaluated at the corners and centre of a typical grid. The distance between these points remains d, so the stability criterion remains

$$\frac{\kappa \Delta t}{d^2} \leq \frac{1}{2}.$$

(3) The LTE are

$$\frac{\partial u}{\partial t} - fv = -g\frac{\partial \eta}{\partial x},$$

$$\frac{\partial v}{\partial t} + fu = -g\frac{\partial \eta}{\partial y},$$

$$\frac{\partial \eta}{\partial t} + \frac{\partial}{\partial x}(hu) + \frac{\partial}{\partial y}(hv) = 0.$$

One possible implementation using the Eliason scheme would be

$$\frac{u_{i,j}^{s+1} - u_{i,j}^{s-1}}{2\Delta t} - fv_{i,j}^{s+1} = -g\frac{\eta_{i+1,j}^{s} - \eta_{i-1,j}^{s}}{2\Delta x},$$

$$\frac{v_{i,j}^{s+1} - v_{i,j}^{s-1}}{2\Delta t} + fu_{i,j}^{s+1} = -g\frac{\eta_{i,j+1}^{s} - \eta_{i,j-1}^{s}}{2\Delta y},$$

$$\frac{\eta_{i,j}^{s+1} - \eta_{i,j}^{s-1}}{2\Delta t} + \frac{(hu)_{i+1,j}^{s} - (hu)_{i-1,j}^{s}}{2\Delta x} + \frac{(hv)_{i,j+1}^{s} - (hv)_{i,j-1}^{s}}{2\Delta y} = 0.$$

It is known that Coriolis at the later time enhances stability. Centred differences have been implemented throughout, but there are choices for the time levels of the space derivatives of u and v. The above choice preserves symmetry and is likely to minimise truncation error. The Fourier method could be used to ascertain whether stability has been achieved, though the algebra is likely to be heavy. It might be more practical to perform numerical experiments.

(4) As in the last question, we start with the LTE

$$\frac{\partial u}{\partial t} - fv = -g\frac{\partial \eta}{\partial x},$$

$$\frac{\partial v}{\partial t} + fu = -g\frac{\partial \eta}{\partial y},$$

$$\frac{\partial \eta}{\partial t} + \frac{\partial}{\partial x}(hu) + \frac{\partial}{\partial y}(hv) = 0.$$

The finite difference version can vary depending on the time level the right-hand sides are evaluated, however, one version is

$$u_{i,j}^{s+1} - u_{i,j}^{s-1} - 2f\Delta t v_{i,j}^{s} = -\frac{g\Delta t}{\Delta x}(\eta_{i+1,j}^{s} - \eta_{i-1,j}^{s}),$$

$$v_{i,j}^{s+1} - v_{i,j}^{s-1} + 2f\Delta t u_{i,j}^{s} = -\frac{g\Delta t}{\Delta y}(\eta_{i,j+1}^{s} - \eta_{i,j-1}^{s}),$$

$$\eta_{i,j}^{s+1} - \eta_{i,j}^{s-1} - 2H\Delta t \left[\frac{u_{i+1,j}^{s} - u_{i-1,j}^{s}}{2\Delta x} + \frac{v_{i,j+1}^{s} - v_{i,j-1}^{s}}{2\Delta y}\right] = 0.$$

(5) This is mostly bookwork; but one advantage of differences is the ease of application (no bookkeeping problems associated with labelling elements). One advantage of elements is the ability to concentrate the grid in places of interest (high variability) and conversely to use very large elements where nothing is happening.

(6) The word *kriging* refers to the use of filtering and associated interpolation and regression, which are applied to the first few EOFs of multivariate data. Two problems in using kriging in oceanography are (1) the sheer volume of the data and (2) the time dependence of the phenomena we are trying to model. Kriging can help validate a numerical model, but only if the stability properties of the numerical scheme are not in doubt.

14.6   Chapter 6

(1) Take the first equation, multiply it by dx and add it to the second multiplied by dy to obtain

$$g\frac{\partial \eta}{\partial x}dx + g\frac{\partial \eta}{\partial y}dy = -\frac{\partial \Omega}{\partial x}dx - \frac{\partial \Omega}{\partial y}dy - \frac{1}{\rho}\frac{\partial p_a}{\partial x}dx - \frac{1}{\rho}\frac{\partial p_a}{\partial y}dy.$$

So, using that

$$dF = \frac{\partial F}{\partial x}dx + \frac{\partial F}{\partial y}dy$$

for any reasonable function $F = F(x, y)$ yields

$$gd\eta = -d\Omega - \frac{1}{\rho}dp_a,$$

integrating gives

$$g\eta = -\Omega - \frac{p_a}{\rho} + K,$$

where K is a constant of integration. So, to answer (a), for pure tide

$$g\eta = -\Omega + K$$

for (b)

$$p_a = -\rho g\eta + K,$$

which for a point z below the sea surface becomes

$$p = p_a + \rho g(\eta - z),$$

which is the standard hydrostatic balance (z points down here).

(2) The Kelvin wave has equation

$$\eta = C_1 e^{-fx/\sqrt{gh}} \cos\left(\frac{\omega y}{\sqrt{gh}} + \omega t + \phi_1\right).$$

The exponential term has a factor $-fx/\sqrt{gh}$ so, in order to assume the value of around unity (order of magnitude), then

$$x \approx \frac{\sqrt{gh}}{f} = a.$$

The length a is called the barotropic radius of deformation or, in other books, the Rossby radius of deformation. For the North Sea, the values $h = 200$ m, $g = 10$ m s^{-2} and $f = 1.2 \times 10^{-4}$ s^{-1}. Giving a value $a \approx 40$ km. This is why the superposition of Kelvin waves works as a reasonable model for the northern North Sea whose width is ten times this distance, so the UK waves travelling south interfere very little with the Norwegian wave travelling north.

(3) A "cold" start means all variables are set to zero. The tide is imposed from an open boundary and the model is allowed to settle down over ten days or so. This is done principally because it works best. Starting from some kind of guess values will work equally well but inputting any values other than zero into all the matrices is more difficult and not worth doing. Also, tide does propagate from the neighbouring ocean via the open boundaries, so starting a model this way mirrors reality.

(4) The wind is towards the north east, therefore the surface Ekman flow will be 45 degrees to the right of this, i.e. easterly. The surface current can be deduced from the formula

$$V = \frac{\pi}{\rho D_0 |f|} \sqrt{2(\tau^x)^2 + 2(\tau^y)^2} = \frac{\pi}{\rho D_0 |f|} \tau \sqrt{2}.$$

Inserting the values given in the question, we get $V \approx 0.3$ m s^{-1}. The net flux is given by τ/f and is 2.68×10^3 Ns. The density is not required as *flux* denotes mass transport, so the second of the definitions of flux, whereby

$$(U, V) = \int_{-\infty}^{0} (\rho u, \rho v) dz = \left(\frac{\tau^y}{f}, -\frac{\tau^x}{f} \right),$$

applies.

(5) Storm surges are, by and large, well represented by models, however, short term fluctuations (see Figure 6.10) tend to be smoothed out by linear models. Storm surges as a name can also be (mis)used by the media, in particular to describe local flooding at coasts caused by the interaction of a flooding river with the incoming tide. These are sometimes not well predicted as floods in terms of both quantity and timing can be hard to get right.

(6) Using the data, there are 150 km for the tsunami to travel over the deep 3 km water then 50 km to travel over the shallower 200 km. The speed everywhere is given by $\sqrt{gh}$ and so over the deep water it is $\sqrt{30000} = 173$ m s^{-1}, so the time taken is $150000/173$ s $= 867$ s. For the shallower water, the speed is the much slower $\sqrt{2000} = 44.7$ m s^{-1}, so the time taken is the longer $50000/44.7 = 1118$ s. This gives the total time as

1985 s = 33 mins. Comments: the huge height stemming from the deep water wave "catching up" the slower part over the continental shelf means that the depth over the continental shelf will be slightly deeper than 200 m, perhaps 230 m. This corrected value increases the speed to 47.9 m s^{-1} leading to the second time of 1044 s, so the tsunami arrives 74 seconds earlier. This is not significant, and stems from the square root dependence of the time on wave height. The precise forecasting of the actual height of the tsunami is hard and unnecessary; it is the timing that is vital as the earthquake itself will probably be a surprise. 33 minutes is not long for a full scale evacuation.

(7) Using Merian's formula with the numbers given gives a period of 4.5 hours, which is a 3% error.

(8) For a harbour, one end is open, so instead of nodes at both ends, there will be a node at the closed end and an antinode at the open end. Thus, the length of the harbour will be half the wavelength of the seiche. Hence, Merian's formula will be

$$\frac{4a}{\sqrt{gh}}.$$

14.7   Chapter 7

(1) The proportions are found by using

$$\alpha 35 + \beta 32 = 34.5 \quad \text{where } \alpha + \beta = 1.$$

Hence, α (the proportion of S_1) is 0.83 and β (the proportion of S_2) is 0.17.

(2) This follows from the linearity of the Fickian diffusion model — even the advection parts are pseudo linear as it is only the *mean* of the flow that multiplies the spatial gradients in contaminant. If the pollutants interact, this will no longer be the case and the actual interaction will have to be incorporated into the modelling. If the reaction time is shorter than the diffusion time, then the new substance so formed can be modelled using a third diffusion co-efficient appropriate to this new pollutant.

(3) Qualitatively, homogeneous turbulence in two dimensions leads to diffusion on a t^2 timescale whereas three dimensions lead to diffusion on a t^3 time scale. In the ocean, the third dimension is restricted by the sea bed, the sea surface and the thermocline, therefore a $t^{2.3}$ law seems reasonable. As the new ideas on stratified turbulence emerge, there may be a more precise interpretation — see the introduction by Paul Linden to the book by von Larcher and Williams (2015).

(4) The details of the calculation are omitted, but

$$F(z) = \int_0^z \int_0^b (u - \langle u \rangle) d\zeta \, db = \int_0^z \int_0^b \frac{1}{2}(1 - 3\zeta^2) d\zeta \, db$$

and performing the integration gives

$$F(z) = \frac{1}{4}\langle u \rangle (2z^2 - z^4).$$

So it is the mean value of $(u - \langle u \rangle)F(z)$ that needs evaluating and this gives rise to the integral

$$\frac{1}{8}\langle u \rangle \int_0^1 (2z^2 - z^4)(1 - 3z^2) dz$$

and this computes to $-4\langle u \rangle / 105$. This gives the result.

(5) In a turbulent fluid, the transfer of energy per unit volume is

$$\rho \nu_z \left(\frac{\partial u}{\partial z}\right)^2.$$

The time rate of increase of potential energy will be $g\rho w$, which when averaged over turbulent fluctuations, is

$$g\langle \rho' w' \rangle = -g\kappa_z \frac{\partial \rho_0}{\partial z}.$$

The ratio of these terms is the flux Richardson number, R_f, which if less than one means that kinetic is larger than potential turbulent energy and the turbulence will persist (stability). Lewis (1997) says that usually R_f is less than about 0.2. The inequality $R_i < 1$ is necessary for overall stability and so, since

$$R_f = R_i \frac{\kappa_z}{\nu_z} < 1 \quad \text{and} \quad R_i < 1,$$

it has to be the case that $R_i < \nu_z / \kappa_z$. There is more discussion of this in Lewis (1997), page 78.

(6) These models are a useful adjunct to understanding. Problems include: the accurate modelling of the front, although there are good schemes now that seem to work; the presence of shear and density gradients over a variety of length scales are still difficult to model with confidence as some of the underlying processes are still poorly understood. Such a model can be useful to an engineering company that has to assure regulating bodies that production and processing are not harmful to the environment, provided that the model is validated and accepted as accurate.

14.8   Chapter 8

(1) This is routine integration.

(2) As

$$w(k) = w(k_0) + \frac{dw}{dk}(k - k_0) + \cdots$$

we have that

$$\frac{w(k) - w(k_0)}{k - k_0} = \frac{dw}{dk} + \cdots .$$

So, since

$$e^{wt - kx} = e^{(w - w_0)t - (k - k_0)x} e^{w_0 - k_0}$$

it is the ratio

$$\frac{w(k) - w(k_0)}{k - k_0} .$$

that is the speed of travel of the wave group as w approaches w_0 (simultaneously k approaches k_0, of course, as w is a continuous and differentiable function of k). Hence, the group velocity is dw/dk.

(3) The SMB and PNJ methods are very different; the PNJ method is based on fixed relations and takes the form of using charts and reading off wave heights given wind. The SMB method is based upon spectra and can change as the knowledge of these spectra change. This is the reason for its longevity; it is adaptable, whereas the PNJ method is not.

(4) Directional wave theories are less important for coastal regions simply because, due to Snell's law, the waves line themselves to the coast, rendering the spectrum largely one dimensional. Longshore waves are different in structure and are analysed separately.

(5) When viscosity is included in the model, the boundary conditions change from slip (normal flow is zero) to no flow adjacent to the boundary. The viscous terms in the equations contain more derivatives so extra integration is involved in solving them. If the viscosity is allowed to shrink to zero, the problem does not revert to the inviscid problem (mathematicians say that there is a singularity at the boundary — in reality, the gradients in the flow get extremely large). Hence, inviscid interior mass transport is generated by these boundary driven viscous conditions. Books on boundary layer theory (e.g. van Dyke, 1964) say a lot more.

(6) The equation for the amplitude of the edge wave is

$$(hF')' + \left(\frac{\omega^2}{g} - hl^2\right) F = 0$$

so, with $h(x) = \alpha x$,

$$(xF')' + \left(\frac{\omega^2}{g\alpha} - l^2 x\right) F = 0.$$

Now write

$$F(x) = e^{-lx} L(2lx).$$

Differentiating gives

$$F' = -le^{-lx} L + 2le^{-lx} L',$$

where the dash represents differentiation with respect to $2lx$. So

$$(xF')' = -le^{-lx} L + l^2 xe^{-lx} L - 2l^2 xe^{-lx} L' \\ + 2le^{-lx} L' - 2xl^2 e^{-lx} L' + 4xl^2 e^{-lx} L''$$

or

$$(xF')' - l^2 xF = -le^{-lx} L - 4l^2 xe^{-lx} L' \\ + 2le^{-lx} L' + 4xl^2 e^{-lx} L''.$$

Rearranging this gives

$$(xF')' + \left(\frac{\omega^2}{g\alpha} - l^2 x\right) F = 2le^{-lx}[2lxL'' + (1 - 2lx)L']$$

$$+ e^{-lx}\left(\frac{\omega^2}{g\alpha} - l\right)L = 0,$$

which on division by $2l$ is Laguerre's equation

$$XL'' + (1 - X)L' + \gamma L = 0,$$

where $X = 2lx$ and

$$\gamma = \frac{1}{2}\left(\frac{\omega^2}{gl\alpha} - 1\right),$$

as required.

(7) The PM spectrum is for a fully developed sea and its analytical form is consistent with that required using dimensional analysis for high frequency and decays exponentially for small ω. The JONSWAP spectrum preserves these aspects, but also is more peaked in the centre as the spectrum is not fully developed and these waves have not had the length of time or fetch to be anything other than near sinusoidal. In order to match these not-fully-developed conditions there are more free parameters that have to be matched to different seas. The JONSWAP spectrum is not popular in the USA, for example.

(8) The calculation is straightforward using $p = 1/60$ and the formula under Figure 8.7, viz. $\ln(\ln(1/p)) = 0.945\ln(H_s - 0.95) + 0.278$. The one year return period ($p = 1/60$) gives the wave $H_s = 4.26$ m. The 50 year return period ($p = 1/3000$) gives $H_s = 7.68$ m and the 100 year return period ($p = 1/6000$) gives $H_s = 8.30$ m.

(9) If

$$u \approx u_*\left(\frac{z}{h}\right)^\alpha,$$

then firstly the dimensions are correct and secondly one has a free parameter α to fit to data. As so often a log–log plot is used to fit experimental data, the parameter α is simply the slope of the regression line fit. (In reverse, suppose $y = x^k$, then taking logs gives $\ln y = k\ln x$, so plotting $\ln y$ against $\ln x$ gives a straight line of slope k.)

(10) GENESIS uses the bulk equations of conservation and a great deal of empiricism. MIKE 21 is securely based on current finite difference methods using discretised equations of motion and continuity in flux form.

14.9 Chapter 9

(1) The principal properties of the Michaelis–Menten relationship are: (1) No growth when $N = 0$,

(2)
$$\frac{dN}{dt} \to \alpha \ (\text{constant})$$

for large N, and

(3)
$$\frac{dN}{dt} = \frac{1}{2}\alpha \ \text{ at } N = k,$$

so in a sense $N = k$ is analogous to the "half-life" of the nutrient N.

(2) Inserting the values $P = 89$ at $t = 10$ and $P = 103.64$ at $t = 20$ and solving the simultaneous equations gives $a = 0.0193$ and $b = 3.59 \times 10^{-7}$. This very small value of b means that early on the model is virtually linear. Using the iterative formula (based on Euler's method)

$$P(t + \Delta t) = P(t) + \Delta t(aP(t) - bP^3(t)),$$

gives the following table:

Year	t	$P(t)$ (approx)	$P(t)$ (measured)
1900	0	76.10	76.1
1910	10	89.00	92.4
1920	20	103.64	106.5
1930	30	119.64	123.1
1940	40	136.59	132.6
1950	50	153.80	152.3
1960	60	170.42	180.7
1970	70	185.55	204.9
1980	80	198.42	226.5
1990	90	208.67	259.6
2000	100	216.32	281.4
2010	110	221.73	308.7
2020	120	225.39	?

The fact that the nonlinear term increases as the cube of P rather than the square of P also means less growth because of the negative sign in the formula for dP/dt. The steady state value for P is given by setting the right-hand side of

$$\frac{dP}{dt} = aP - bP^3$$

to zero and so is $P = \sqrt{a/b} = 231.86$, which is a lot less than the model in the text. This model is inferior as it does not model the limitation correctly according to the data. However, the model in the text was very accurate by pure chance.

(3) Adding the equations gives

$$\frac{d}{dt}(N + P + Z) = 0,$$

hence $N + P + Z = $ constant. This is a constraint and means that if the nutrient has a given value as P increases, Z decreases and vice versa. Solving for zero right-hand side

$$b_2 PZ - dZ = 0,$$
$$aNP - bPZ = 0,$$
$$-aNP + b_1 PZ + dZ = 0,$$

which gives the non-zero solution $P = Z = d/b_2$ and $N = bd/ab_2$. This solution tends to be the one numerical methods pick out as the one the variables edge towards as time progresses. The more trivial solution is: $P = Z = 0$ and N can be any value. This corresponds to no plants and animals at all, just an arbitrary amount of nutrient.

(4) For this model, $x = x_0, y = 0$ is perfectly allowable and corresponds to no predator and a fixed amount of prey. Nutrient is not in the model (it only has two variables) so one assumes there are enough nutrients to keep the prey fed. The state of peaceful coexistence is given by setting the left-hand side to zero, which implies

$$x = \frac{c}{b_2}, \quad y = \frac{a}{b_2}\left(1 - \frac{c}{x_0 b_2}\right).$$

If the model is started with some arbitrary positive values of x and y, then they oscillate and tend to this state of peaceful coexistence. The "phase portrait" (plot of x against y) is a spiral with this point in the centre. The fixed point is one of stable equilibrium.

(5) The food web depicted in Figure 9.6 will apply to a wider class of ecosystem problems, but the details of the model presented will be different. For example, the sediment interaction for mud or sand will differ; the water may be more or less nutrient rich, and the chemistry and biology of how nutrient is absorbed will differ with climate and other aspects of the environment. This is a complex question about which a lot more can be written.

(6) The obvious next step to ERSEM and POLCOMS is to improve the physics as any animal and plant migration is crucially dependent on a correct representation of local currents.

(7) The main worry is that the particular organ that is modelled in this way may not always be an exemplar for the entire animal. It is for the marine mussel, but modelling fish or marine mammals is more of a challenge and more internal biology may have to be modelled in order for these kinds of models to be useful.

14.10 Chapter 10

(1) The derivation of the Sverdrup balance follows by obeying the instructions in the question. Since

$$\frac{\partial w}{\partial z} = -\frac{\partial u}{\partial x} - \frac{\partial v}{\partial y},$$

the divergence produces a negative value of

$$\frac{\partial w}{\partial z} = -9 \times 10^{-9}.$$

Inserting the rest of the figures into the formula

$$v = \frac{f}{\beta} \frac{\partial w}{\partial z}$$

gives $v = -10$ cm s^{-1}. This is a southerly flow over the interior of the ocean, and represents the "return flow" for the Gulf Stream or Kuroshio, although the reality is a little more complex.

(2) This needs breaking down. The first term is

$$\frac{\partial}{\partial x}\left(\frac{Dv}{Dt}\right) = \frac{\partial}{\partial x}\left(\frac{\partial v}{\partial t} + u\frac{\partial v}{\partial x} + v\frac{\partial v}{\partial y}\right)$$

$$= \frac{\partial}{\partial t}\left(\frac{\partial v}{\partial x}\right) + \frac{\partial}{\partial x}\left(u\frac{\partial v}{\partial x}\right) + \frac{\partial}{\partial x}\left(v\frac{\partial v}{\partial y}\right)$$

$$= \frac{\partial}{\partial t}\left(\frac{\partial v}{\partial x}\right) + \frac{\partial u}{\partial x}\frac{\partial v}{\partial x} + u\frac{\partial^2 v}{\partial x^2} + \frac{\partial v}{\partial x}\frac{\partial v}{\partial y} + v\frac{\partial^2 v}{\partial x \partial y}.$$

The second term is

$$\frac{\partial}{\partial y}\left(\frac{Du}{Dt}\right) = \frac{\partial}{\partial y}\left(\frac{\partial u}{\partial t} + u\frac{\partial u}{\partial x} + v\frac{\partial u}{\partial y}\right)$$

$$= \frac{\partial}{\partial t}\left(\frac{\partial u}{\partial y}\right) + \frac{\partial}{\partial y}\left(u\frac{\partial u}{\partial x}\right) + \frac{\partial}{\partial y}\left(v\frac{\partial u}{\partial y}\right)$$

$$= \frac{\partial}{\partial t}\left(\frac{\partial u}{\partial y}\right) + \frac{\partial u}{\partial y}\frac{\partial u}{\partial x} + u\frac{\partial^2 u}{\partial x \partial y} + \frac{\partial v}{\partial y}\frac{\partial u}{\partial y} + v\frac{\partial^2 u}{\partial y^2}.$$

Subtracting these gives

$$\frac{\partial}{\partial x}\frac{Dv}{Dt} - \frac{\partial}{\partial y}\frac{Du}{Dt} = \frac{\partial}{\partial t}\left(\frac{\partial v}{\partial x} - \frac{\partial u}{\partial y}\right) + [\text{eight terms}]$$

and the eight terms referred to are

$$\frac{\partial u}{\partial x}\frac{\partial v}{\partial x} + u\frac{\partial^2 v}{\partial x^2} + \frac{\partial v}{\partial x}\frac{\partial v}{\partial y} + v\frac{\partial^2 v}{\partial x \partial y} - \frac{\partial u}{\partial y}\frac{\partial u}{\partial x} - u\frac{\partial^2 u}{\partial x \partial y} - \frac{\partial v}{\partial y}\frac{\partial u}{\partial y} - v\frac{\partial^2 u}{\partial y^2}.$$

Rearranged these are

$$\left(\frac{\partial u}{\partial x} + \frac{\partial v}{\partial y}\right)\left(\frac{\partial v}{\partial x} - \frac{\partial u}{\partial y}\right) + \left(u\frac{\partial}{\partial x} + v\frac{\partial}{\partial y}\right)\left(\frac{\partial v}{\partial x} - \frac{\partial u}{\partial y}\right)$$

that, when combined with the pervious equations, give the required result.

(3) The conservation of planetary vorticity equation is

$$\frac{D}{Dt}\left(\frac{\zeta + f}{H}\right) = 0,$$

which is

$$\frac{D}{Dt}(\zeta + f) = 0$$

for constant H. Expanding the total derivative gives

$$\left(\frac{\partial}{\partial t} + u\frac{\partial}{\partial x} + v\frac{\partial}{\partial y}\right)(\zeta + f) = 0;$$

neglecting nonlinear terms and remembering that $f = f_0 + \beta y$, the β-plane approximation gives

$$\frac{\partial \zeta}{\partial t} + v\frac{\partial f}{\partial y} = 0.$$

Now

$$\zeta = \frac{\partial v}{\partial x} - \frac{\partial u}{\partial y} = -\nabla^2 \psi,$$

where

$$u = \frac{\partial \psi}{\partial y} \quad \text{and} \quad v = -\frac{\partial \psi}{\partial x}$$

and $\beta = \frac{\partial f}{\partial y}$, so

$$-\frac{\partial}{\partial t}\nabla^2 \psi - \beta\frac{\partial \psi}{\partial x} = 0$$

which, multiplying by -1, is the classic Rossby wave equation.

(4) Substituting the wave into the (free) Rossby wave equation gives

$$k^2 + l^2 + \beta\frac{k}{\omega} = 0,$$

which implies

$$k = -\frac{\omega}{\beta}(k^2 + l^2).$$

Everything on the right of the minus sign is positive, therefore $k < 0$ and the wave has a westerly component.

(5) Adaptive grids will concentrate nodes where there are gradients in current. If the algorithm had to take account of an increased number of nodes to trigger closer nodes, the mechanism for adaptation would (probably) be more sluggish. One suspects, therefore, that to have a higher order scheme would act against the advantages of adaptive grids and should thus be avoided.

(6) Climate change models are models of long term changes. Adaptive grids are used for modelling flows (or, in general, *fields*) that exhibit high structure. Some climate change models do this, albeit over long time scales. Therefore adaptive grids could be used, although it does depend on what field is being modelled.

14.11   Chapter 11

(1) Estuaries are not one-dimensional (like the rivers that flow into them); they have the sea on the other side, so there is a salt–fresh water interface. They are (probably) tidal. All estuaries are different from one another.

(2) (a) Box models are useful for assessing overall balances and getting an overall picture of the long term circulation. The flushing time is the difference between the time a particle enters an estuary and the time it leaves the estuary and is an indicator of the ability of an estuary to clean itself of pollution. Box models can provide a rough estimate of such parameters. (b) Dimensionless numbers (Froude number, Richardson number) provide a global classification of an estuary. They will help the modeller to know whether or not it is stable to disturbances, whether it can support a tidal bore, but no detail. They can also provide estimates of terms in the equations and can help the modeller to decide what terms to omit and what terms to include. (c) Analytical models help in the understanding of the dynamic processes of an estuary, but do not give details of specific currents, elevations or density. They can be very helpful to modellers by going further than dimensional analysis in the assessing in both a spatial and temporal manner where and when certain variables are important. (d) For real answers, however, after these three approaches,

a numerical model usually has to be built. Numerical models give accurate answers to estuary specific questions.

(3) Non-dimensionalising, as suggested,

$$\frac{U^2}{L}\left[u\frac{\partial u}{\partial x} + v\frac{\partial u}{\partial y}\right] = \frac{\nu U}{L^2}\frac{\partial^2 u}{\partial x^2} + kg\Delta T$$

dividing by U^2/L throughout gives

$$u\frac{\partial u}{\partial x} + v\frac{\partial u}{\partial y} = \frac{\nu L}{U}\frac{\partial^2 u}{\partial x^2} + \frac{L}{U^2}kg\Delta T.$$

The combination $Re = UL/\nu$ is the Reynolds number, so this equation is

$$u\frac{\partial u}{\partial x} + v\frac{\partial u}{\partial y} = \frac{1}{Re}\frac{\partial^2 u}{\partial x^2} + \frac{1}{Re^2}\frac{L^3}{\nu^2}kg\Delta T.$$

The combination

$$Gr = \frac{L^3}{\nu^2}kg\Delta T$$

is the Grashof number, and this multiplies the thermal term. However, the Reynolds number is also important as the combination is Gr/Re^2. It is not unusual for Re to be constant for a given type of problem or regime, so it is the magnitude of Gr that governs the importance of the thermal term.

(4) This is a sort of trick question as the Grashof number is not often important by itself in an estuarial model. However, it can be important locally in estuaries linked with models of sedimentation.

(5) The advantages are: An accurate-as-can-be model of a particular estuary, dynamics on many scales, most (every?) mechanism is included in the model. Disadvantages: Can be over complex and mask understanding of underlying dynamics. There are aspects it cannot model well, like the formation and migration of fronts. (These seem to be absent from this particular inlet.) Expensive modelling, and some of it might be unnecessary even though the structured grid is efficient. If there is one variable that does not need resolving (e.g. stratification in the Yellow Sea) whereas

the velocity does, the grid cannot vary between variables. These, however, are quibbles; it's a very good model.

(6) This is book work. Langmuir circulation is the name given to helical roll vortices that occur near the surface of the sea, see Figure 11.11. They are formed by the interaction between the Stokes drift of the surface waves and the wind-driven current.

(7) The next step in the Firth of Forth model would probably be to include diffusion. Sea bed friction in particular. Interface friction is probably less important.

14.12   Chapter 12

(1) Two effects that help to trap waves are the Coriolis acceleration (see, for example, Kelvin waves) and changes in depth (see continental shelf waves). Stratification hinders trapping by allowing baroclinic modes that leak energy out from the coast.

(2) Substituting the formula

$$H = H_0 e^{-2\lambda x}$$

gives

$$\frac{1}{2H} \frac{dH}{dx} = -\lambda,$$

so the equation

$$\frac{d^2\phi}{dx^2} + \left\{ \frac{d}{dx} \left(\frac{1}{2H} \frac{dH}{dx} \right) - \left(\frac{1}{2H} \frac{dH}{dx} \right)^2 - l^2 - \frac{fl}{\omega} \frac{1}{H} \frac{dH}{dx} \right\} \phi = 0$$

becomes the much simpler

$$\frac{d^2\phi}{dx^2} + \left[-\lambda^2 - l^2 + \frac{2fl\lambda}{\omega} \right] \phi = 0.$$

So, assuming the solution

$$\phi = e^{ikx}$$

so that

$$\frac{d^2\phi}{dx^2} = -k^2\phi,$$

gives

$$-k^2 - \lambda^2 - l^2 + \frac{2fl\lambda}{\omega} = 0$$

or

$$\omega = \frac{2fl\lambda}{\lambda^2 + k^2 + l^2},$$

as in the text. Given that

$$\psi(x, y, t) = e^{-\lambda x} \sin kx \cos(ly - \omega t),$$

straightforward differentiation gives the velocity (u, v) as

$$\left(\frac{l}{H_0} e^{\lambda x} \sin kx \sin(ly - \omega t), \frac{l}{H_0} e^{\lambda x} \cos kx \cos(ly - \omega t) \right).$$

This decays exponentially for large $|x|$ (remember x is negative) and also $H \to \infty$, so the solution is idealistic.

(3) The horizontal derivatives will have dimensions pfL^{-2} and the other terms have dimensions $f^2 H^{-2} N_0^{-2} pf$, where p is pressure which still has dimensions, of course. Comparing these two expressions gives

$$S = \left(\frac{N_0 H}{fL} \right)^2.$$

As $S \to 0$ this is, in effect,

$$\frac{\partial}{\partial z} \left(\frac{1}{N^2} \frac{\partial^2 p}{\partial z \partial t} \right) = \rho_0 \frac{\partial w}{\partial z} = 0$$

using the energy equation. So, w does not vary with depth and the situation is barotropic.

(4) Starting with the given equation

$$\frac{(\omega^2 - f^2)}{g} \eta + \left[\frac{\partial}{\partial x} \left(H \frac{\partial \eta}{\partial x} \right) \right] + \left[\frac{\partial}{\partial y} \left(H \frac{\partial \eta}{\partial y} \right) \right] + \frac{if}{\omega} \left| \frac{\partial(H, \eta)}{\partial(x, y)} \right| = 0,$$

noting that

$$\eta(x, y, t) = F(x) e^{ily - i\omega t},$$

$$\frac{\partial \eta}{\partial y} = ily, \quad \text{so} \quad \frac{\partial}{\partial y} \left(H \frac{\partial \eta}{\partial y} \right) = -Hl^2 \eta$$

and

$$\frac{if}{\omega}\left|\frac{\partial(H,\eta)}{\partial(x,y)}\right| = \frac{if}{\omega}\left(\frac{\partial H}{\partial x}\frac{\partial \eta}{\partial y} - \frac{\partial H}{\partial y}\frac{\partial \eta}{\partial x}\right) = -\frac{fl\eta H'}{\omega},$$

as H does not depend upon y. Cancelling the exponential factor thus gives the required equation

$$\frac{d}{dx}\left(H\frac{dF}{dx}\right) + \left(\frac{lf}{\omega}\frac{dH}{dx} + \frac{\omega^2 - f^2}{g} - l^2 H\right)F = 0.$$

Before answering the specific questions, it is worth noting that the last two terms on the left-hand side inside the parentheses,

$$\frac{\omega^2 - f^2}{g} - l^2 H$$

represent gravity waves modified by rotation (f).

(a) If $lf > 0$, then in addition to the gravity waves there is a Kelvin wave (see Chapter 6) trapped against the coast with maximum elevation at the coast itself but decay offshore without reaching zero.

(b) If $lf < 0$, the solutions are edge waves, which were discussed briefly in Chapter 8, section 8.3. Edge waves are a general term for waves that travel parallel to the coast; they technically include Kelvin waves but are usually restricted to smaller scale waves that have offshore nodes and are responsible for the longshore transportation of sediment (see Chapter 8). It is the term

$$\frac{lf}{\omega}\frac{dH}{dx}$$

that is responsible for classical continental shelf waves that depend not on gravity but vorticity conservation for propagation. They are meanders in current up and down a continental slope.

(c) With no slope, there are no continental shelf waves and no edge waves, only pure inertia-gravity waves remain. These are gravity waves that are large enough to be influenced by the Earth's rotation.

(d) the expression $\frac{\omega^2 - f^2}{g} - l^2 H$ governs whether or not the waves are trapped, if the expression is positive, then trapping is possible. If it is negative it is not.

(5) The extra condition that can be imposed is that

$$F(r, \theta) = F(r, \theta + 2\pi),$$

where $F(r, \theta)$ is any variable such as surface elevation or current. The main advantage is that all variables have to be periodic in θ, which enables analytical progress and compensates a little for the equations in cylindrical polar co-ordinates being more awkward to solve than the equivalent equations in Cartesian co-ordinates.

(6) The presence of r does change the trapping criteria, at least in theory. As mentioned in the text, it is theoretically possible for waves to be trapped by the rapid increase of H with r. If this rate is greater than $1/r^2$, which is a parabolic profile, then trapping is assured. However, on a practical level, the criteria for trapping are not very different from the straight coastline when considering depth changes offshore. The smaller the island, the tighter the curvature and the less significant Coriolis effects. Therefore, for small enough islands there is no trapping, simply wave diffraction, explored a while ago by the civil engineering community under the heading of the diffraction of waves by vertical piles and applied to offshore piers and other cylindrical structures. There are some other differences due to the curvature of the coast that have not been explored, and cannot be explored with models of exactly circular islands. If the curvature changes as waves travel along the coast, it is entirely possible for abrupt changes such as capes to "throw the waves off", rather like a wake. Numerical experiments seem to indicate this, though definitive theoretical criteria remain the subject of current research. The Coriolis effect always promotes trapping.

(7) With mode $n = 0$, the velocity components in complex form are

$$u = u(r, t) = U(r)e^{-i\omega t}; \quad v = v(r, t) = V(r)e^{-i\omega t},$$

where now, of course, $U(r)$ and $V(r)$ can be complex (in fact $V(r)$ is stubbornly real, but $U(r)$ is pure imaginary). Using the results

$$u = \frac{g\omega}{\omega^2 - f^2} \frac{\partial \eta}{\partial r},$$

$$v = -\frac{gf}{\omega^2 - f^2} \frac{\partial \eta}{\partial r}$$

in the definition of Stokes drift

$$\mathbf{U}_S = (U_S, V_S) = \overline{\left(\int \mathbf{u}dt \cdot \nabla\right)\mathbf{u}}$$

or

$$\mathbf{U}_S = (U_S, V_S) = \overline{\frac{1}{-i\omega}\left(u\frac{\partial u}{\partial r}, u\frac{\partial v}{\partial r}\right)}$$

gives

$$\mathbf{U}_S = \frac{1}{-i\omega}\frac{g^2}{(\omega^2 - f^2)^2}(-\omega^2\eta_r\eta_{rr}, -i\omega f\eta_r\eta_{rr})$$

or

$$\mathbf{U}_S = \frac{1}{\omega}\frac{g^2}{(\omega^2 - f^2)^2}(-i\omega^2\eta_r\eta_{rr}, \omega f\eta_r\eta_{rr})$$

using the suffix derivative notation as defined in the text. We now use the rule for time averaging complex quantities and note two things, firstly that the first component of Stokes drift U_S is entirely imaginary and so has no real part, so must be zero. Secondly, that using the relation

$$\Re\{z_1\}\Re\{z_2\} = \Re\left\{\frac{1}{2}z_1^*z_2\right\}$$

results in

$$\overline{\eta_r\eta_{rr}} = \Re\left\{\frac{1}{2}\eta_r^*\eta_{rr}\right\} = \frac{1}{2}F_rF_{rr}$$

and so

$$V_S = \frac{fg^2F_rF_{rr}}{(\omega^2 - f^2)^2},$$

precisely the same as the result derived in the text using real quantities.

14.13   Chapter 13

(1) This is a completely open question; however, one area might be flooding. There are no models at present for immediate forecasts

of flooding due to sudden rainfall, like the one that occurred in Cornwall, UK in 2004, devastating the village of Boscastle. These kind of immediate models may well use methods such as ARMA or ARMAX (auto regressive moving average without or with exogenous input). These are time series-based models that are extremely data hungry, but could be designed to respond to very short term input and data assimilation. Interfacing these with existing models could be difficult.

(2) Taking the lead from the previous answer, perhaps more statistically based modelling.

(3) As the knowledge of stratified turbulence increases, it will be essential for these to replace the more empirically based aspects of the $k - \epsilon$ model. It is not possible at the moment to give any detail on how this might develop.

Bibliography

Accad, Y. and Pekeris, C. L. (1978) Solution of the tidal equations for the M_2 and S_2 tides in the world oceans from a knowledge of the tidal potential alone, *Sect. A Philos. Trans. R. Phil. Soc. London*, **A290**, 235–266.

Acheson, D. J. (1990) *Elementary Fluid Dynamics*, Oxford University Press, Clarendon Press, Oxford, U.K., pp. 397.

Aliabadi, M. H. (2002) *The Boundary Element Method; Part 2, Applications in Solids and Structures*, John Wiley & Sons, Hoboken, New Jersey, USA, pp. 580.

Asselin, R. (1972) Frequency filter for time integrations, *Mon. Weather Rev.*, **100**, 487–490.

Augier, P., Chomaz, J.-M. and Billant, P. (2012) Spectral analysis of the transition to turbulence from a dipole in stratified fluids, *J. Fluid Mech.*, **713**, 86–108.

Backhaus, J. O. (1982) A semi-implicit scheme for the shallow water equations for application to shelf sea modelling, *Cont. Shelf Res.*, **2**(4), 343–354.

Baines, M. J. (1994) *Moving Finite Elements*, Oxford Scientific Publications, Oxford University Press Clarendon Press, Oxford, UK.

Baines, P. G. (1974) The generation of internal tides over steep continental slopes, *Phil. Trans. R. Soc. London*, **A277**, 27–58.

Baines, P. G. (1995) *Topographic Effects in Stratified Flows*, Cambridge University Press, Cambridge, CB2 8BS, United Kingdom, 482.

Banks, J. E. (1974) A mathematical model of a river-shallow sea system used to investigate tide, surge and their interaction in the Thames-southern North Sea region, *Phil. Trans. R. Soc.*, **A275**, 567–609.

Baretta, J. W., Ebenöh, W. Ruardij, P. (1995) The European regional seas ecosystems model, a complex marine ecosystems model, *Neth. J. Sea Res.*, **33**, 233–246.

Bartello, P. and Tobias, P. M. (2013) Sensitivity of stratified turbulence to the buoyancy Reynolds number, *J. Fluid Mech.*, **725**, 1–22.

Batchelor, G. K. (1953) *The Theory of Homogeneous Turbulence*, Cambridge University Press, Cambridge, CB2 8BS, United Kingdom, pp. 197.

Bertino, L., Evensen, G. and Wackernagel, H. (2003) Sequential data assimilation techniques in oceanography, *Int. Stat. Rev.*, **71**(1), 223–242.

Blackford, J. C., Allen, J. I. and Gilbert, F. J. (2004) Ecosystem dynamics at six contrasting sites: a generic modelling study, *J. Mar. Syst.*, **52**, 191–215.

Blumberg, A. F. and Mellor, G. L. (1987) A description of a three-dimensional coastal ocean circulation model, In: N. S. Heaps (Ed.), *Three-Dimensional Coastal Ocean Models*, Coastal and Estuarine Sciences, Series 4, A.G.U. pp. 1–16.

Boris J. P. and Book D. L. (1973) Flux corrected transport SHASTA 1. A fluid transport algorithm that works. *J. Computational Physics*, **11**, 38-69.

Brethouwer, G., Billant, P., Lindborg, E. and Chomaz, J.-M. (2007) Scaling analysis and simulation of strongly stratified turbulent flows, *J. Fluid Mech.*, **585**, 343–368.

Brink, K. H. (1982) The effect of bottom friction on low-frequency coastal trapped waves, *J. Phys. Oceanogr.*, **12**, 127–133.

Brink, K. H. (1991) Coastal-trapped waves and wind-driven currents over the continental shelf, *Annu. Rev. Fluid Mech.*, **23**, 389–412.

Brink, K. H. (1999) Island-trapped waves, with applications to observations off Bermuda, *Dyn. Oceans Atmos.*, **29**, 93–118.

Buchwald, V. T. and Adams, J. K. (1968) The propagation of continental shelf waves, *Proc. R. Soc. London, Ser. A*, **305**, 235–250.

Budd, C. J. and William, J. F. (2009) Moving mesh generation using the parabolic Monge-Ampère equation, *SIAM J. Sci. Comput.*, **31**, 3438–3465.

Bunk, S. (2003) The alpha project, *The Scientist*, **29**, 24–25.

Chatwin, P. C. and Allen, C. M. (1985) Mathematical models of dispersion in rivers and estuaries, *Annu. Rev. Fluid Mech.*, **17**, 119–149.

Chen, H. and Dyke, P. P. G. (1998) Multivariate models for suspended sediment concentration, *Cont. Shelf Res.*, **18**, 123–150.

Chen, C., Liu, H. and Beardsley, R. (2003) An unstructured grid, finite-volume, three-dimensional, primitive equations ocean model: Application to coastal ocean and estuaries, *J. Atmos Oceanic Technol.*, **20**, 159–186.

Craik, A. D. D. and Leibovich, S. (1976) A rational model for langmuir circulations, *J. Fluid Mech.*, **29**, 337–347.

Crank, J. (1975) *The Mathematics of Diffusion,* Oxford Science Publications, Oxford University Press, Clarendon Press, Oxford, U.K., pp. 414.

Dale, A. C. and Sherwin, T. J. (1996) The extension of baroclinic coastal-trapped wave theory to superinertial frequencies, *J. Phys. Oceanogr.*, **26**, 2305–2315.

de Kok, J. M. (1994) Tidal averaging and models for anisotropic dispersion in coastal waters, *Tellus*, **46A**, 160–177.

Dingemans, M. W. (1997) *Water Wave Propagation Over Uneven Bottoms, Part 1 Linear Wave Propagation,* World Scientific Press, Singapore, pp. 471.

Durran, D. R. (1999) *Numerical Methods for Wave Equations in Geophysical Fluid Dynamics,* Springer, pp. 465.

Dyer, K. R. (1997) *Estuaries: A Physical Introduction,* 2nd edn. John Wiley & Sons, Hoboken, New Jersey, USA, pp. 195.

Dyke, P. P. G. (1977) A simple ocean surface layer model, *Riv. Ital. Geofis.*, **4**, 31–34.

Dyke, P. P. G. (1980) On the Stokes' drift induced by tidal motions in a wide estuary, *Estuarine Coastal Mar. Sci.*, **11**, 17–25.

Dyke, P. P. G. (1987) Water circulation in the Firth of Forth, Scotland, *Proc. R. Soc. Edinburgh*, **93B**, 273–284.

Dyke, P. P. G. (1996) *Modelling Marine Processes*, Prentice-Hall, New Jersey, USA, pp. 158.

Dyke, P. P. G. (2001) *Coastal and Shelf Sea Modelling*, Kluwer Academic Press, pp. 264.

Dyke, P. P. G. (2005) Wave trapping and flow around an irregular near circular island in a stratified sea, *Ocean Dyn.*, **55**, 238–247.

Dyke, P. P. G. (2014) *An Introduction to Laplace Transforms and Fourier Series*, 2nd edn. Springer, pp. 381.

Dyke, P. P. G. and McVeigh, A. (2009) A first model of the biological functions of a simple animal: The marine mussel, *Appl. Math. Modell.*, **33**, 783–796.

Dyke, P. P. G. and Robertson T. (1985) The simulation of offshore turbulence using seeded eddies, *Appl. Math. Modell.*, **9**(6), 429–433.

Edwards, A. M. (2001) Adding detritus to a nutrient–phytoplankton–zooplankton model: a dynamical-systems approach, *J. Plankton Res.*, **23**(4), 389–413.

Elliott, A. J. (1991) EUROSPILL: Oceanographic processes and NW European shelf databases, *Mar. Pollut. Bull.*, **22**, 548–553.

Erdogan, M. E. and Chatwin, P. C. (1967) The effects of curvature and buoyancy on the laminar dispersion of solute in a horizontal tube, *J. Fluid Mech.*, **29**, 465–484.

Fasham, M. J. R., Ducklow, H. W. and McKelvie S. M. (1990) A nitrogen-based model of plankton dynamics in the oceanic mixed layer, *J. Mar. Res.*, **48**, 591–639.

Fischer, H. B. (1972) Mass transport mechanisms in partially stratified estuaries, *J. Fluid Mech.*, **53**, 672–687.

Flather, R. A. and Heaps, N. S. (1975) Tidal computations for morecambe bay, *Geophys. J. R. Astron. Soc.*, **42**, 489–517.

Gent, P. R., Bryan, F. O., Danabasoglu, G., Doney, G. C., Holland, W. R., Large, W. G. and McWilliams J. C. (1998) The NCAR Climate system model global ocean component, *J. Clim.*, **11**, 1287–1306.

Gill, A. E. (1982) *Atmosphere — Ocean Dynamics*, Academic Press, London, UK; New York, USA (now Elsevier), pp. 662.

Gill, A. E. and Clarke, A. J. (1974) Wind induced upwelling, coastal currents and sea-level changes, *Deep-Sea Res.*, **21**, 325–345.

Gill, A. E. and Schumann, A. H. (1974) The generation of long shelf waves by the wind, *J. Phys. Oceanogr.*, **4**, 83–90.

Glover, D. M., Jenkins, W. J. and Doney, S. C. (2011) *Modeling Methods for Marine Science*, Cambridge University Press, Cambridge, CB2 8BS, United Kingdom, pp. 571.

Gray, W. G. and Lynch, D. R. (1977) Time-stepping schemes for finite element tidal computations, *Adv. Water Resour.*, **1**(2), 83–95.

Greenspan, H. P. (1968) *The Theory of Rotating Fluids*, Cambridge University Press, Cambridge, CB2 8BS, United Kingdom, pp. 328.

Gurney, W. S. C. and Nisbet, R. M. (1998) *Ecological Dynamics*, Oxford University Press, Clarendon Press, Oxford, U.K., pp. 355.

Haidvogel, D. B. and Beckmann, A. (1999) *Numerical Ocean Circulation Modelling*, Imperial College Press, London UK, pp. 320.

Hansen, D. V. and Rattray, M. Jr. (1965) Gravitational circulation in straits and estuaries, *J. Mar. Res.*, **23**, 104–122.

Hansen, D. V. and Rattray, M. Jr. (1966) New dimensions in estuary classification, *Limnol. Oceanogr.*, **11**, 319–326.

Heaps, N. S. (1984) Tides, storm surges and coastal circulations, In *Offshore and Coastal Modelling*, Dyke, P. P. G., Moscardini, A. O. and Robson, E. H. (eds), pp. 2–54. Lecture Notes in Coastal and Estuarine Studies, volume 12, Springer-Verlag, Heidelberg, London, New York. Dordrecht.

Hemmings, J. C. P., Srokosz, M. A., Challenor, P. and Fasham, M. J. R. (2004) Split-domain calibration of an ecosystem model using satellite ocean colour data, *J. Mar. Syst.*, **50**, 141–179.

Hinze, J. O. (1975) *Turbulence*, McGraw-Hill, New York, USA, pp. 790.

Holloway, G. (1986) A shelf wave/topographic pump drives mean coastal circulation. Part (i), *Ocean Modell.*, **68**, 12–15.

Holloway, G. (1992) Representing topographic stress for large scale oceanographic models, *J. Phys. Oceanogr.*, **22**, 1033–1046.

Hunter, J. R. (1980) An interactive computer model of oil slick motion, *Oceanol. Int. Session M*, Brighton, UK, pp. 42–50.

Hunter, P. J., Kohl, P. and Noble, D. (2001) Integrative models of the heart: Achievements and limitations, *Phil. Trans. R. Soc. London*, **A359**, 1049–1054.

James, G. (ed.) (2015) *Modern Engineering Mathematics*, 5th edn. Pearson, New York, USA, pp. 1025.

James, I. D. (1996) Advection schemes for shelf sea models, *J. Mar. Syst.*, **8**, 237–254.

James, I. D. (2002) Modelling pollution dispersion, the ecosystem and water quality in coastal waters: A review, *Environ. Modell. Software*, **17**, 363–385.

Johnson, R. S. (1997) *A Modern Introduction to the Mathematical Theory of Water Waves*, Cambridge University Press, Cambridge, CB2 8BS, United Kingdom, pp. 445.

Jones, P. D., Briffa, K. R., Osborn, T. J., Lough, J. M., van Ommen, T. D., Vinther, B. M., Luterbacher, J., Wahl, E. R. *et al.* (2009) High-resolution palaeoclimatology of the last millennium: a review of current status and future prospects, *The Holocene*, **19**, 3–49.

Jones, J. E. and Davies, A. M. (2005) An intercomparison between finite difference and finite element (TELEMAC) approaches to modelling west coast of Britain tides, *Ocean Dyn.*, **55**, 178–198.

Kaufmann, S. A. (1969) Metabolic stability and epigenesis in randomly constructed genetic nets, *J. Theor. Biol.*, **22**, 437–467.

Kellog, O. D. (1929) *Foundations of Potential Theory*, Barman Press, pp. 400.

Klein, P. and Steele, J. H. (1985) Some physical factors affecting ecosystems, *J. Mar. Res.*, **43**(2), 337–343.

Komen, G. J., Cavaleri, L., Donelan, M., Hasselmann, K., Hasselmann, S. and Janssen, P. A. E. M. (1994) *Dynamics and Modelling of Ocean Waves*, Cambridge University Press, Cambridge, CB2 8BS, United Kingdom, pp. 532.

Kowalik, Z. and Murty, T. S. (1993) *Numerical Modeling of Ocean Dynamics*, World Scientific, Singapore, pp. 481.

Leblond, P. H. and Mysak, L. A. (1978) *Waves in the Ocean*, Elsevier Oceanography Series 20, Elsevier Press, Amsterdam, The Netherlands, pp. 602.

Leclair, M. and Madec, G. (2009) A conservative leap-frogging time stepping method, *Ocean Modell.*, **30**, 80–94.

Lesieur, M. (1997) *Turbulence in Fluids* 3rd edn. Kluwer Academic Publishers, P.O. Box 17, 3300 AA Dordrecht, the Netherlands, pp. 515.

Lewis, R. (1997) *Dispersion in Estuaries and Coastal Waters*, John Wiley & Sons, Hoboken, New Jersey, USA, pp. 312.

Longuet-Higgins, M. S. (1953) Mass transport in water waves, *Phil. Trans. R. Soc. London*, **A345**, 535–558.

Lynch, D. R. and Gray, W. G. (1979) A wave equation model for finite element tidal computations, *Comput. Fluids*, **7**, 207–228.

Magazenkov, L. (1980) Trudi Glavnoi Geofizicheskoi Observatorii (Transactions of the Main Geophysical Observatory) **410**, 120–129.

Mann, M. E., Bradley, R. S. and Hughes, M. K. (1998) Global-scale temperature patterns and climate forcing over the past six centuries, *Nature*, **392**, 779–787.

Marsden, J. E. and Tromba, A. J. (1988) *Vector Calculus*, W H Freeman & Co, New York, pp. 655.

Masselink, G., Hughes, M. G. and Knight, J. (2014) *Introduction to Coastal Processes and Geomorphology*, Routledge, pp. 416.

Masselink, G., Hughes, M. and Knight, J. (2011) *Introduction to Coastal Processes and Geomorphology*, (Second Edition) Routledge, pp. 413.

Mei, C. C. (1989) *The Applied Dynamics of Ocean Surface Waves*, World Scientific, Singapore, pp. 740.

Mellor, G. and Blumberg, A. (2004) Wave breaking and ocean surface layer thermal response, *J. Phys. Oceanogr.*, **34**, 693–698.

Mellor, G. L. and Yamada, T. (1974) A heirarchy of turbulence closure models for planetary boundary layers, *J. Atmos. Sci.*, **31**, 1791–1896.

Mesinger, F. and Arakawa, A. (1976) *Numerical Methods used in Atmospheric Models*, GARP Publications Series No. 17, pp. 64.

Miller, K. (1981) Moving finite elements II. *SIAM J. Num. Anal.* **18**, 1033–1057.

Mitsuyasu, H., Tasai, F., Suhara, T., Mizuno, S., Ohkusu, M., Honda, T. and Rikiishi, K. (1975) Observations of the directional spectrum of ocean waves using a cloverleaf buoy, *J. Phys. Oceanogr.*, **5**, 750–760.

Moberg, A., Mohammad, R. and Mauritsen, T. (2008) Analysis of the Moberg *et al.* (2005) hemispheric temperature reconstruction, *Climate Dyn.*, **31**, 957–971.

Mohammadi, B. and Pironneau, O. (1994) *Analysis of the k − ε Turbulence Model*, John Wiley & Sons, Chichester, pp. 196.

Moore, D. (1970) The mass transport velocity induced by free oscillations at a single frequency, *Geophys. Fluid Dyn.*, **1**, 237–247.

Moore, M. N. and Allen, J. I. (2002) A computational model of the digestive gland epithelial cell of the marine mussel and its simulated response to pollutants, *Mar. Environ. Res.*, **54**, 579–584.

Moore, M. N. and Willows, R. I. (1998) A model for cellular uptake and intracellular behaviour of particulate-bound micropollutants, *Mar. Environ. Res.*, **46**, 509–514.

Munk, W. H. (1950) On the wind-driven ocean circulation, *J. Meteorol.*, **7**, 79–93.

Murdock, J. and Barnes, J. A. (1974) *Statistical Tables*, Macmillan, London and Basingstoke.

NEMO (2012) The NEMO Ocean Engine manual version 3.4. Go to http://www.nemo-ocean.eu for the latest version.

Oey, L.-Y. (1984) On Steady Salinity Distribution and Circulation in Partially and Well Mixed Estuaries, *J. Phys. Oceanogr.*, **14**, 629–644.

Okubo, A. (1971) Oceanic diffusion diagrams, *Deep Sea Res.*, **18**, 781–802.

Okubo, A. (1974) Some speculations on oceanic diffusion diagrams, *Rapp. P.-V. Cons. Int. Explor. Mer.*, **167**, 77–85.

Ortega, J. M. and Poole, W. G. (1981) *Numerical Methods for Differential Equations*, Pitman Publishing, pp. 329.

Pain, C. C., Umpleby, A. P., de Oliveira, C. R. E. and Goddard, A. J. H. (2001) Tetrahedral mesh optimisation and adaptivity for steady-state and transient finite element calculations, *Comput. Methods Appl. Mech. Eng.*, **190**, 3771–3796.

Pedlosky, J. (1986) *Geophysical Fluid Dynamics*, 2nd edn. Springer, pp. 710.

Piggott, M. D., Gorman, G. J., Pain, C. C., Allison, P. A., Candy, A. S., Martin, B. T. and Wells, M. R. (2008) A new computational framework for multi-scale ocean modelling based on adapting unstructured meshes, *Int. J. Numer. Methods Fluids*, **56**, 1003–1015.

Pond, S. and Pickard, G. L. (1991) *Introductory Dynamical Oceanography*, 2nd edn. Pergamon, Oxford, pp. 329.

Pritchard, D. W. (1955) Estuarine circulation patterns, **81**, *Proc. ASCE*, Sept. No. 177.

Proctor, R., Flather, R. A. and Elliott, A. J. (1994) Modelling tides and surface drift in the Arabian Gulf — application to the Gulf oil spill, *Cont. Shelf Res.*, **14**, 531–545.

Proudman, J. (1953) *Dynamical Oceanography*, Methuen, pp. 409.

Pugh, D. and Woodworth, P. (2014) *Sea-Level Science (Understanding Tides, Surges, Tsunamis and Mean Sea-Level Changes)*, Cambridge University Press, Cambridge, CB2 8BS, United Kingdom, pp. 395.

Reeve, D., Chadwick, A. and Fleming, C. A. (2004) *Coastal Engineering: Processes, Theory and Design Practice*, Spon Press, pp. 461.

Reeve, D., Li, B. and Thurston, N. (2001) Eigenfunction analysis of decadal fluctuations in sandbank morphology at Gt Yarmouth, *J. Coastal Res.*, **17**(2), 371–382.

Reynolds, R. W. (1978) Some effects of an elliptic ridge on waves of tidal frequency, *J. Phys. Oceanogr.*, **8**, 38–46.

Rhines, P. B. (2006) Sub-arctic oceans and global climate, *Weather*, **61**(4), 109–118.

Robert, A. J. (1966) The integration of a low order spectral form of the primitive meteorological equations, *J. Meteorol. Soc. Jpn.*, **44**, 237–245.

Robinson, A. R. (1964) Contintental shelf waves and the response of sea level to weather systems, *J. Geophys. Res.*, **69**, 367–368.

Roulstone, I. and Norbury, T. (1994) On the Hamiltonian structure of the semi-geostrophic equations, *Geometrical Methods in Geophysical Fluid Dynamics. Proceedings of Woods Hole Summer Programme 1993*, WHOI Tech Rep. WHOI-94-12.

Ruddick, K. G., Deleersnijder, E., de Mulder, T. and Luyten, P. J. (1994) A model study of the Rhine discharge, *Tellus*, **46A**, 149–159.

Schlichting, H. (1975) *Boundary Layer Theory*, 7th edn. McGraw-Hill, New York, USA, 817 pp.

Semtner, Jr. A. J. Jr. (1974) An oceanic general circulation model with bottom topography, In: *Numerical Simulation of Weather and Climate*. Technical Report No. 9, UCLA Department of Meteorology.

Shields, A. (1936) Application of similarity principles and turbulence to bed-load movement, *Mitteilunger der Preussischen Versuchsanstalt für Wasserbau und Schiffben*, **26**, 5–24.

Shields, A. F. (1936) Application of similarity principles and turbulence research to bed-load movement, vol 26. Mitteilungen der Preussischen Versuchsanstalt für Wasserbau und Schiffbau, Berlin, Germany, pp. 524

Sleath, J. R. A. (1984) *Sea Bed mechanics*, John Wiley & Sons, Hoboken, New Jersey, USA, pp. 335.

Smith, G. D. (1965) *Numerical Solutions of Partial Differential Equations*, Oxford University Press, Clarendon Press, Oxford, U.K., pp. 179.

Smith, R. (1978) Asymptotic solutions of the Erdogan–Chatwin equation, *J. Fluid Mech.*, **88**, 323–337.

Smith, R. (1979) Buoyancy effects upon lateral dispersion in open channel flow, *J. Fluid Mech.*, **90**, 761–779.

Smith, R. (1980) Buoyancy effects in longitudinal dispersion in wide well-mixed estuaries, *Phil. Trans. R. Soc. London Sect. A*, **296**, 467–496.

Soulsby, R. (1997) *Dynamics of Marine Sands. A Manual for Practical Applications.* Thos. Telford, Chiswick, London UK, pp. 249.

Stoker, J. J. (1958) *Water Waves*, Wiley-Interscience, Hoboken, New Jersey, USA (Classics Edition published 1992), pp. 567.

Stommel, H. (1948) The westward intensification of wind driven ocean currents, *Trans. Am. Geophys. Union*, **99**, 202–206.

Stommel, H. (1965) *The Gulf Stream,* 2nd edn. University of California Press, Oakland, California, USA, pp. 202.

Summerfield, W. C. (1969) *On the trapping of wave energy by bottom topography*, Horace Lamb Centre for Oceanographic Research, Research paper 30.

Tkalich, P., Huda, M. K. and Gin, K. Y. H. (2003) A multi-phase oil spill model, *J. Hydraul. Res.*, **41**(2), 115–125.

van Dyke, M. (1964) *Perturbation Methods in Fluid Mechanics*, Applied Mathematics and Mechanics 8, Academic Press, London, UK; New York, USA (now Elsevier), pp. 229.

Vlasenko, V., Stashchuk, N. and Hutter, K. (2005) *Baroclinic Tides*, Cambridge University Press, Cambridge, CB2 8BS, United Kingdom, pp. 372.

von Larcher, T. and Williams, P. D. (2015) *Modeling Atmospheric and Oceanic Flows: Insights from Laboratory Experiments and Numerical Simulations*, John Wiley & Sons, Hoboken, New Jersey, USA, pp. 368.

Waite, M. L. (2013) The vortex instability pathway in stratified turbulence, *J. Fluid Mech.*, **716**, 1–4.

Walters, R. A. (1983) Numerically induced oscillations in finite element approximations to the shallow water equations, *Int. J. Numer. Methods Fluids*, **3**, 591–604.

Walters, R. A. and Carey, G. F. (1983) Analysis of spurious oscillation modes for the shallow water and Navier–Stokes equations, *Comput. Fluids*, **11**, 51–68.

Walters, R. A. and Carey, G. F. (1984) Numerical noise in ocean and estuary models, *Adv. Water Resour.*, **7**, 5–20.

Wang, D.-P. and Mooers, C. N. K. (1976) Coastal trapped waves in a continuously stratified ocean, *J. Phys. Oceanogr.*, **6**, 853–863.

Watson, G. N. (1922) *A Treatise on the Theory of Bessel Functions*, Cambridge University Press, Cambridge, CB2 8BS, United Kingdom, pp. 804.

Webb, A. J. and Metcalfe, A. P. (1987) Physical aspects, water movements and modelling studies of the Forth Estuary, *Proc. R. Soc. Edinburgh.*, **93B**, 259–272.

Westerink, J. J., Luettich, R. A. and Muccino, J. C. (1994) Modelling tides in the western North Atlantic using unstructured graded grids, *Tellus*, **46A**(2), 178–199.

Widdows, J. and Donkin, P. (1992) Mussels and environmental contaminants: bioaccumulation and physiological aspects. In: E. Gosling (Ed.), *The Mussel Mytilus: Ecology, Physiology, Genetics and Culture*. Elsevier Science Publishers, Amsterdam, pp. 383–424.

Williams, P. D. (2009) A proposed modification to the robert–asselin time filter, *Mon. Weather Rev.*, **137**, 2538–2546.

Wolanski, E. (1988) Circulation anomalies in tropical Australian estuaries. In: B. Kjerfve (Ed.), *Hydrodynamics of Estuaries*, CRC Press, pp. 53–59.

Wrobel, L. C. (2002) *The Boundary Element Method. Part 1: Applications in Thermofluid and Acoustics*, John Wiley & Sons, Hoboken, New Jersey, USA, pp. 451.

Yanuma, T. and Tsuji, Y. (1995) Observation of edge waves trapped on the continental shelf in the vicinity of makurazaki harbor, Kyushu, Japan, *J. Oceanogr. Soc. Jpn.*, **54**, 9–18.

Zalesak S. T. (1979) Fully multidimensional flux-corrected transport algorithms, *J. Computational Physics*, **31**, 335–362.

Zienkiewicz, O. C. and Taylor R. L. (2000a) *The Finite Element Method — The Basics*, Butterworth-Heinmann, Amsterdam, The Netherlands, pp. 689.

Zienkiewicz, O. C. and Taylor, R. L. (2000b) *The Finite Element Method — Volume 3 Fluid Mechanics*, Butterworth-Heinmann, Amsterdam, The Netherlands, pp. 334.

Index

Printed in the United States
By Bookmasters